*Management
Systems
for Profit
and Growth*

RICHARD F. NEUSCHEL
Director, McKinsey & Company, Inc.

Management
Systems
for Profit
and Growth

McGraw-Hill Book Company
New York   St. Louis   San Francisco   Auckland   Düsseldorf
Johannesburg   Kuala Lumpur   London   Mexico
Montreal   New Delhi   Panama   Paris
São Paulo   Singapore   Sydney
Tokyo   Toronto

Library of Congress Cataloging in Publication Data

Neuschel, Richard F
 Management systems for profit and growth.

 First ed. published in 1950 under title:
Streamlining business procedures.   Second ed.
published in 1960 under title: Management by system.
 Includes index.
 1. System analysis. 2. Management. I. Title.
HD20.5.N47   658.4   75-37661
ISBN 0-07-046323-9

1234567890   KPKP   785432109876

*The editors for this book were W. Hodson Mogan and Ruth L. Weine,
the designer was Naomi Auerbach, and the production
supervisor was Teresa F. Leaden. It was set in
VIP Baskerville by University Graphics, Inc.*

It was printed and bound by The Kingsport Press.

The previous edition of this book was published under the
title *Management by System.*

To Jean, Robin, and Debra

Contents

Preface   ix

PART ONE   *The Case for a "Systematic*
            Ordering of Operations"   *1*

1. The Role of Systems in the Business of Management   *3*
2. The Pay Dirt in Systems Analysis   *15*

PART TWO   *The Ingredients of a Successful*
            Systems Improvement Program   *35*

3. Barriers to Systems Improvement   *37*
4. Organizing and Staffing the Systems Improvement Program   *49*
5. The Top-Management Approach to Systems Analysis   *72*
6. Gaining Acceptance of the Systems Improvement Program   *86*
7. Managing the Systems Staff   *108*

PART THREE   *Systems Analysis and*
              Improvement Techniques   *125*

8. Planning and Controlling the Individual Systems Project   *127*
9. Gathering the Facts   *145*
10. Improving the Performance of Line Operating Functions   *180*
11. Cutting Clerical and Other Administrative Costs   *206*

12. Improving Top-Management Effectiveness Through Better Decision-Making and Control Systems   *250*

13. Strengthening Management Information Systems   *270*

14. Optimizing Development of Computer-Based Systems   *293*

PART FOUR   *Cashing In On Systems*
                Improvement Recommendations   *311*

15. Presenting and Selling Recommendations   *313*

16. Installing Approved Systems and Effecting Change   *327*

    Appendix   350

    Index   354

Preface

Thirty years have passed since Elton Mayo—one of the all-time great teachers of management—wrote:

> In modern, large-scale industry the three persistent problems of management are:
> 1. The application of science and technical skill to some material good or product
> 2. The systematic ordering of operations
> 3. The organization of teamwork—that is, of sustained cooperation.[1]

Among students and practitioners of present-day management, there is little doubt that the problems Professor Mayo cited then are still very much with us today.

MANAGEMENT SYSTEMS FOR PROFIT AND GROWTH is concerned with the development of a logical, organized, top-management approach to the second of these "persistent problems of management." Its principal theme is that the continuous simplification, strengthening, and adaptation of general-management and operating systems are major requisites for achieving a "systematic ordering of operations."

Much of the book is aimed at answering the "why" and "how" implicit in this central theme.

In its answer to the first question, the book seeks to spread an understand-

---

[1] Elton Mayo, *The Social Problems of an Industrial Civilization,* Division of Research, Graduate School of Business Administration, Harvard University, Boston, 1945, pp. 69–70.

ing of the vital part that systems play in the management of all organizations of any significant size. It treats systems not as mere matters of paperwork mechanics or clerical detail, but as one of the basic administrative techniques through which the complex organization acts and by which it controls and coordinates its actions. This broad perspective emphasizes the far-reaching benefits that can be realized from an intelligent and continuing effort to improve management systems.

In answering the "how" of systems improvement, the book discusses the principles of organizing, planning, and administering a comprehensive systems analysis and development program as well as the methods of attacking individual systems studies.

In light of this statement of scope, the book is designed for three groups of readers.

First are the *top-management executives* of all types of business and public-sector organizations. They are the ones that must take the lead in identifying and capitalizing on opportunities to improve significantly the general-management systems through which they carry out their overall planning and control responsibilities—both strategic and operational.

Second are the *functional executives and supervisors* of an organization. For this group, the book spells out a detailed approach to developing improved systems that cut the costs and increase the effectiveness with which it carries out its function—whether it be sales or production, finance or personnel, accounting or distribution, or any other.

Third are the *systems and procedures staff analysts* that can be found in most of today's organizations. For them, the book treats all the key steps in a dynamic, profit-oriented systems improvement program. These include identifying the major systems problems and improvement opportunities in the organization, gathering the facts, applying a fundamental top-management approach to the analysis of findings and the development of recommendations, presenting and selling recommendations to top management and to operating personnel, installing approved systems, and effecting change.

Notwithstanding its title, the book is by no means limited to profit-making enterprises. The approaches and techniques it sets forth are equally applicable to all types of public-sector organizations, and much of the case material presented is drawn from my experience and that of my colleagues in serving such organizations. The use of the words PROFIT AND GROWTH in the title, rather than being restrictive, underscores the point that the true test of well-designed management systems is whether they help the organization achieve some significant, quantifiable improvement in its *end-result effectiveness*.

Throughout my professional career, I have seen a steady increase in the number of women who hold executive and supervisory positions or who are involved in systems and procedures work. This fact is implicitly recognized in many of the case examples contained in the book. But, even with this recognition, I have frequently used the pronouns "he" and "his" in a purely generic sense to accommodate to the limitations of our language and avoid the awkwardness of a "she/he" or "her/his" construction.

The thinking and experience of many of my associates in McKinsey & Company are reflected in this book. Among them I acknowledge particularly the substantial help that my former partner, Harvey Golub, gave me in preparing the chapter, "Optimizing Development of Computer-Based Systems."

In addition, a number of other colleagues made constructive critiques of parts of the manuscript or gave me valuable substantive inputs. For these sorts of contributions, I am especially grateful to Michael Abrams, Robert Alexander, Warren Cannon, Robert Slane, John Soden, and Peter Walker.

Others to whom I am deeply indebted are those who typed countless drafts of the manuscript, drew the graphic presentations, and proofread the typed and printed versions. They include Gladys Cazakoff, Annabelle Devine, Elvira Hawkins, and Delio Silva.

Finally, my warmest thanks go to my secretary, Denise Tufaro, who served so ably as the "beachmaster" of the project, and to Mary Perkins who did such superb work in editing the manuscript and preparing the Index.

Richard F. Neuschel

*Management
Systems
for Profit
and Growth*

PART ONE

The Case for a "Systematic Ordering of Operations"

The Role of Systems in the Business of Management

The success of any large-scale, organized human endeavor—public or private—depends, more than anything else, on two critical factors.

First is the organization's entrepreneurial drive—its strategic thrust—its capacity to identify and capitalize on major new areas of need or opportunity. That is what this book is concerned with, only in part.

Second is the effectiveness and efficiency with which the organization carries out its present basic mission. That is what this book is concerned with primarily.

These two key ingredients of success are, in turn, shaped and controlled by a host of subelements, the most important of which are:

1. The quality and motivation of the organization's human resources—among which, its leadership talent and its reservoir of technological competence are two of the most telling

2. The management tools that it employs; these consist of:

   a. The way in which the enterprise is organized.

   b. The basic objectives and policies it has adopted—instruments that determine its character and govern its recurring decisions of every kind.

   c. The network of systems—both operational and managerial—through which it plans, executes, and controls not only its repetitive day-to-day activities but also its long-term, more fundamental moves.

SYSTEMS IN THE PRIVATE AND PUBLIC SECTORS

With this overall framework in mind, we can deepen our insight into the role and widespread application of systems as a tool of management by reflecting on some representative examples from the real-life world.

The Brokerage Industry's Backroom Foul-Up

During the latter years of the 1960s, the nation's business press was flooded with articles about the nearly disastrous "back office" problems of the security brokerage houses. The critical tenor of these stories was typified by such stinging assessments as "a nightmare of bureaucratic chaos," "Wall Street drowns in paper," "the industry's operational collapse," "hopeless bungling of customers' accounts," and "ballooning costs and financially crippled brokerage firms."

The root cause of this highly publicized problem—as everyone with the benefit of hindsight readily conceded—was that the systems by which brokerage firms processed their basic transaction, the security sale, were just not geared to handle the sharp increase in trading volume that occurred in 1968 and 1969. During the preceding several years, a number of the so-called retail brokerage firms had pushed hard for volume growth—opening additional offices, training more sales representatives and otherwise expanding their capacity to generate new business. But they had given little if any attention to improving their paperwork techniques for handling the increased volume that would inevitably result.

This lack, in turn, led to the widespread occurrence of errors and "fails"—a term applied to any trading transaction in which the securities involved are not delivered to the purchaser within 5 days. Customers' monthly statements of account were shot through with mistakes that seemingly defied correction. In addition, the volume of fails skyrocketed, reaching at its highest point almost $6 billion in value, an amount twice as large as the industry's total capital.

The full impact of this systems breakdown was by no means limited to the industry's tarnished image with the investing public. For many firms, it included a heavy—indeed, even in a few instances a ruinous—economic penalty. At least three companies (one of them among the industry's top ten) were put out of business by the paperwork snarl and related problems. Countless others paid a great—though fortunately not fatal—price in the form of loss of income or impairment of net capital resulting from the following conditions and corrective measures:

1. In the second half of 1968, the markets were closed on Wednesdays to curtail business volume. In early 1969, the 5-day week was restored, but the trading day was shortened to 4 instead of 5½ hours.

2. The exchanges imposed substantial percentage cuts in the number of trading transactions that each of some fifty firms could execute in a single day.

3. Payroll costs increased dramatically as back office workers labored overtime to clean up backlogs.

4. Cash flow became erratic. Because firms did not receive—and hence could not deliver—traded securities anywhere near on time, they encountered corresponding delays in collecting their commissions.

5. The Securities and Exchange Commission imposed a requirement that if a brokerage firm's volume of fails exceeded a prescribed level, it must make a deduction from its net worth in determining the amount of its capital to meet exchange requirements. The deduction ranged from 10 to 30 percent of the value of each trade, depending on the age of the fail.

Cost Overruns on Major Defense Systems Programs

One of the most challenging management jobs of our time is that of the U.S. Air Force Systems Command (AFSC), which is charged with the basic responsibilities of "advancing aerospace technology, adapting it into operational aerospace systems, and acquiring qualitatively superior aerospace systems and material needed to accomplish the Air Force mission."[1]

In carrying out these responsibilities, the AFSC depends heavily, of course, on the aerospace industry, with which it contracts for a tremendous variety of research, development, production, and spares-provisioning programs. Thus, in the simplest context, the key AFSC management task is to plan and control the individual contractor's execution of Air Force programs in terms of three parameters: meeting technical performance objectives, adhering to delivery schedules, and staying within cost limits. What makes this straightforward and obvious task so complex and difficult are these inescapable facts of life:

1. Most AFSC development and acquisition programs extend over a long span of years.

2. The technology of most programs is typically so advanced that the AFSC can find little if any prior experience to tap when plans and estimates are developed.

3. Throughout the life of the program—from the earliest conceptual phase to final production contract deliveries—countless changes are made in the system's performance specifications, engineering design, delivery schedules, and quantities procured.

Against this background, the news media reported on several occasions during the early to middle 1960s that the Department of Defense was experiencing massive cost overruns on certain Air Force development and acquisition contracts—most notably on the Minuteman missile program. What was more, according to the news accounts, these overruns—in the magnitude of hundreds of millions of dollars—often came as a surprise long after the relevant early-warning signals should have been detected.

Why this shortfall? The AFSC Systems Program Office responsible for

---

[1]Air Force Regulation No. 23-8.

monitoring each program was doing an effective job of tracking actual costs incurred on the program against a time-phased budget or spending plan, and actual work progress against a program schedule. But the monitoring systems then employed did not make provision for relating these two elements of performance to each other; cumulative costs were not reviewed in terms of the *value of work actually performed.*

In light of this experience, the AFSC launched a major project to develop, pilot-test, and install an improved cost planning and control system that has come to be known as the "cost/accomplishment system." The key objectives of this new cost management approach are to give aerospace contractors and the AFSC

1. A reliable measure of actual costs incurred versus the value of work performed (that is, current overrun-underrun status)

2. Earlier visibility of potential cost growth at contract completion

With this improved information system, the AFSC is able not only to avoid surprise overruns at higher-management levels in the Department of Defense, but also to act much earlier itself in making tradeoff decisions among cost-savings options, and in stimulating contractors to take timely corrective action on incipient cost growth.

Service to Independent Agents as an Insurance Company Competitive Weapon

The great bulk of property and casualty insurance coverage, at least in the United States, is placed by (or distributed through) so-called independent agents—the local agents who sell us our automobile and homeowner's insurance policies. Today the large independent agents (together with the marketing outlets referred to as insurance brokers) also place most commercial and industrial insurance. Virtually none of these independent agents has an exclusive arrangement with any one carrier. Rather, each typically seeks to maximize his ability to place business of widely varying kinds and degrees of risk by representing some three to five different insurance companies.

Given this type of product distribution system, one of the large property and casualty insurance companies recently found that it was slowly but steadily losing its share of the market in many parts of the country. To learn the reasons for this decline, the company surveyed a large sample of its independent agents. The findings were sobering. Although the company was rated average or better in its general standing within the industry, it was—without exception—at or near the bottom of the list on the speed and quality of its service to agents. Typical of the comments gathered during the survey was this: "Your branch and home office service has become so slow, and the volume of errors has grown so large, that I usually wind up trying to place a piece of business with your company only when I can't place it anywhere else."

These findings led the company to take three actions:

1. Completely overhaul its system for processing each of the major transactions making up its business—e.g., issuing new policies, renewals, and endorsements

2. Inaugurate more effective error-detection systems

3. Sharpen the skills of its front-line supervisors in managing the work flow.

As a result of these measures, the company today is giving competitively superior service to its agents without any significant increase in its operating expense ratio. Moreover, it seems likely to continue this level of performance in the future. For one thing, it has developed a more realistic comprehension of the significance of service to agents as a competitive weapon in its industry. In addition, it has established an ongoing system by which it routinely measures the quality of each key element of its service from the agent's point of view.

Logistics Systems in an International Petroleum Company

Beginning with the closing of the Suez Canal in 1967, the large international oil companies have had to adapt to a host of new developments that have enormously complicated their operations in all areas of the world. These forces at work include the burgeoning demand for petroleum products, particularly in Europe and Japan; the endless succession of constraints imposed by oil-producing countries, culminating in the Arab oil embargoes of 1973 and 1974; the advent of the supertanker, which can carry up to 3.5 million barrels of crude oil; the ecological pressures against power-generating plants that burn fuel oil with a high sulfur content; the substantial increase in the variety of petroleum products and product grade specifications demanded of the oil companies; and the rapid growth of the industry's fixed investment in refining, storage, and transportation facilities.

These and similar developments have confronted each of the international oil companies with a compelling need to develop and implement new systems for making major operating decisions. The systems that are required not only must be responsive to changing external forces but also must continuously balance a company's far-flung logistics network—from the extraction of crude oil to the marketing of finished products—in a way that achieves maximum efficiency and effectiveness.

Developing these systems is an uncommonly difficult task because of the large number of recurring decisions that are involved and the many variables or constraints affecting those decisions.

For example, decisions must be made monthly or more often on which types of crude oil from what sources should go to which refineries by what means; which products each refinery should make, and in what quantities; and how much crude oil or finished product should be sold to, purchased from, or exchanged with other oil companies.

Reaching decisions like these is all the more difficult because of the wide variations that exist within each of the elements making up a company's logistics system. These include variations in:

1. Finished-product yield among types of crude. (In the distillation process, different crude oils yield substantially different proportions of end product—for example, fuel oil versus gasoline.)

2. Overall capacity and the amount and type of flexibility in yield among refineries. (Most refineries can be operated in a variety of ways that result in changes in the proportion of products obtained from a given crude oil.)

3. The mix of products sold among the markets in which the company operates.

4. Elapsed time and cost involved in transporting oil from each source of crude to each refinery.

5. Cargo capacity and speed of the tankers making up the company's marine fleet.

6. Storage capacity (*a*) for each type of crude among producing fields and refineries, and (*b*) for each type of finished product among refinery and sales terminals.

7. Port facilities (for example, docks and pumping capacity) among refinery and sales terminals.

Notwithstanding the difficulty of keeping pace with the growing complexity of this logistics management job, the oil companies have had no alternative. They have had to develop and continuously upgrade advanced information and decision-making systems, simply because the penalty costs of not meeting this need are so very high. To illustrate: Systems that result in less-than-optimal decisions at any point in the long process from wellhead to consumer can produce the following types of profit erosion:

1. Excessive refining costs. These might be caused by:
   a. Crude-oil shortages or product buildups, which force the company to operate the refinery at less than capacity.
   b. Allocation of the wrong crude oil to a given refinery, a combination which reduces product yield.

2. Excessive transportation costs, which may arise from
   a. Poor scheduling of tanker movement—for example, delivering crude oil to refineries, or products to sales terminals, more frequently than storage facilities at those destinations require.
   b. Delays in port, perhaps because there are too many tankers in port at the same time, or because tankers must wait for a cargo of finished product to be produced.
   c. Excessive spot-chartering of tanker capacity to accommodate sudden disruptive changes in the movement schedule.
   d. Failure to identify short-term excesses in tanker capacity which might have been chartered out to other companies.

3. Losses incurred from overproduction of a given product at any one refinery. These losses might result from:

    a. Dumping the product excess at less-than-normal margins.

    b. Storing it at high cost (in inventory investment and rented storage space).

    c. Blending the excess into a more readily salable but lower-quality product (that is, making a "product quality giveaway").

    4. Losses incurred from underproduction of a given product at any one refinery. In this case, the company must forgo the sale and resulting profit or, more likely, meet the shortage by having products shipped over longer distances from another less efficient refinery.

    5. Losses incurred in trying to balance the logistics system by making unfavorable crude-oil or product purchases from, sales to, or exchanges with other oil companies.

Strategic Planning Systems in a Multibusiness Enterprise

In a cover story entitled "GE's New Strategy for Faster Growth," the July 8, 1972, issue of *Business Week* reported on General Electric's development of a rigorous analytical and decision system for planning and executing the company's far-ranging strategic moves. In describing the details of this new system, the article stated: "Strategic planning has become GE's route to faster corporate growth." Basically, it is "a technique that treats the company's vast array of ventures as an investment portfolio, sorting out the winners and losers through systematic analysis. The aim is to decide which ventures deserve the most investment capital and which should be dumped."

The approach, "which has become a buzzword in company circles since its adoption nearly two years ago, involves a continuing in-depth analysis of market share, growth prospects, profitability, and cash generating power of each venture. The probing goes far deeper than any planning operation GE has conducted before."

At another point—in underscoring the importance of this development in management methodology—the article stated: "Reg Jones [then the newly elected GE president], who is now charged with applying the system on a continuing basis, calls it an 'extremely significant' move for GE. 'Its impact on the company,' says Jones 'will be as profound as Ralph Cordiner's decentralization in the early 1950s.'"

THE FUNCTIONS OF SYSTEMS

These vignettes from the worlds of business and government help to highlight the multipurpose role that systems play in the management of every type of organization of any significant size.

Systems as an Operations Catalyst

First, systems—and their component subsystems or "procedures"—are the means by which all the recurring actions of an enterprise are initiated, carried forward, controlled, and stopped. In a manufacturing company, for example, it is through forecasting and budgeting procedures that financial and operating plans are developed; through purchasing procedures that materials are procured; through employment, transfer, and work order procedures that the labor force is recruited and deployed; through order processing procedures that the manufacturing process is initiated or finished goods are shipped from stock; through production planning, scheduling, and dispatching procedures that workers, materials, and machines are coordinated; through material control procedures that investment in inventories and losses from obsolescence are minimized; through inspection procedures that qualitative standards are applied to the product and its components; and through accounting procedures that the information required for control of costs, profits, and assets is gathered, summarized, and reported.

These are but a few of the activities performed day after day in a make-and-sell business. They are sufficient, however, to show that the systems and procedures carried out by nonproduction personnel are among the basic tools that make it possible for management to manage. They are the mechanism through which action and control are achieved. They are the traffic laws and signals governing the form, the direction and rate of movement, the stopping places, and the ultimate destination of each instruction and piece of information through which today's complex enterprise functions. They make control possible by providing for inspection and recording of performance, and by specifying the kinds of performance data that are to be submitted to the managerial group.

Because these procedural activities are performed by clerical personnel and involve paperwork rather than physical operations, a pervasive misconception exists that procedures are peculiar to the office. This, of course, is not true. The notion persists, however, with the result that too many companies have taken an unrealistic and ineffective approach to the task of systems and procedures improvement.

Perhaps the most marked characteristic of systems and procedures as a management tool is their widespread application. They are not purely an office phenomenon in terms of either participants or areas affected. They are no more restricted to the office than are policies, organization charts, or any other administrative tool. They play a vital role in the operation of every department and in the performance of every function of an organization. We have seen, for example, that paperwork is an integral part of the production process—that by governing the flow of materials and tools, it determines how effectively workers and machines are used. The forms prepared and records maintained by shop clerks, schedulers, storekeepers, or receiving clerks are just as much a part of the company's action and control patterns as are sales orders, cost sheets, help requisitions, or budget reports.

It should be recognized, however, that the role of systems and procedures in carrying on the activities of a department varies considerably from function to function. For example, within a manufacturing company the number of employees engaged in procedural routines is generally smaller in the sales and industrial relations departments than it is in the accounting and production control departments. In fact, in accounting and production control, procedures and record keeping are such a large part of the whole that these functions become known as the accounting *system* and the production control *system*. But whatever the department—however much judgment, professional skill, or technical knowledge its members must have, and however variable and nonrepetitive its activities are—a significant part of its work will always require established routines for getting things done in a prompt, well-ordered, efficient way.

The same is true of repetitive operating or decision-making processes in any other type of organization, whether they be premium accounting and billing in an insurance company; check processing, stock transfer, or money transfer in a commercial bank; value analysis of purchased components in an automobile company; merchandise control in a department store; worldwide project management in the United Nations Development Program; or capital budgeting and control in an international airline.

One of the most marked differences that one encounters among the other businesses just named is that, in some of them, paperwork becomes an even greater part of the whole than it is in manufacturing companies. Indeed, in industries such as banking and insurance, clerical labor is very nearly the entire means for processing all the business of the enterprise.

Systems and Top Management

Thus one of the major functions of systems is to facilitate and control the work of the various functional groups that make up an enterprise. But, in addition, systems and procedures play an equally significant part—though perhaps a less perceptible one—in the affairs of top management.

1. *The formulation and continuous reappraisal of major plans and policies* are well-recognized top-management responsibilities. But plans and policies are rarely self-executing. Their implementation generally requires the careful development and installation of systems and procedures that will ensure timely and effective execution. Let us suppose, for example, that it is the policy of a given company to compete on a specialized-merchandise, customer-service basis. Any such policy would be meaningless unless the company's scheduling and expediting procedures and its production progress records were all closely related to customers' orders. A personnel policy to promote from within the organization and to achieve the best use of the existing work force could hardly be administered effectively without the development and maintenance of an inventory record of employees' skills and experience.

Yet many soundly conceived policies and programs fail in execution because members of top management tend to regard the promulgation of a policy as the fulfillment thereof. Top executives often overlook the size of the gap that lies between decision and implementation. Many a program of annual profit planning and control, for example, has failed to improve its company's profit performance because of either one of two shortcomings—one an error of omission, the other of commission. In the first instance, adequate instructions are not issued to translate top management's philosophy and objectives into operating reality. More specifically, the underlying procedures that are devised may fail to (1) provide for development of the basic information needed for intelligent planning decisions; (2) specify any qualitative standards governing the plans submitted; or (3) make any provision for regular evaluation of each department's performance against agreed-upon plans. In the error of commission, the opposite extreme is encountered—too much system rather than too little. In this case, management's desire for dynamic profit planning is subverted by a cumbersome, excessively expensive planning system that comes to be regarded as a fruitless, time-consuming exercise instead of a unifying force that helps the whole organization to consistently plan and strive for peak performance.

Thus, in a very practical sense, the full effectiveness of key executives is realized only to the extent that relevant and efficient procedures are developed through which operating personnel can convert static plans and policies into a living management force.

2. *The exercise of control* to ensure that the organization achieves its objectives and plans is another traditional responsibility of top management. Here again systems and procedures play a supporting role. They specify the form, substance, and frequency of information essential to the exercise of control, and they provide the channels through which that information flows to the top-management group.

3. The final function performed by systems as an instrument of top management is that of *facilitating coordination of effort among functional groups* so that the best interests of the enterprise as a whole are served. The development of every procedure requiring participation by more than one department or division is fundamentally a means of promoting coordination. The full extent to which systems and procedures serve this purpose will be covered in succeeding chapters. It will suffice to point out here that procedures achieve coordination among departments by prescribing the nature and timing of each group's participation in a common process.

Summary

Among organizations in every field, systems and procedures play a major role in the planning, execution, coordination, and control of all recurring activities. In this respect, they are one of the generic, constitutive parts of the management process. How well they are designed will have a very real impact on operating costs, production speed, quality of service to customers, and overall effectiveness of the enterprise. Soundly conceived and continuously adapted, they will do much to ensure the "systematic ordering of operations" of which Mayo writes. Improvidently developed and permitted to grow unattended, they will produce only waste, confusion, and delay. As another writer states, it is the difference between "lifeblood" and "sleeping sickness."

TYPES OF SYSTEMS

Added understanding of the role of systems and their relationship to other parts of the management process can be gained by classifying them into categories that delineate their basic purposes. Although different types of classifications have been advanced, the following breakdown, in our experience, is the most useful—particularly as an aid in setting the objectives of an ongoing systems improvement program.

Most if not all of the systems covered by this book can be divided into two major groups: operational systems and general-management systems. These groups, in turn, can be subdivided and defined as follows:

1. *Operational systems*
   a. Transaction-processing systems, which are networks of routine procedural steps for processing the recurring transactions of the organization. Typical examples are:
      (1) A manufacturing company's payroll system, sales order processing system, accounts payable system, and fixed-assets record-keeping system
      (2) An airline's reservation processing system
      (3) A department store's complaint handling system
      (4) A utility company's repair order system
   b. Operational decision-making systems, which are concerned with day-to-day decisions on operating matters that typically involve an aggregation of many transactions. For example:
      (1) A production planning and scheduling system
      (2) A cost accounting system
      (3) A system for planning and controlling field sales representatives' use of time
      (4) A commercial bank's money-management decision system (covering its short-term acquisition and disposition of funds)
      (5) The international petroleum company's logistics decision system that was described earlier in the chapter.

These two subclasses of operational systems (*a* and *b*) are largely, but not entirely, exclusive. They overlap slightly, in the sense that transaction-processing systems often involve some decision making. But where this is true, the decisions are typically simple, straightforward "unit transaction" ones. For example: deciding whether to accept a risk as a step in the insurance application processing system, or deciding what credit terms to extend to a customer in a sales order processing procedure. In contrast, the decisions that are the end product of an operational decision-making system are typically of a higher order of complexity and importance, and represent nearly the whole purpose for which that system exists.

Another useful distinction between the two subclasses lies in the fact that transaction-processing systems must be carried out well to avoid a negative impact, but doing so is the normal expectation and, therefore, does not produce any positive gain. In contrast, the quality of performance in executing an operational decision-making system can have a major impact on the organization's operating results. For example, it may result in cutting unit costs, reducing inventories, speeding up service, or optimizing the use of facilities.

2. *General-management systems.* As the words imply, the systems that make up this second broad classification are not concerned exclusively with a function or combination of functions. They apply across the board to all divisions and departments as well as to the management of the organization as a whole. Typical general-management systems are concerned with:

 a. Strategic planning.

 b. Operational planning and control

 c. Manpower management—that is, the system for planning and development of human resources

 d. Capital budgeting and control

TWO

The Pay Dirt
in Systems Analysis

Some years ago, Alfred P. Sloan, then president of General Motors Corporation, wrote in a statement to stockholders:

> Experience has demonstrated that the day-to-day problems that confront the Corporation's operating executives are so absorbing in their demands of time that too little opportunity is afforded for . . . the development of a better operating technique, which requires much study and research. The importance of research work as contributing to the advancement of industry through scientific study is well established. The marvelous contribution that it has made to the progress of industry and the advancement of the standard of living is universally accepted. It is not so generally recognized, however, that research may be equally well applied—and it is important that it should be applied—to all functional activities of a business.

It is a fundamental premise of this book that the application of organized, intelligent, continuing "research" to operating and management systems offers an almost unparalleled opportunity to improve the effectiveness of any large-scale organization—whether business or nonbusiness. This is held to be true for two reasons:

1. The fruits of systems research can be extraordinarily abundant and varied. They cannot be weighed solely in terms of clerical cost reduction, as is so commonly supposed. Their full measure must include the increased effectiveness of the entire organization that results from faster, more reliable, more purposeful service by the clerical group. The widespread application of systems produces equally widespread benefits from their systematic improvement.

2. The gains to be realized from upgrading operational and management systems still represent a vast, relatively unexploited opportunity in most organizations. In the industrial world, the application of scientific management techniques to factory operations has long been commonplace. What began over 60 years ago as elementary motion analysis at the workplace has expanded in most companies to a full-fledged program for continuous and systematic analysis of manufacturing processes, tools and machines, flow of work, layout, and material handling methods.

But the efforts to improve paperwork systems have not advanced nearly so far. To be sure, in recent years many organizations have launched formal systems and procedures programs. Most of these programs, however, have only begun to tap the profit potential that lies in the paperwork area. And in other companies that do not yet have formal improvement programs, procedural processes receive only sporadic attention, and that is ordinarily given only under the stimulus of computer or other office equipment salesmen, cost reduction drives, or a breakdown in the services rendered.

WHY SYSTEMS ANALYSIS PAYS OFF

If any organization is to capitalize fully on the foregoing opportunities, its key people must first have a keen awareness of the diverse, high-impact benefits that can be realized from a soundly conceived, skillfully executed program of systems development and improvement. These benefits can be grouped into the following end-result categories:

1. Significantly improving the organization's operational performance
2. Increasing top-management effectiveness
3. Cutting clerical and other paperwork costs
4. Compensating for clerical labor shortages
5. Capitalizing on the full potential of the computer
6. Developing teamwork among departments
7. Improving policy execution
8. Raising employee morale
9. Coping with complexity
10. Faciliating change

Improving Operational Performance

By far the most substantive contribution that a systems improvement program can make to the vitality of an organization is helping it accomplish its basic mission in a competitively superior way. In a manufacturing organization, for example, this contribution can take the form of cutting the direct costs of making or distributing the company's product, reducing inventories, improving service to customers, or increasing the effectiveness of the sales force and of engineering, research and development, or other technical personnel. Following are some specific ways in which systems analysis can produce these gains.

Cutting direct costs   Manufacturing cost reduction is normally considered the province of methods men, tool designers, and industrial engineers. But their work involves the physical aspects of handling and processing materials. They are concerned with reducing and simplifying worker motions; developing faster, more automatic, or larger-capacity machines; selecting better tools; simplifying product design; or substituting materials that require less processing time.

It must be remembered, however, that the availability of materials and tools at the right time, in the right quantities, and at the right place is governed by paperwork. Thus, in large measure, the effectiveness with which workers and machines are utilized is determined by the adequacy, accuracy, and usefulness of production and material control records and the forms supporting them.

Here are some of the ways in which development of better records and paperwork can reduce direct costs:

1. They can eliminate idle worker and machine time that results from poor scheduling or unavailability of materials.

2. They can reduce the need for overtime work by direct labor.

3. They can minimize investment in inventories and losses from obsolescence or shrinkage.

4. They can achieve tighter control over scrap and rework costs.

The experiences of the following companies testify to the size of the direct-cost penalty that lack of control imposes.

Specialty equipment company.   The production plans of a specialty equipment manufacturer were based on contract commitments calling for the completion and shipment of sixty-five units each month. Adherence to this schedule required full utilization of the available labor supply on a fully manned first shift and a partially manned second shift.

The manufacturer had trouble with fulfilling contracts because shortages of detail parts frequently interrupted assembly operations, particularly during the first half of each month. In some instances, inventory records showed that the required parts were available when actually none were in stock. In other instances, the supply of certain parts simply did not keep up with assembly demand.

Investigation revealed that the fundamental cause of this condition was lack of accurate and adequate control information of four types:

1. Lists of parts shortages caused by work spoilage, rejection of incoming shipments, and design changes. Without this information, parts replacement was haphazard and incomplete.

2. Records covering the status of casting orders issued to the foundry and showing quantities ordered, in process, and delivered to stock.

3. Records covering flow of work and performance in the machining department.

4. Stock records of detail parts. The discrepancy between stock records and physical inventories was caused by defects in the procedure for issuing parts from stock.

In spite of these handicaps, the company came close to meeting its commitments, but only at the penalty of a 25 percent increase in production costs. The reasons for the cost increment are evident in Fig. 1.

During the first half of each month, there was considerable idle time in the assembly department because of the depletion of parts stock. Workers had to move back and forth from one assembly job to another. Machines that were almost completed stood for days awaiting one or two missing parts. Toward the end of each month, all orderly processes of manufacturing control were abandoned under the pressure to make up lost time. Both the machining and assembly departments operated on a heavy overtime schedule. Parts were "borrowed" from machining lots to keep assembly operations going. Factory supervisors, shop clerks, and production control personnel dropped their normal functions to expedite the flow of parts.

The impact of this month-end scramble on the following month's production is obvious. The cycle simply repeated itself: stock depletion, idle time, and hand-to-mouth assembly, followed by overtime, high expediting costs, and exhaustion.

Defense systems hardware company. A company engaged in the development and manufacture of defense systems hardware had a large number of military equipment contracts. The company made several types of defense systems, it had more than one contract on each, and it subcontracted substantial parts of the work.

Because of technical advances in this rapidly changing field, some of the contracts were periodically canceled. Cancellation notices required that the company not only stop its own work on the contracts affected but also promptly notify subcontractors to stop their work on specific parts and subassemblies. Failure to act quickly involved a serious financial risk, for the company could

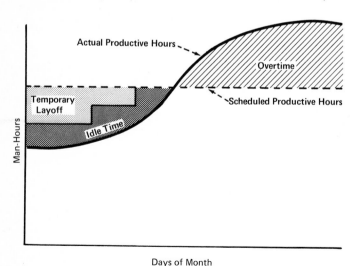

FIG. 1 *Inadequate control information caused idle time and overtime.*

not usually recover the cost of work done more than a few days after receipt of the cancellation notice. Yet, the company's material control and contract status records were so inadequate and poorly maintained that it took an average of 3 weeks to identify the subcontract parts affected, determine the quantities to be canceled, and issue instructions to suppliers.

Other systems-poor companies. Not all opportunities to cut direct costs by developing better control records are so spectacular as these, but there are few businesses where some opportunities do not exist. One company, for example, found that it was losing discounts on purchased materials because its receiving, incoming inspection, and invoice verification procedures had become so time-consuming. Another discovered that its outstanding purchase commitments were $600,000 in excess of requirements because of a faulty procedure for converting its sales forecasts and production plans into a schedule of material requirements. A third company found that tighter production control of the manufacturing cycle made it possible to cut in-process inventories by 20 percent.

Reducing inventories   In a large chemical company the systems staff was asked to determine whether inventory records on maintenance, repair, and operating supplies should be converted to a computer-based application.

In evaluating the purposes that might be served by this method of record keeping, the project team found that among the 37,000 items in stores there were a great many similar items that served the same purpose. It also found that the average turnover rate on stores investment had declined in the previous 3 years from 2.3 times to 1.4 times per year. In addition, over 20 percent of the items in inventory had not been active (that is, no withdrawals had been made) in the past 2 years. These discoveries led to development of the following organizational and procedural changes:

1. Inventory records were reorganized so that departmental responsibility for the size of the investment in each inventory class was clearly established.

2. Planned dollar levels were set for each inventory class.

3. An inventory standardization program was introduced to eliminate all unimportant or unnecessary variations in type, form, grade, size, shape, or composition of materials.

4. A continuing program was established for the detection and disposition of obsolete or surplus stocks.

Improving customer service   This study was carried out by a company in the fastener industry. The firm manufactures an extensive line of nuts, bolts, rivets, and similar items, most of which are not protected by patents. Moreover, a company in that industry cannot maintain a significant product quality or price advantage, nor can it develop a brand-name franchise. Hence, two major factors affecting competitive strength are the quality and speed of service that a firm provides for its customers.

Against this background, the amount of overtime work in the company's order, traffic, and shipping departments had increased sharply. The systems analysts who were assigned to the problem recognized at the outset that any reductions made in operating costs must not be achieved at the expense of service to customers. This meant that they would first have to determine the minimum standards of customer service that the company must provide to remain competitive.

In its initial inquiry, the study team found that over the past 2 years there had been a steady increase in (1) the average time necessary to process and ship a customer's order, and (2) the percentage of total orders that had to be filled by a series of partial shipments because some items were temporarily out of stock. Moreover, they found that a major reason for the increase in order filling costs was the addition of three order expediters who were needed to make up for the deficiencies caused by cumbersome procedures and by an organizational arrangement that involved six different departments in the order filling cycle.

In light of these findings, the study team was able to bring about fundamental changes in organizational structure and order processing and inventory control procedures. The new procedures (1) decreased the average time required to fill an order by 55 percent, (2) cut down the volume of back orders by 68 percent, and (3) reduced the volume of customer order status inquiries and complaints about delivery service by 82 percent.

Increasing the effectiveness of the sales force   In the face of an accelerating cost squeeze, a large insurance company undertook an extensive study to simplify paperwork in its field sales offices across the country.

The first function that the study task force attempted to eliminate led to widespread and anguished protests by the company's agents—that is, its front-line sales representatives. As a result, the task force backed off momentarily to think through more clearly what the real objective of its study ought to be.

In making this reassessment, the task force recognized that the clerical units under study had the basic mission of rendering service to two groups: agents and policyholders. From this recognition evolved the realization that before the task force could make worthwhile and enduring changes in the character of the work being performed, it had to understand the real marketing significance of the services being rendered to these two groups. Specifically:

1. What was the relative importance of each type of service in increasing the selling effectiveness of the company's agents? What types were most productive? What types were least productive and might be considered expendable? What wholly new services might be added to increase the productivity of the sales force beyond the cost of such services?

2. Similarly, what were the real nature and significance of services to policyholders? That is, what effect did each type of service have on their receptivity to purchasing additional insurance or to keeping their current insurance in force?

From the fieldwork undertaken to develop answers to these questions, the company was able, for the first time, to reach fact-founded decisions on which parts of the service were profitable and which were not. As a result, it eliminated services that had a relatively low payout and added some new ones that materially helped agents without greatly increasing costs. The net effect was that the company substantially reduced the cost of running its field offices and, at the same time, maintained a high level of morale among its agents.

Summary These several case studies show how great an opportunity the systems analyst has to improve the operational performance of a *business* enterprise. But these opportunities are by no means limited to business. They can be found in every kind of organization and in every aspect of management—whether it be the operating costs of a hospital, facilities and manpower utilization of a major city government, research project planning and control within the American Cancer Society, or program evaluation and resource allocation in the British Post Office.

Increasing Top-Management Effectiveness

Some ways in which systems analysis and improvement programs can help increase top-management effectiveness were suggested in Chap. 1. They include improved implementation of the organization's top-level policies as well as substantially better decision making and control in such areas as strategic and operational plans, and capital or other resource commitments.

In addition, skillfully designed systems and procedures contribute in still another way to top-management effectiveness: *They conserve management time.* For example, soundly conceived control procedures on matters requiring top-management surveillance permit the chief executive to delegate decision-making authority to subordinates. Thus, bottlenecks are eliminated and action facilitated, while adequate control is retained by the delegant to ensure that performance is satisfactory. As a case in point, a multinational manufacturer and marketer of industrial machinery recently established a corporate information system that enables top management to control its worldwide operations effectively while delegating major responsibilities to regional management.

As Holden, Fish, and Smith have pointed out in their classic study:

> Each time a well-planned control procedure is substituted for personal approvals, additional time is made available to the top executives for the far more important overall planning and control of the enterprise, which is aimed at securing maximum effectiveness of the company as a whole, rather than control of any individual problem.[1]

---

[1]Paul E. Holden, Lounsbury S. Fish, and Hubert L. Smith, *Top-Management Organization and Control*, Stanford University Press, Stanford, Calif., 1948, p. 78. (Reprinted with the permission of the authors and of the publishers.)

Incidentally, departmental executives and supervisors can utilize the same techniques to save their own managerial time. As patterns are established for carrying on repetitive tasks, the need for close personal direction of subordinates is reduced. The simpler the pattern, the more effective the operating executive becomes.

Cutting Clerical and Other Paperwork Costs

Of all the persons who are gainfully employed today, one of the largest occupational groups neither produces nor sells the products or services of its employers. This is the army of administrative labor performing the paperwork of business and government—an army made up of clerks, stenographers, and keypunch operators; accountants, computer operators, and statisticians; schedulers, expediters, checkers, messengers, and so on.

Back in the 1870s, which mark the beginning of this country's shift from an agricultural to an industrial economy, about 80,000 workers were engaged in clerical occupations. In the century since then, perhaps one of the most significant changes in the socioeconomic distribution of workers has been the constant and rapid increase in the ratio of clerical employees to the total work force. While the number of gainfully employed civilians was increasing about 6½ times, the clerical group multiplied over 175 times.

The 80,000 clerical employees in the 1870s represented less than 1 percent of the nation's workers. By the turn of the century, the percentage had risen to 2.5, by 1910 to 4.6, by 1930 to 8.3, and by 1950 to 11.6. Nor does this phenomenal growth of clerical personnel—both in numbers and in ratio to production labor—show any signs of slowing down. Figure 2 contrasts the relatively modest growth over the past 20 years in the economy's population of "money makers" (people who make a product) with the sharp increase in "money spenders" (people who, for the most part, generate and process paperwork).

The group of 14 million people engaged in clerical occupations is difficult to visualize, particularly as a segment of all those gainfully employed and not as a segment of our total population. For that reason, let us think of the relative importance of this occupational sector in these terms: 14 million people are just about 8 times the number of persons of all occupations working in the nation's food-processing industries; they are over 15 times the total work force in the textile industry; they are over 5 times those employed in banking and insurance; they are almost 3 times the total employed in transportation and communications; and they are over 125 percent of all the workers in our metal manufacturing and fabrication, machinery, automotive, and aircraft industries combined.

Because clerical labor is such a large and growing part of the total cost of operating any enterprise, all sorts of approaches have been taken on many fronts to reduce—or at least slow the growth of—this continuously escalating

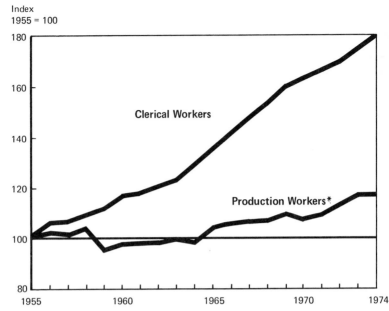

Index
1955 = 100

* Includes blue-collar (factory) workers, service workers and farm workers.

FIG. 2 *Clerical workers account for an ever-increasing proportion of the total labor force in the United States.*

element of expense. But despite these efforts, adequate understanding still does not exist of the magnitude of the potential saving, the diverse ways in which it can be obtained, and its full implication for profit building. Here are some of the facts that must be spotlighted if anything approaching the full, realizable saving is to be achieved:

1. The size of the potential clerical labor saving through systems analysis is enormous. During a sharp cyclical decline in business activity, one industrial equipment manufacturer with annual sales of about $45 million experienced an abrupt drop in profits as the effect of overcapacity in the industry began to make itself felt. At that point, the company undertook a fundamental, profit-conscious attack on its overhead costs, 80 percent of which were accounted for by the salaries of administrative, technical, and indirect factory personnel. It was not a disorderly, panic-driven effort, but a thoughtful, businesslike evaluation of what activities were clearly essential or profitable and how many people were needed in each of them to achieve the best balance between results produced and the cost of getting them. Eighteen months after the study was begun, the company had cut its manpower in these overhead areas from 700 to 450 employees, a reduction of over 35 percent. This saving alone contributed almost $2,250,000 to profits before taxes—which is more than the company had earned in either of the two preceding years. Today, the company's volume has risen from $45 million to $60 million, but it is still operating with an overhead complement of fewer than 500 persons.

Another company, an electrical equipment and supplies manufacturer, was confronted 2 years ago with the necessity of adding substantially to its office space. The firm had been growing steadily for several years, and the acquisition of more space for its administrative personnel seemed a normal and inevitable part of the total growth pattern. But when the board of directors was asked to approve a capital expenditure of roughly $1 million for a new wing to the office building, it demurred and asked that a careful study first be made "to determine whether existing facilities and personnel are being employed to the best advantage." One of several projects that grew out of this request was an intensive review of the company's order filling activities, including sales order processing, credit, invoicing, accounts receivable, and finished-goods inventory record keeping. As a result of the study, the company was able, over a year's time, to reduce the personnel engaged in these activities from 260 to 155, or by just about 40 percent.

A corollary point that should be made here is that, in many make-and-sell companies, opportunities for labor saving through work elimination are greater in the clerical field than in any other part of the business. Work performed by *direct* labor is specified, limited, and purposeful. What is to be done and where it is to be done are controlled by manufacturing orders or operation sheets. The way it is to be done and even the time required are controlled by setup instructions, by jigs and fixtures, by automatic machines, and by standard time allowances.

Such preciseness as this does not exist in clerical operations, and no one should have any contrary illusions. The determination of what clerical tasks are to be performed is the product of a hundred minds—executives, supervisors, and operators—all with widely varying sensitivities to value-cost relationships. Within most organizations of any size, requests for information, desire for protection, ignorance of what is done elsewhere, worker preference, and extemporizing to meet changing conditions—all work to produce a monumental structure of uncoordinated activities, forms, records, and reports. And concealed within this complex structure, duplication, excessive control, and unnecessary operations go unnoticed. Nor does this process of accretion ever stop if left unattended. It continually feeds on itself as its very complexity produces the need for more and more control.

2. *Some companies do a vastly better job than others in reducing and controlling clerical and other overhead costs.* As a concrete example, Fig. 3 shows the composite performance of four life insurance companies that are winning the expense battle and four others that appear to be losing it. The lesson to be learned from this graphic representation is that there are always some companies that manage to outperform others in their industry, whether the general trends of the industry and the forces affecting it are favorable or unfavorable. But the sort of superior expense performance illustrated in Fig. 3 never just *happens*. It

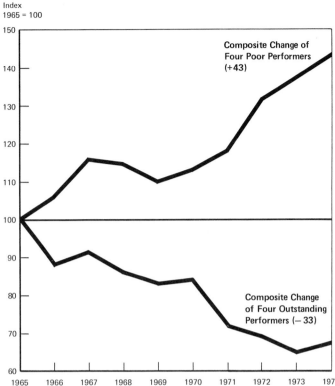

Index
1965 = 100

Composite Change of
Four Poor Performers
(+43)

Composite Change
of Four Outstanding
Performers (− 33)

FIG. 3. *One group of life insurance companies proved to be vastly more effective than another group in controlling its general and administrative expenses as a percent of premium income.*

inevitably results from an intelligent recognition of need and opportunity and a systematic, aggressive, and skillful effort toward substantial, continuing cost reduction.

3. *Savings in the clerical field are not limited to a reduction of the work force.* Systems studies can also produce substantial—though perhaps not spectacular—savings by eliminating the chronic need for overtime work, by reducing equipment or space requirements or increasing the utilization of these resources, by cutting communications costs or other nonpeople expenses, by making possible the use of lower grades of labor, or by significantly reducing errors through the gearing of paperwork routines to the skills of the available clerical labor supply.

4. *Clerical cost savings are a direct contribution to company profit.* Too often managers take a passive attitude toward reduction of paperwork costs because they unconsciously tend to measure savings against gross sales instead of against net profit. A $10,000 reduction in office expense causes little excitement when it is compared with annual sales of $50 million. But what is frequently overlooked is that this saving represents a 100 percent contribution to a company's

profits, whereas only a very small part of each sales dollar ever reaches the profit point. In a company that earns 4 percent on its sales dollar, the elimination of just one $8,000-a-year clerical position contributes as much to net profits as sales totaling $200,000. Moreover—and this, too, is often overlooked—the saving in operating expense automatically repeats itself year after year, whereas the $200,000 worth of finished goods must be resold each year.

5. *The economics of overhead expense produces a compelling incentive to dig for these savings.* In most companies, administrative expenses are many times greater than net profits. Because such expenses consist mostly of salaries, they are relatively insensitive to fluctuations in business volume. For this reason, they constitute a continuous and, in many companies, a serious threat to profits. Even in the face of a moderate decline in sales, their inflexibility can increase unit costs to a point where profits are seriously eroded. Certainly much of the responsibility for this jeopardy to profits must be borne by clerical costs, since they constitute a major part of administrative expenses.

Thus necessity is added to opportunity. A positive approach to profit protection requires the reduction and control of overhead costs that can be achieved through continuing systems analyses.

In recent years, these facts of business life have attracted an increasing amount of top-management attention. In his company's annual report to stockholders, the chairman of one large corporation stated, "The company's earnings improved in 1968 despite a dip in sales volume. . . . The increase over 1967 reflects our aggressive companywide program to reduce general, administrative, and distribution expenses."

In still another company, the president, in inaugurating a profit improvement program, pointed out; "A reduction of 15 percent in our clerical, EDP, and related costs [a goal that is certainly not unrealistic] would be equivalent to the total annual dividend paid in recent years to our stockholders."

6. *The very steps that cut clerical costs will also produce a better clerical product.* If a company reduces its advertising appropriation or sales promotional expense, it normally expects a lower volume of sales. To be sure, the planned reduction in expense may be justified on the ground that it will have a relatively small effect on volume. But some decline in sales performance is usually anticipated.

Quite the reverse is true of clerical expense. By reducing the cost of getting the clerical job done, a company is very likely to realize improved clerical performance. The work simplification measures that produce clerical savings also eliminate the most prolific source of clerical error: complexity. The volume of error is further decreased by providing adequate checking points in the streamlined procedure.

But improvement in performance is not limited to higher quality of work. As needless operations and backtracking are cut out and work scheduling improves, backlogs vanish and the whole procedural cycle is accelerated.

These two factors—greater accuracy and faster execution—play a leading role in the achievement of most of the other benefits enumerated in this chapter.

Compensating for Clerical Labor Shortages

A few years ago, one large military equipment manufacturer grew so rapidly that in a relatively short time it became the largest employer in its geographic area. As a result it almost wholly exhausted the supply of clerical labor in the two communities where its plants were located.

Progressive managements are not content to regard such circumstances as an excuse for high clerical costs or low productivity. Instead, they consider it all the more imperative that systems and procedures be streamlined to reduce personnel requirements in numbers as well as in the level of skill and experience necessary to do the job.

Capitalizing on the Full Potential of the Computer

No development in the past 25 years has had so great an impact on management as the advent and widespread—indeed, almost universal—application of the computer. In the 17-year period 1958 to 1975, for example, the number of general-purpose computers installed in the United States grew from about 3,800 to over 72,000. As a major new tool of management, in business and the public sector alike, the computer has clearly become a key element of operating expense. In fact, the total cost represented by the 72,000 general-purpose computers in use in 1975 (including hardware, operating personnel, software, and other directly associated costs) is estimated at almost $33 billion, or just about 30 percent more than the amount spent by United States industry in 1975 on its total research and development work.

But all costs, of course, are relative. And, for that reason, the critical question is whether these mounting computer expenditures are clearly being matched by comparable economic returns.

Partly in answer to this question, a McKinsey research study made a few years ago came to this conclusion: "In terms of technical achievement, the computer revolution in U.S. business is outrunning expectations. In terms of economic payoff on new applications, it is rapidly losing momentum."[2] One of the basic findings leading to this conclusion was that a majority of enterprises had not yet learned how to use the computer and related tools of management so skillfully that they had become an integral part of the way in which the enterprise was being run.

On this score, David Hertz states:

> The real reason why the computer *must* be used—and used effectively—is as simple as this: if *anyone* uses the computer effectively, then everyone will have to use it or fall behind. In a way, the computer is very much like a new and more efficient production tool. If no one has it, then no one needs it. But if one company in an

[2]*Unlocking the Computer's Profit Potential—A Research Report to Management,* McKinsey & Company, Inc., New York, 1968.

industry gets it, then all the rest are at a disadvantage until they follow suit—even though the cost of the new tool may be very steep in terms of investment, training, manpower, or management education.[3]

Yet, in spite of the urgency of the need, our experience has been that many of the largest enterprises in the world have still not come close to capitalizing on the real profit potential of these new management tools. True, companies are investing in faster, costlier, more sophisticated hardware; they are continuously enlarging the size of their computer staffs; and they are developing increasingly more ingenious and elegant computer programs. But with all these ostensible advances, and the burgeoning costs they entail, the flow of profitable end results is still only a trickle.

As one of the underlying causes of this disappointing result, it must be recognized that no development in business history has so exhausted the lexicon of catchword superlatives as the computer has. For years, and from all sides, we have been lured by the promise implicit in such terms as "the electronic frontier," "instant information," "push-button management," "automatic decision making," and "the management revolution." And, from time to time, we have been given further assurances that the computer will have "a profound impact on management effectiveness"; that by failure to use it, a company "condemns itself to competitive catastrophe"; that the computer "breaks through the barrier of human limitations"; and that, therefore, it may well "control your business future and the future of your company."

There is little doubt whatever that magic words such as these have produced a sort of electronic hypnosis among many managements and instilled in them the blind faith that they can keep pace with competition simply by importunate spending on computer hardware and software systems and by entrusting the company's use of these tools almost entirely to computer specialists.

Fortunately, this sort of witless abdication did not get many organizations into deep trouble so long as their computer development activities were focused on converting routine accounting and administrative systems. But the design and installation of such applications have now been completed in virtually all organizations.

In the face of this fact, the critical questions that must now be answered are: "What next? Where do we go from here in the use of these new management tools?" These questions simply cannot be left for the EDP specialist to answer. The reason is that the technical capability of the hardware can no longer be controlling. From now on, the competitive leaders in the use of these tools will be those who have the wisdom, the business judgment, and the combination of skills needed to build their computer applications around (1) the operational and management decision-making needs, and (2) the economic characteristics of their particular business or form of endeavor. Achieving this goal requires systems design talent of the highest order—talent that includes not only in-

[3]David B. Hertz, *New Power for Management,* McGraw-Hill, New York, 1969, p. 180.

depth knowledge of what the computer can do, but more important, a sophisticated understanding of the operations of the enterprise and the key factors critical to its success.

Developing Teamwork Among Departments

The specialization of function, which has been the natural consequence of growth in business size, has tremendously increased the burden of coordination. Urwick attempted a measure of this burden when he stated, "It seems incontestable that the most urgent need in modern . . . business is a rapid evolution of new machinery and methods designed to improve coordination."[4] Despite departmentalization, the interaction and interdependence of the organizational units making up a business require that there be unity of the whole if the company is to achieve its objectives promptly and economically—or indeed, if it is to achieve them at all.

It is not enough to rely on an organizational manual to supply the mechanics or framework of coordination. Organizational definitions of responsibility delineate the boundaries of departmental function. Systems and procedures, however, tie departments together in an active, living relationship. They specify how departments serve or are served by each other in performing the recurring work of the enterprise. Complete, soundly developed systems will do much to minimize the uncertainty and the areas of controversy that lead to "buck passing" on the one hand and to conflict, ill will, and ungainful competition on the other.

Aside from the contribution that established systems make to the development of teamwork, the very process of formulating or of studying and improving systems accomplishes a similar purpose. Through interchange of ideas and information, executives become more familiar with each other's activities and acquire a more sympathetic comprehension of each other's problems.

Improving Policy Execution

Chapter 1 mentioned that systems and procedures are among the principal means by which policies are implemented. How effectively policies are administered will be determined, to a large extent, by the thought and skill applied to developing mechanics for their execution. Credit policies cannot be carried out satisfactorily unless records are adequate to do the job. The seniority clause in a union contract will be a fertile source of grievances unless the procedures for handling layoffs and transfers are carefully tied in with it. A specialty store's policy of always offering a wide selection of fashion merchandise cannot be adhered to without serious inventory risk unless the clerical organization is geared to furnish buyers with up-to-the-minute sales statistics by item, style, color, and size.

[4]L. Urwick, *The Elements of Administration,* Harper & Brothers, New York, 1944, p. 74.

Raising Employee Morale

Among the basic ingredients of sustained employee morale are understanding of one's work, interest in it, pride of accomplishment, and confidence in the leadership of the group. Few workers can long maintain their interest and loyalty if they do not feel that their work is necessary and fruitful—that they are making a perceptible and worthwhile contribution to the affairs of the organization. Moreover, they want to perform their work in an atmosphere of calm efficiency. Although occasional rush jobs may help develop team spirit, continuously hectic operations and recurring crises quickly induce lunchroom speculation about "whether the boss knows what he's doing." There just cannot be much job enthusiasm or confidence among an embattled force of clerks struggling with red tape, antiquated methods, or complexity that defies comprehension.

Yet many organizations seem to rely on other forces to sustain employee interest—on the romance of the industry, on dynamic growth, on size or publicity or leadership in the public mind. Although these factors are unquestionably attractive to new employees, their influence soon diminishes as the newcomers begin to associate themselves more with their own jobs and their immediate environment than with the enterprise as a whole.

Here, then, is one of the most significant intangible benefits of an organized systems improvement program. By developing simple, effective, purposeful clerical routines, it eliminates confusion and promotes understanding. It ensures that all the work being performed is necessary and useful. It removes backlog pressures, error hazards, and frequent overtime work. It eliminates the necessity for constant interruptions of normal routine by special expediters trying to pry something loose or to find something that is lost. It substitutes orderliness and dispatch for disorder and delay. It fosters tidy thinking and work habits. It facilitates clerical job classification and, therefore, makes possible the development of a more equitable compensation plan.

Nor is the program's effect on employee morale confined to clerical workers. A large manufacturing company having three plants within a 15-mile radius frequently transferred employees within each plant and between plants. Weekly pay envelopes of factory workers were delivered by pay wagons to each department and distributed to the workers. But the procedure for transferring payroll and personnel records was incredibly slow, with the result that transferred employees were seldom paid in the normal manner until they had been at their new location for several weeks. If the pay wagon contained no envelope for a worker, he or she had to get the supervisor's permission to leave the department, go to the cashier's cage, prove that payment had not been made, and collect an advance against wages from petty cash funds.

It is not hard to imagine the cumulative impact that this condition had on morale throughout the factory. Workers mentally refuse to ally themselves with a management for which they have no respect. And the measure of their respect depends on such small daily occurrences that affect them directly and personally.

Coping with Complexity

In all probability, no management task ever undertaken is so immensely complex as that of running the U.S. Department of Defense (DOD).

This is true, first, because of the sheer size of the military establishment. No other organization in the free world has so many millions of employees, so many billions of dollars to spend, or so gravely important a mission to fulfill. Almost certainly, an organization as vast as this cannot be comprehended as a whole.

But size alone is not the only determinant; to it must be added a number of other causes of management complexity. These include:

1. An almost unbelievable diversity of activities, varying all the way from operating a system of supply depots, test ranges, SAC wings, and military academies to negotiating and monitoring the execution of a research and development project for an advanced supersonic bomber or the Trident submarine.

2. Long lead times on the development and production of defense hardware systems.

3. Rapidly changing technology, particularly in the field of aerospace and ballistic missiles.

4. The intricate interrelationships of the basic military missions for different weapon systems—a condition that calls for a continuum of large-scale tradeoff analyses and decisions.

5. The unusual mobility of Department of Defense personnel. Military officers and civilian managers of the Department move rapidly from job to job. While perhaps desirable from a career point of view, extensive mobility almost certainly diffuses the continuity of professional management in the Department.

Because of this combination of factors, the United States military establishment is very much a multidirected organization. No one individual, however skillful, can make more than a tiny fraction of the important decisions that face the Department every day. Yet somehow the organization must go forward toward its goals with coherent direction and with a sound balance of effort between short- and long-term needs and programs.

One of the principal steps taken by the Department to provide a clear sense of direction in the face of unparalleled complexity has been the development of the so-called planning, programming, and budgeting system (PPB). Among the many contributions this system has made, two are of special significance:

1. PPB has extended DOD's planning horizons. Recognizing that military preparedness requires a long lead time for both equipment and trained manpower, the system calls for the annual development of so-called force plans,

which extend 8 years into the future, while their financial impact is estimated 5 years ahead.

2. In addition, PPB requires that each proposal be critically analyzed, not just in terms of the resources needed, but also in the light of alternative courses of action and the mission to be accomplished. The focusing of decision making on programs and program elements, organized by mission, has resulted in the integrated analysis of related issues and thus has reduced the danger of making suboptimal decisions. Stated another way, the system forces decision makers to look at the *balance* among the elements making up the nation's defense capability.

Although the DOD is perhaps the most dramatic example that might be cited, we should not assume, of course, that the challenges it faces are peculiar to management in the governmental arena. Complexity has become a dominant force in business management also as companies have grown in size, diversified into new and unrelated industries, extended the multinational scope of their operations, and encountered increased regulatory constraints.

Because of these conditions, the executive team has had to grapple with more products to be managed, an increased number of markets with different requirements for success, new geography and cultures, and exploding technological change. These developments, in turn, have greatly increased the dependence of management—and particularly of top-management executives—on more formal communication, information, and decision-making systems as a means of institutionalizing the management processes that, in less complex times, were adequately served by the personal abilities, informal contacts, and intuitive judgments of a few key people.

Facilitating Change

Continuous adaptation of its operational and management systems is one of the means by which a dynamic organization keeps in step with changing conditions.

As one example, in response to growing consumerism, the automobile manufacturing companies have developed warranty and defect recall systems that keep track of each individual car from the time it is entered into the production schedule until the warranty runs out several years—and often several owners—later.

Another example of adaptive change in operational systems can be found in the commercial banking industry, which, in recent years, has experienced a sharp change in the proportions of its "free money" supply (demand deposits) that have come from two major sources: businesses and individual checking account customers. Specifically, during the period from 1959 through 1974, demand deposits from households increased 143 percent while those from all United States nonfinancial businesses (corporate and noncorporate) increased by only 16 percent (Fig. 4). This change has required the typical commercial bank to go far beyond its traditional corporate lending function and develop the sort of consumer-marketing-management systems that would ena-

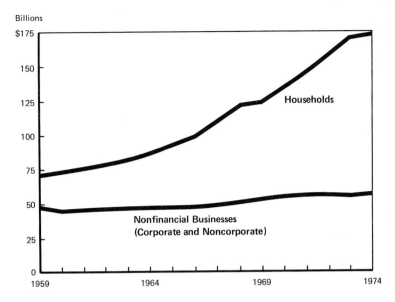

Source: Flow of Funds Accounts 1959-1974, Board of Governors of the Federal Reserve System.

FIG. 4 *Individual depositors account for an increasing proportion of the funds acquired by commercial banks.*

ble it to obtain its full share of the growing "raw material" supply provided by individual depositors.

As a final example, accelerating technological change and shortening product life cycles have required the development of much more effective approaches to strategic analysis and decision making and to management of the new-product-development-and-introduction process.

Other types of changes make the need for a thoroughgoing overhaul of operational and management systems equally evident. Company operations are decentralized, departments performing related functions are combined, mergers take place, new companies are acquired and others divested. Changes also take place, at home and abroad, in the character of competition, in markets served, in consumer behavior and patterns of demand. Mounting labor costs, severe price competition, or growth of overcapacity in an industry require development of tighter financial control systems. New legislation or regulatory decisions create new record-keeping and reporting requirements. Changes in the educational and skill levels of the available clerical labor force increase the need for streamlined, error-resistant procedures. Shorter decision lead times mean that the flow of management information must be accelerated. At the same time, the enormous growth in our capacity to generate information requires a high order of systems design skill to keep the organization from being glutted with data.

Too often these and other kinds of changes provide an excuse rather than a reason for complexity, duplication, or breakdown of operational and management systems. A conscious, continuing effort to adapt systems to new requirements will greatly facilitate change; without it, change will be grievously impeded.

SUMMARY

To supplement the statement of Alfred Sloan which appears at the beginning of this chapter, we might regard the scientific or technical research and development that is carried on by most modern businesses as a type of insurance policy—insurance against product obsolescence—and the cost of doing that work as the premium paid on the policy. Similarly, organized "research" work on operational and management systems is a kind of insurance policy that can be written against other competitive factors that loom large on the horizon—the competitive factors of organizational effectiveness and efficiency.

Few enterprises—whether in the public or the private sector—can afford *not* to carry this type of insurance, for there is probably not a single system or procedure in any organization anywhere that cannot be performed more simply, more effectively, and more economically. The savings and other improvements realized must, of course, pay for the effort expended in securing them. But modern-day enterprise is still far from reaching that point of diminishing returns. The pay dirt for profit building through systems and procedures improvement is a long, long way from being exhausted.

PART TWO

The Ingredients of a Successful Systems Improvement Program

THREE

Barriers to Systems Improvement

The first ingredient of a dynamic, profit-oriented systems improvement program is clear-cut recognition of the reasons why so many such programs fail. The word "fail" is not used here to imply that a large number of formal systems and procedures programs die out or are abandoned. In fact—once inaugurated—most of them are carried along indefinitely. But, despite this fact, the accomplishment record and the operating climate of these staff groups all too often provoke questions such as these: Why are so many systems and procedures staffs bogged down in the minutiae of forms control, procedures patchwork, or computer reprogramming—without ever producing significant, measurable savings or helping to improve the management process in their companies? Why do so many conscientious but frustrated systems and procedures staff members chronically complain about lack of top-management support and operating-management cooperation? What are the real barriers to achieving all the potential benefits of systems analysis that were cited in the preceding chapter? Why are these improvement opportunities still awaiting their fullest exploitation in most companies?

The starting point in developing realistic answers to such questions lies in recognizing that the need, in this instance, is not merely one of overcoming normal resistance to change. The problem is more subtle and complex than that. Specifically, it flows from a number of basic conditions and attitudes peculiar to the office area which make systems and procedures improvement work a good bit more difficult than staff work in the factory. These conditions and attitudes are:

1. Lack of top-management understanding
2. Lack of a strong, continuing improvement orientation among most departmental managers
3. The functional compartmentation of most organizations
4. Lack of acceptance of the systems staff by line managers and supervisors
5. Failure to tackle systems improvement projects in a broad-gauged, fundamental way

Lack of Top-Management Understanding

Top executives seldom fully appreciate the role of systems in the management process, the obstacles that stand in the way of systems improvement, and the price that must be paid to make the improvement program succeed.

This lack of appreciation is certainly understandable, for the line of progression to top management is generally through key positions in the sales, production, or financial branches, or, in nonbusiness organizations, through comparable operating functions. Rarely, if ever, do top executives come up through the channel of paperwork or office management. Another factor conditioning top-management attitudes is that the most dynamic growth in business history has traditionally been associated with entrepreneurial moves and shrewd, intuitive decision making. As a result, the important contribution that better administrative management can make to the success of an enterprise has not yet fully impressed itself on top management's consciousness.

But whatever the cause, the result is the same. In most instances top management has not done the things it must do—and that only it can do—to ensure the success of the systems improvement program.

Lack of a Strong, Continuing Improvement Orientation

It is unfortunately true that many well-established organizations in both the public and the private sectors are much more concerned with *maintaining* operations—with getting the job done, turning out the work, handling problems, and keeping the show going—than with *changing, upgrading,* and *improving* operations. Most managers tend to think in terms of administering their routine activities—filling orders, handling personnel problems, answering mail, and the like. Few are inclined to accept the idea that substantially changing and improving operations is an obligatory part of their jobs, ranking as high as, if not higher than, day-to-day management.

Even forms typically used for appraising managers' and supervisors' performance tend to focus on innate personal qualities, on how skillfully they use management tools, and on how well they have kept things going—rather than on what they have achieved in the way of significant improvement. These forms cover such attributes as cooperativeness and ability to get along, flexibility, human relations skills, and similar qualities. But one has to look hard to find passing reference, if any, to what the individual has done to upgrade, not just sustain, the operation he or she has been running.

One of the identifying characteristics of an organization that lacks strong improvement orientation is that its members have too much reverence for tradition and authority, and not enough for knowledge, concept, and principle. They are personality-oriented, instead of problem-oriented. They tend to be more concerned with *who* is right that with *what* is right. But the most telling characteristic of such an organization is that its value system and its approach to decision making are dominated by experience and preconception rather than by analysis and experimentation.

That this condition should be such a widespread organizational affliction was well explained by Everett Smith in *The Wall Street Journal:*

> Men who began their careers with an organization decades ago, and who have focused their skills and experience ever since on immediate functional and operating needs, cannot usually be expected to be *au courant* with changes taking place outside their sphere of interest and expertise, however relevant these might be to the future. That explains why so few industries seem capable of major innovations from within. . . .
>
> Moreover, the gap between the demands of change and the ability of organizations to respond to change is growing. The reason is simple enough: the more complex organizations become, the more they have to depend on established patterns of behavior to maintain stability. To populate organizations with people who have mastered the requisite skills requires some orderly basis of tenure and promotion. Yet these very systems of order and stability reduce the organization's ability to identify and deal with change.[1]

The painful relevance of this fact of organizational life can be found in the experience of a major company that for many years had operated successfully under conditions of great stability, homogeneity, and predictability. During this time, it had developed and codified its policies, practices, operating procedures, and approaches to decision making in minute detail. In such an environment, the accolade, "He is a good company man," or "She is an outstanding manager," meant—for all practical purposes—that the person so identified was one who had learned "the system" thoroughly and administered it faithfully.

But within a decade, vast changes occurred in the external forces affecting this company—changes that made many of its approaches to running the business completely obsolete. In face of these changes, the company was slow to react for the basic reason that its management people had been trained as administrators, not *problem solvers*. Confronted with uniqueness—with conditions that the company had never before experienced—most of them simply lacked the ability to analyze and innovate at anywhere close to the level of need.

This inability to meet change can, of course, affect all aspects of an organization's operations. But, to bring the point into the focus that concerns us, we

---

[1]E. Everett Smith, "The Goldberg Dilemma: Directorships," *The Wall Street Journal,* Feb. 7, 1973, p. 16.

need to ask: What are the special factors or causes that act as inhibitors to the conduct of a strong, continuing improvement program in the *systems* arena? In our experience, these barriers include the following:

1. The hidden character of systems improvement opportunities

2. Lack of hard-headed profit consciousness among most of the managers of functional segments of an organization

3. The desire for protection among supervisors of paperwork activities.

4. The illusion that administrative and other paperwork costs are controlled through budgets

The hidden character of systems improvement opportunities   Much of the waste and inefficiency in the office activities of modern-day enterprise is hidden. It is seldom as perceptible or dramatic as the scrap pile or a shut-down machine. It is not evident from casual observation or from traditional accounting information.

As a result, there is widespread failure—at all levels of management—to appreciate how much pay dirt lies in this area. Most managements would be shocked at the suggestion that reductions of 25 to 30 or even 40 percent could safely be made in the cost of their paperwork activities. Hence, they have generally been content to achieve marginal gains in these areas and to leave things much as they are, as long as they seem to be operating smoothly. Yet, in business enterprises, we have seen the cost of operating purchasing departments range from $7 to over $40 per purchase order issued; the cost of the payroll function range from $1 to almost $5 per payroll check prepared; and the cost of a single-item billing and collection operation vary from $1.25 to $6.50 per collection. In all these cases, the high-cost companies have no way of knowing that they are high-cost. The plain fact is that there are no cost standards for comparison and very few other guidelines to indicate how many people are actually needed to run an office activity. And without some sort of revealing cost comparison, managements are in the dark on two counts: first, they have no reliable yardstick of worker productivity; second, they seldom realize how many of the activities being performed in these paperwork areas can safely be eliminated.

The same is true of other kinds of potential improvements in operational and management systems—whether they be in the sophistication of computer applications, in the optimization of cash flow, in the quality of information for management decision making, or in any other benefit-producing area. Almost invariably, management's limited knowledge of what has been achieved elsewhere blinds it to the size of the opportunity that exists in its own operations.

Lack of profit consciousness   A direct outgrowth of the condition discussed above is that supervisors of office activities do not often have a highly developed profit consciousness. This does not necessarily mean that they are poor managers; they may be doing a conscientious, highly competent job. At the same time, they tend—as we all do—to concentrate on those parts of their job on which they are most frequently checked or judged. And because there are so few cost standards for measuring efficiency or optimizing performance in paperwork functions, the persons who supervise those areas are most frequently judged by the quality and speed of the service they render. It is not uncommon, therefore, to find departments overstaffed so that they can handle rush jobs or work peaks as promptly as possible without interrupting the regular routine. Since the supervisors usually cannot control these peaks and special demands, their natural inclination is to be prepared for every contingency. Thus, consciously or unconsciously, they will often resist any effort that might impair their flexibility for rendering this kind of service.

In addition, supervisors of paperwork activities typically do not have a highly developed sense of value-cost relationships. The more conscientious they are, the more they tend to be perfectionists and to regard as fundamental the activities that happen to be their own peculiar daily concern. As a result, they usually discount—often vehemently—the likelihood that what they are doing does not need to be quite as thorough and precise as they have tried to make it.

But this lack of profit consciousness is not limited, by any means, to down-the-line supervisors or managers. It is just as likely to exist among higher-level executives heading up major functions of the business.

An example can be found in the experience of a company that some time ago launched a 3-year "administrative improvement program" aimed at both cutting costs and strengthening procedures in virtually all of the company's major systems areas. Recently, the head of the administrative improvement staff met with his superior, the company president, to review progress on the program. Although the two men agreed that some real accomplishments had been made, they both felt that, on the whole, they were a good bit behind schedule and that parts of the program were dragging badly. In trying to analyze the reasons for the generally unsatisfactory progress, the two turned their attention to the sort of support that the whole program had been receiving from key people in the organization. In this connection, the staff head said to the president:

> As I see it, there are seventeen executives in our company on whom the success of this program depends. They are the five operating executives reporting directly to you and the twelve major executives at the next lower level, who report to some of the first five. Now, in order to get a true picture of how much support our program is getting, let's try to rate these seventeen executives in two different ways—first, according to how interested they are in making improvements and, next, how interested they are in cashing in on them. In other words, how much attention have they, as individuals, given to straightening out their plan of organization, streamlining their operating systems, eliminating work, and adopting better methods and equipment? And, how interested is each one in cashing in on these improvements—that is, in converting statistical savings into realized savings?

The two men agreed that within each of the two categories they would rate each executive as being either strongly interested, moderately interested, or little interested. As a result of this process they found that, in all honesty, they could rate only two of the seventeen executives as being strongly interested both in making improvements and in cashing in on them. In the face of this finding, the president said: "I guess this means that I haven't done my part of the job. The real job for me now is to see that the other fifteen get the same sort of 'religion' that the top two have got."

The desire for protection   Closely related to lack of profit consciousness is the desire for protection, which inflates clerical costs by building up excessive controls and increasing the work force.

Once an organizational unit has been accused of delaying an activity, making a mistake, or losing a document, it has a compelling incentive to create records that will either prevent a repetition of the error or prove the group's innocence if the error does recur. It is an unusual organization that does not have a host of registers, double checks, extra copies, follow-up records, and additional reports exclusively for this reason. Yet the cost of these controls is often many times greater than the cost of the errors they protect against. Their value for blame fixing is, of course, even more questionable.

A second form of desire for protection might be labeled the "protection of numbers." In our competitive economy, the number of one's subordinates has somehow become a symbol of importance among managerial personnel, particularly among front-line and middle-management supervisors. Their set of values is pervaded by the belief that the size of one's unit is a measure of one's responsibility in the eyes of both one's associates and one's superiors. Many feel that this factor affects their compensation and the consideration they are given when important vacancies occur. Where this belief persists, it discourages voluntary efforts to develop systems improvements that will bring about any sizable reduction in the work force.

The illusion that administrative costs are controlled through budgets   The income, cost, and expense goals making up a budget are aimed at achieving a predetermined profit objective—but not necessarily the maximum profit, over either the short or the long run. The reason is that, outside the production area, the level of expenditures provided for in a budget is essentially an arbitrary rather than a "scientifically" determined amount. For most activities, the budget usually amounts to little more than a reflection of past performance, eloquently defended. It is arrived at by a process of negotiation in which the most articulate participant is motivated by personal incentive and a host of purely departmental considerations rather than by the goal of maximum contribution to company profits. In other words, expense budgets are usually subjectively determined by the very persons for whom they serve as standards of measurement.

In any event, the illusion that paperwork costs are effectively controlled by the budgetary process is one of the obstacles that keep management from taking more fundamental steps to achieve such control.

Functional Compartmentation

The third major barrier to systems improvement is the rigid compartmentation of function that exists in many organizations. This condition is, at the same time, both a strength and a weakness. First, we all recognize that, in part at least, many enterprises have grown large and strong because of specialization—that is, the continuous breaking down of the total job into smaller and smaller pieces so that a high degree of proficiency can be developed in the performance of each piece. But it is equally clear that we have not gained the benefits of specialization without paying a very real price for them. That is, organizational compartmentation or departmentalization has created a situation in which the objectives or standards of the functional segments of an enterprise not only may fail to add up to, but may sometimes even *conflict* with, the goals of the organization as a whole.

To be more specific: The success of most business enterprises is judged by their long-term profit results—for example, by their rate of return on investment. Yet it is unfortunately true that throughout the functional divisions of a business hundreds of decisions are made each day without any consideration of their effect on profit. Left to their own devices, management personnel all up and down the line are motivated by the surroundings and standards of the specialized worlds in which they operate—that is, within a manufacturing company, for example, by the sales, production, or engineering world; or, within an insurance company, by the actuarial, underwriting, or agency world. To illustrate:

> *In a make-and-sell business, the credit manager may be driving to cut the company's bad-debt losses in half because of the belief that this achievement will reflect favorably upon him not only within the company but also within the professional credit circles in which he moves. Yet, from a maximum-profit-contribution point of view, it may well be that the company should double its bad-debt losses as a stimulus to sales, rather than seek to cut them in half.*

The significance of these observations in the present context lies in the fact that systems, by definition, are networks of clerical, administrative, and managerial activities that cut across various functional segments of an organization. In so doing, they tie those segments together in the performance of major recurring processes. In this light, optimization of any system means, perforce, that tradeoff decisions must be made not just among the interests of individual departments but, more important, between them and the needs of the organization as a whole. These tradeoffs, in turn, often require altering the role or responsibilities of a given department, changing its methodology, imposing constraints or new requirements on it, and so forth. Moves such as these are not easy to make in an environment where individual departments are preoccupied with their own functional pursuits and fail to give adequate attention to the best interests of the total organization or to the influence of their activities on the operating costs or effectiveness of other departments.

Lack of Staff Acceptance

Still another major barrier is the fact that systems improvement staffs are not so readily accepted among managers of office activities as staff units are in the factory or in the equivalent physical operations of other enterprises. In industrial companies, staff specialization in the manufacturing area has become traditional and well accepted. Over the years it has gradually relieved factory supervisors of the responsibility for planning facilities, developing production processes, and even, in many cases, determining the size of the work force. These are the activities that have been taken over by such specialists as the process engineer, the methods engineer, and the industrial engineer.

No such development, however, has taken place in the office. Hence, office executives and supervisors still consider themselves responsible for developing their own objectives, standards, systems, and procedures—for deciding what will be done, how it will be done, and how many people will be needed to do it. Thus, they view any attempt to change this process of decision making as an intrusion on their domain, or at least as something of a reflection on their managerial record.

Lack of a Fundamental Approach

The barriers already discussed have one element in common: they all reflect attitudes or conditions that exist among operating management. In contrast, the fifth and final obstacle is created by the systems staff itself. It grows out of the common failure of such staffs to approach the systems improvement program in a broad-gauged, fundamental way.

The plain truth is that many so-called "systems and procedures" staff members are technical specialists rather than analysts and problem solvers. For this reason, they tend, in their work, to apply suboptimizing approaches that all too often result in perfecting only a small piece of the total systems and procedures structure—even to the point of harming the structure as a whole.

More specifically, these narrow, specialized approaches can take the following forms:

1. The fire-fighting or trouble-shooting approach
2. The accounting-systems approach
3. The manual-writing approach
4. The forms-control approach
5. The computerization approach

The fire-fighting approach   The staff that plays a fire-fighting role is concerned with solving paperwork problems as they arise, usually in the form of requests from operating personnel. Although this sort of approach has the lure of being service-oriented, it usually amounts to little more than an expedient patching up of weak spots in the routines of the organization. It keeps systems staff members so busy handling the urgent that they have no time for the important. With their attention absorbed in tracking down the causes of a rash of billing errors, straightening out work-scheduling practices in the stenographic pool, or evaluating the need for the proposed purchase of a piece of office equipment, they never find the time or develop the ability to (1) design systems that enable management to run the organization better, or (2) bring about major reductions in clerical, administrative, and other overhead costs.

The accounting-systems approach   As the words imply, the whole focus in this approach is on perfecting the accounting system, with particular emphasis on seeing that tight protective control exists over all transactions affecting company assets.

One newly formed systems staff undertook as one of its first projects the development of better methods for maintaining the company's records on fixed assets. The staff spent two worker-years on the effort studying different record-keeping practices in the company's several divisions, developing and obtaining agreement on recommended changes, and directing the changeover to the new system.

As a result of this project, the company converted its fixed-asset records to a computer system that had the advantages of (1) being uniform among all divisions, (2) being continuously more up to date, and (3) providing the means for extracting summary information from the records more quickly than was possible under the former system.

Over a period of time, however, these features have turned out to be more in the nature of niceties than true benefits. The reason is that the new system has produced no saving in clerical costs, no better information for reaching capital appropriation decisions, no greater protection against loss of assets, and no greater accuracy in calculating the company's overall depreciation expense.

The manual-writing approach   In still other instances, organizations that have launched formal systems and procedures programs have made the mistake of assuming that the job consists of developing and maintaining manuals of standard practice instructions or standard operating procedures. But manual preparation, by itself, does *not* constitute systems improvement. It is only the recording and communication phase. It makes no provision for the primary creative functions of investigation, analysis, and design.

The forms-control approach   Proponents of this field of specialization assert that, since forms are the raw material of office production, forms control is the key to administrative cost control. But although a sensible degree of control over forms should be included in every well-balanced systems program, forms control by itself is an extremely limited, "in-the-back-door" effort to solve the broad problem of administrative costs and effectiveness by picking away at one of its minor segments. In a very real sense, the claim of these proponents is very much like saying that the key to making profits in the automobile industry lies in developing tighter control over raw material specifications and inventories.

An example that puts both the forms-control and manual-writing approaches in perspective can be found in the experience of one of the nation's largest companies in the financial services field.

A major operating department of this company employed over 2,500 clerks in its office operations spread across the country. With paperwork volume of this magnitude, the department had set up its own so-called methods and procedures staff in the late 1950s. In subsequent years this staff implemented a highly developed manual-writing and forms-control program. All recurring procedures of the department—both at headquarters and in field branches—were reduced to detailed standard practice instructions. The resulting set of three loose-leaf manuals was distributed to every manager and supervisor of the department. In addition, form-numbering and functional classification schemes were adopted for all printed and duplicated forms. Overlapping forms were combined. Physical specifications were established on the size and grade of paper, form layout, and the like. Printing standards were spelled out, the most economical ordering quantities were prescribed, and the rule was laid down that no new form could be ordered without approval of the departmental methods and procedures staff.

Almost 10 years after the departmental staff had been created, members of another staff—the corporate organization and systems group—were brought in to tackle the problem of spiraling clerical costs in the department. By applying a meaningful and imaginative approach to the problem, this staff was able to develop changes that cut the department's clerical costs by $4 million a year and, at the same time, substantially improved the usefulness of the department's management reports as tools for decision making and control.

The basic reason why these gains took 10 years to achieve is that the department's own staff had let itself become preoccupied with the trappings of systems and procedures instead of with the fundamentals.

The computerization approach   Of all forms of suboptimization that characterize systems improvement work, an intense preoccupation with computerization can be the most pernicious. It has its roots in two illusions that pervade today's large-scale organizations. The first is that the improvement of office operations by and large means computerization of the office. The second is that if a given operation can be done on computers, then that per se must be the best way of doing it. Yet some of the highest-cost clerical operations exist in organizations that are using the most sophisticated computer hardware in most of their operations. This does not mean, of course, that the costs in these offices are high because of the computer. It simply means that the managements of these offices have fallen into the trap of assuming that computerization is synonymous with efficiency and in so doing have overlooked the other, more fundamental ways to reduce paperwork costs.

Much of the blame for this preoccupation must be borne by staff specialists who have become captives of the new EDP technology. With eyes bright but out of focus, more than a few of them have lost their ability to be objective and tough-minded about the savings or other benefits that EDP will really produce. In this respect, they have metamorphosed from systems analysts into computer advocates. They no longer have an impartial professional concern for improving administrative processes—by any or all means that can be profitably employed—but a partial concern for justifying or finding new applications for a particular piece of hardware.

The danger implicit in this suboptimizing approach is underscored by the following experience:

Five years ago, a company with a clerical force of 3,400 undertook what it called a basic long-term study aimed at bringing its control over office costs up to the level of its control over factory costs. But the task force that was charged with this mission limited its efforts to evaluating the feasibility of converting from what was basically a "first-generation" computer installation to a much more advanced EDP hardware-software configuration.

Not surprisingly, the team's evaluation led, in time, to conversion. When completed, the new installation replaced some 150 of the company's 3,400 clerical employees, with a consequent payroll saving somewhat higher than the incremental cost of the upgraded computer operation. By the application of this cost-versus-saving criterion, the venture was considered successful. Yet the company's clerical cost control problem was still far from solved because less than 5 percent of its clerks were covered by the study. Moreover, one wonders how much greater savings might have been achieved had the task force applied to all 3,400 clerical positions an amount of imaginative work-simplification effort equivalent to the 36 worker-years of analytical talent expended on the electronics conversion study.

The true role of computerization (or any other form of mechanization) in the systems improvement program can best be summarized by thinking in terms of *all* the steps that make up a dynamic approach to (1) cutting administrative costs, and (2) strengthening management information. These two end-result objectives are selected because they represent the areas of justification on which recommendations for conversion to computer applications are most frequently based.

Here are the fundamental steps in cutting paperwork and other administrative costs:

1. Motivate line managers to *want* to cut their costs.

2. Reevaluate the work that *must* be done or that "earns a profit."

3. Modify the indirect factors, such as policies and organization structure, that affect the volume or complexity of clerical work.

4. Select the best method—manual, mechanical, or electronic—of performing the essential work.

5. Apply objective measurement techniques for determining how many people are really needed to do the required work by the methods selected.

As Chap. 13 will spell out more fully, the steps in a dynamic approach to improving management information are these:

1. Identify the critical aspects of the business—that is, the end results and other elements of performance—that must be planned and controlled.

2. Determine what specific information is needed for decision making and control for each of these factors.

3. Train and then motivate line managers to use the new information tools.

4. Determine the best method of preparing the required information within essential time limits.

The point to be stressed in thinking about these two lists is this: Although methods selection is admittedly an essential step in each approach, it is the last or nearly the last step to be taken. In our experience, it is often the least important of the steps in terms of its contribution to the end result. In any event, there can be no doubt whatever but that computerization by itself can never eliminate the need for, or excuse managements from, taking the other steps listed. In the area of paperwork and administrative cost reduction, unless the other four steps are taken, no amount of office automation will produce real efficiency. Similarly, in the area of decision making and control, increasing an organization's information-generating capacity or speed is no substitute for fresh, disciplined thinking about what information is needed or can be profitably used in managing the enterprise.

FOUR

Organizing and Staffing the Systems Improvement Program

The orderly and successful performance of any major undertaking, of whatever character, requires that a definite sequence of steps be followed in its planning and execution. These steps are:

1. Clearly defining the objectives and scope of the undertaking
2. Developing policies necessary to achievement of those objectives
3. Building a logical, workable organization to do the job—which means defining and fixing responsibility as well as staffing the organization with competent people
4. Working out a comprehensive approach and plan of action
5. Applying skilled techniques to the work undertaken
6. Maintaining control through some positive means for measuring results achieved

If a program of systems and procedures improvement is to be purposeful and fruitful, it must evolve according to this conceptual scheme. There is no other quicker or easier path. The field of paperwork is littered with the wreckage of systems and procedures programs that have broken down altogether or are producing only fragmentary results. Almost without exception these failures have occurred because, in the conception or development of the improvement program, one or more of the steps in the foregoing list were ignored or given only perfunctory attention.

The first three steps of this formula will be considered in this chapter, and the remainder in the three succeeding chapters.

SETTING PROGRAM OBJECTIVES

The objectives of a systems program, which were defined in Chap. 2, can be restated here as the *development, implementation, and maintenance of effective and efficient operational and management systems* for the purposes of:
1. Significantly improving the organization's operational performance
2. Increasing top-management effectiveness
3. Cutting clerical, paperwork, and other administrative costs
4. Compensating for shortages of clerical labor
5. Capitalizing on the full potential of the computer
6. Developing teamwork among departments
7. Improving policy execution
8. Raising employee morale
9. Coping with complexity
10. Facilitating change

Mere definition, however, is not enough. To be meaningful, these objectives must be made known to the entire organization and must be rooted in management's conviction that their achievement is both necessary and possible.

ESTABLISHING POLICIES

To be fully effective, the systems program must be based on the following two top-management policies:

1. Systems development and improvement work must be positive and continuous—not passive, intermittent, or expedient. In its positive form, systems work is concerned with continuous observation and analysis aimed at foreseeing operating problems long before their magnitude compels attention and at uncovering new opportunities for improvement that otherwise would remain hidden or obscure. In its passive form, systems work is confined to:

    a. Development of new procedures whenever they are required to implement new policies, carry out new activities, or conform to new legislation

    b. Revision of existing procedures as operating experience reveals the need for change—which is a euphemistic way of saying "putting out operating fires."

If the initiation of systems improvements awaits the sword point of crisis or the compulsion of competition, the progress achieved is likely to take the form of a series of mutations rather than constant, well-balanced evolution. Promising opportunities for cutting costs or improving performance remain neglected, while the changes that are made seldom produce substantial permanent benefits. Systems reforms wrought under the pressure of something gone wrong are apt to do little more than solve the immediate problem at hand. And where the need for action is acute, unsatisfactory compromises must often be made in order to bring the situation under control. But once the patched-up system is functioning again—however inadequate or cumbersome it may be—it quickly becomes assimilated in the day-to-day routine as attention is shifted to other pressure points.

It seems worth emphasizing, therefore, that the first top-management policy essential to controlling the quality of operational and management systems is that the approach to their development and improvement cannot be just passive and reactive. It must be a planned, positive approach involving the continuous review of existing systems and procedures to detect evidence of inadequate information or control, duplication, irrelevance, excessive performance time, or excessive cost.

2. *The systems improvement program must be complete.* It cannot be focused on a few basic transaction-processing routines or on the accounting function, as it often is. It must comprehend the paperwork of every department regardless of who performs that work—whether they be clerks, sales representatives, engineers, purchasing agents, actuaries, underwriters, loan officers, administrators, or anyone else.

DEFINING ORGANIZATIONAL REQUIREMENTS

Given the essential objectives of the systems improvement program and the policies governing it, it follows that the organization within which the program functions must meet some special requirements:

1. It must clearly establish accountability.
2. It must ensure a sustained effort.
3. It must differentiate between the two basic types of systems and procedures—*intra*departmental and *inter*departmental.

Establish Accountability

It is evident, of course, that there cannot be a positive, continuing attack unless responsibility for the systems program is fixed and clearly defined. This means that some specific position or positions must be made accountable not only for carrying out systems studies but also for observing the need for such work—for anticipating problems and isolating opportunities. It is basic to the success of the program that this responsibility be assigned in such a way that accountability is established over these three factors:

1. Completeness of the program—that is, the problems and projects to which effort is applied
2. Progress or rate of accomplishment
3. Quality of the systems and procedures developed

Ensure Sustained Effort

Any attempt at organizing a systems improvement program must quickly come to grips with one of the practical problems involved in ensuring that sustained effort is applied to the program. One of the serious obstacles to developing systems improvements is the pressure of day-to-day operating responsibilities.

This is understandable, for the conduct of a systems study often requires much painstaking research and analysis. The problem must be defined; the facts must be gathered and evaluated; and the new or revised system must be planned, tested, and installed. The line executive or supervisor seldom has adequate time to devote to this process. Although he may recognize problems that need investigation, he often cannot tackle them because of his absorption in regular administrative duties. Should he take the time to do so, his efforts are interrupted so frequently that he must perforce conduct a hasty, superficial study to arrive at any kind of solution within a reasonable period.

These circumstances dictate that systems development and improvement be set up as a staff function and that the persons engaged in the work be assigned no other responsibilities. Only in this way are thoroughness and continuity of effort likely to result.

Aside from having *time* adequate to the task, a special staff can bring at least two other resources to the systems program.

1. Specialized knowledge and skills. Extensive and varied experience in systems work leads to the development of specialized techniques for gathering, organizing, and analyzing data. If skillfully applied by the systems analyst, these techniques help to produce substantially better results in a shorter time. They tend to ensure that the study is carried out in a rigorous and imaginative way, that all pertinent facts are considered, and that no opportunity is overlooked.

Moreover, successful systems studies—particularly those involving analysis of alternative methods—require a fund of information about office machines and equipment, especially EDP and data communications equipment. Although analysts need not be experts in each of the many types of equipment available, they should at least understand its function, costs, and possible applications. Acquiring adequate knowledge in these specialized fields is difficult except through continuous study and on-the-job experience.

2. An objective and critical approach. Staff members are likely to approach a systems problem with more detachment than line personnel. Since they neither perform nor supervise the routine being studied, there is less chance that the thinking they apply to it will be inhibited by tradition, habit, or personal incentive.

In addition to greater objectivity, they bring to the problem a more penetrating and critical approach. Since systems analysis and design are, by their very nature, evaluative and conceptual processes, staff members who are continuously engaged in them acquire a mental predisposition for questioning what is being done and seeking better ways.

These arguments for organizing the systems function as a staff activity need some qualification, however, lest the staff's function be misconstrued. In no sense does the staff usurp the responsibility of the line executive for decision making on systems issues. Nor does existence of the staff diminish the need for the operating supervisor to think creatively about systems and procedures. Indeed, much of the staff's success will depend on how well its members work with and through front-line supervisors, drawing on their firsthand knowledge of operations, stimulating their ideas, and testing proposals against their practical judgment.

Differentiate Between Two Types of Systems

If the organization for a systems improvement program is to meet the test of organizational principles, it must clearly distinguish between two basic types of systems and procedures: those confined exclusively to one department and those involving two or more departments.

Intradepartmental systems   This category consists of detailed work routines that are wholly departmental in scope. Their method of performance or their modification has little or no direct effect on any other department. Their function often, though not always, is to prescribe how the unit should carry out a responsibility assigned to it in an interdepartmental system. For example, an interdepartmental system covering the control of capital expenditures may specify that all requisitions to purchase capital items must be reviewed and approved by the budget director to ensure that adequate funds are available in the capital appropriation. The means by which the budget director carries out this responsibility—the records required, the participating personnel, and the flow of documents within the department—all are matters of intradepartmental procedure.

Interdepartmental systems   These systems are a type of detailed organization plan, in the sense that they assign to specific departments the responsibility for performing each of the major steps in a repetitive process. They also prescribe the sequence and timing of each department's participation; they ensure against avoidable duplication of effort between departments; and they specify the approvals prerequisite to certain types of actions.

One interdepartmental system common to most manufacturing companies is the receipt of incoming shipments. This process generally involves the participation of many departments, including receiving, inspection, material control, storekeeping, purchasing, and accounting. Another such system is that governing the preanalysis and control of engineering or design changes in metalworking and assembly industries. Although this system is not so typical as the first, it is perhaps more useful as an illustration because of the number and diversity of participating departments. In one industrial company, for example, close to a dozen departments participate in some or all of the three phases that make up the system:

1. Gathering, correlating, and presenting the pertinent facts
2. Making the decision
3. Coordinating the preparatory work necessary to putting approved engineering changes into effect

In the fact-gathering phase, the engineering department is responsible for determing the effect of the change on product performance and appearance and on the interchangeability of parts. The methods department shares with purchasing and cost accounting the responsibility for calculating the effect on labor and material costs. Other departments participate in determining the quantity of material that will be rendered obsolete by the change: The material control department compiles this information for raw materials and finished parts in stock, production control for in-process parts, and purchasing for outstanding purchase commitments covering these items; then accounting converts the quantity data to dollar values. Finally, the sales division is responsible for predicting what effect the design change will have on market receptivity.

The same kind of interdepartmental participation characterizes the decision and implementation phases. The system specifies who must approve the design change before it is accepted and under what special conditions top management's approval must be obtained. Once the change has been approved, the cutover in the manufacturing process must be carefully planned. Thus, the system also provides for coordination of the preparatory work so that new tools, jigs, fixtures, patterns, materials, drawings, and operation sheets will all be available at approximately the same time. Finally, the system assigns responsibility for deciding what disposition should be made of obsolete items and for carrying out the decision reached.

From examination of this process, it is apparent that sound decisions and prompt execution require unity of action by all the participants. The primary function of interdepartmental systems is to make such unity possible. On the one hand, they ensure that no essential part of a given process is left unassigned; on the other, they guard against delays or loss of effort due to backtracking, duplication, friction, or dysfunctional competition. In short, one of the principal objectives of interdepartmental systems is to promote coordination of effort on recurring matters so that all branches of the organization function together as a harmonious whole.

With this understanding of the three organizational requirements for systems development and improvement—that we must (1) clearly assign responsibility, (2) ensure a sustained effort through creation of a systems staff, and (3) distinguish between intradepartmental and interdepartmental systems—we are now ready to address the question: To whom should the systems staff report? That is, at what point or points in the company organization structure should responsibility for systems development and improvement be placed?

ASSIGNING RESPONSIBILITY FOR INTRADEPARTMENTAL SYSTEMS

Within business organizations, department heads are normally considered to be responsible for determining the procedures and clerical methods by which the activities assigned to them will be carried out. This responsibility is as intrinsic to their function as that of deciding the organizational structure of the positions under their line authority or of selecting the people to staff that organization. Although they may seek and accept recommendations on intradepartmental procedures from their superiors or from outside specialists, they cannot be relieved of responsibility for decision making in this matter without impairment of their authority. For example, a production control superintendent, accountable for dispatching and following the progress of manufacturing orders through the shop, cannot properly be relieved of the responsibility for determining how these activities will be carried out or what information is needed to control their execution.

This general principle is subject to two qualifications: (1) intradepartmental procedures must not conflict with the activities of other departments or the basic policies of the company, and (2) in any multiple-unit operation, such as a chain of retail stores or a company with several plants or warehouses, standard procedures may be necessary or desirable for the same departmental function in all units. Where this requirement has been made a matter of company policy, the responsibility for procedures design may be appropriately withdrawn from the local heads of like departments and placed with some central planning group.

A corollary conclusion is that staff personnel engaged in the analysis of procedures that are wholly departmental in scope should, in principle, report to the head of the department. In its practical application, however, this principle requires some amplification. One reason is that any one department may not have enough procedures work to keep one analyst continuously busy. Another reason is that the existence of a number of departmental procedures staffs throughout the company (for example, in manufacturing, engineering, sales, and personnel) can create major problems of coordination. Even if everyone agrees that these departmental staffs will work on *intra*departmental procedures only, the line marking this limit cannot always be precisely drawn. As a result, conflicts and overlapping among such staffs are almost certain to occur.

One alternative for overcoming this inherent problem is to have the company's (or division's) *general* systems and procedures staff serve the individual departments on internal procedural matters. Where this alternative is followed, it is important to stress that, on intradepartmental problems, the general staff is strictly a service unit. It is available to render assistance as requested, but no department is obliged to use its services. In addition, when the general staff is asked for assistance on a departmental problem, the staff member assigned to the project should, for the duration of the assignment, be regarded as a staff assistant to the head of that department.

Where this alternative is not followed and, instead, major departments set up their own intradepartmental procedures staffs, special effort should be made to keep the various staffs properly coordinated. One company successfully achieved such coordination by removing all departmental procedures staffs from their respective departmental areas and locating them adjacent to each other in a single area. The natural coordination that grew out of this physical arrangement was supplemented by a regular weekly meeting of the heads of the several procedures staffs.

ASSIGNING RESPONSIBILITY FOR INTERDEPARTMENTAL SYSTEMS IN A SINGLE-ENTERPRISE ORGANIZATION

Within a single-enterprise organization, responsibility for interdepartmental systems may be placed at three different points in the organization structure. Each of these alternatives can be found in actual use among companies that have undertaken an organized program of systems work. They are:

1. Responsibility can be distributed among all departments, with provision made for proper clearance and approval of specific proposals.

2. Responsibility can be assigned to the head of a major function of the business—for example, the controller.

3. Responsibility can be assigned to some member of top management—that is, to some general-management executive rather than a functional executive.

These three alternatives merit thorough exploration because hasty, ill-considered choice among them is perhaps the greatest single cause of ineffective systems and procedures improvement work.

Alternative 1: Assigning Responsibility to All Departments

Where this alternative has been adopted, each department head is considered responsible for undertaking the development or revision of those interdepartmental systems that affect his operations to the greatest extent. This arrangement does not, however, preclude his making recommendations on any other systems or procedures that are of concern to him, no matter to what extent. Along with this division of responsibility, some agency is generally set up to ensure that procedures proposals are reviewed and approved by the heads of all departments involved. This agency may be either a committee of department heads or a staff position reporting to one of the top executives. A final customary feature is the assignment of responsibility for reducing approved procedures to a standard format (usually referred to as "standard practice instructions" or "standard operating procedures") and for distributing copies to the executive and supervisory personnel affected by them.

In large companies where responsibility is divided in this way, it is not uncommon to find a number of divisional or departmental staffs continuously engaged in systems and procedures studies. Departments that do not have a permanent staff generally designate one of their members to work on procedures problems as the need arises. One such company had full-time procedures staffs in four of its divisions or departments: accounting, engineering, production control, and industrial relations. In addition, it had general staff assistants to several other division and department heads who spent varying but considerable amounts of time on procedural matters. This group included assistants to the sales manager, to the director of purchasing, to the traffic manager, and to the inspection superintendent. According to an organization bulletin of the company, the assistant to the president was responsible for "obtaining, reconciling, and compiling the views and recommendations of the executives with respect to new or existing policies and procedures." The extent of this responsibility was amplified by another directive which stated: "Division and department heads requesting adoption or revision of policies and procedures will make their recommendations as complete as possible, leaving as little as possible to be done by the assistant to the president. Before sending in recommendations, the originator should also discuss them with the individuals who will be directly affected."

Arguments favoring departmental responsibility   A number of benefits, claimed to be unobtainable by other means, are credited to this division of responsibility for interdepartmental systems and procedures work. Among the merits most frequently cited are the following:

1. Systems and procedures are likely to be more practical from an operating standpoint when each department head has primary responsibility for developing and applying them in his area. They will be designed and installed by people who have intimate, day-to-day contact with the department's problems and routines.

2. When a procedure is developed by, or under the immediate direction of, the department heads who are to use it, a natural pride of authorship evolves. The department heads understand the procedure and believe in its desirability or necessity. The procedure will be followed not because its use is demanded but because it is accepted as the most practical way of performing the activity it covers.

3. A department will be better served by its own personnel than by an outside group. Not only can the department head supervise his own staff more readily, but also he has the assurance that it is always available to work on the procedural problems affecting his operations. A staff with outside supervision, on the other hand, might be diverted from the department's problems just when they most need attention.

4. Departmental procedures staffs are a source of personnel to handle special, nonrecurring operating jobs that arise from time to time. They also serve as a training ground for supervisory positions within the department.

Basic weakness of departmental responsibility   The importance of these benefits—particularly the first two—seems incontrovertible. Although they can also be realized in full measure through other types of organization, there is little doubt that departmental responsibility for systems work produces them most readily. But their value is largely offset by a fundamental weakness in this division of responsibility: It fails almost completely to meet the three control specifications that were listed earlier as essential to success of the systems and procedures improvement program.

When responsibility for interdepartmental systems is broken up and distributed among all departments, accountability for completeness of the program is wholly lost. No position can be held responsible for observing needs or crystallizing objectives. Nor can there be any assurance than the most promising opportunities are being exploited, particularly since so many of them lie in the "twilight zone" of interdepartmental relationships.

To the same extent, this alternative fails to meet the second specification— that of providing effective means of control over progress or rate of accomplishment. And finally, although it partially meets the third specification by stipulating that new or revised procedures be jointly approved by all department heads affected, such an arrangement is an ineffective means of controlling the quality of procedures adopted.

How does lack of control over these three factors offset the benefits that departmental responsibility produces? The answer lies in the conditions and results arising from an uncoordinated departmental attack on *inter*departmental systems and procedures.

Conditions characteristic of the departmental approach   Three conditions can be expected to emerge under this type of organization.

1. Conflict and duplication of development effort.   Two or more systems staffs are often apt to be working simultaneously on some phase of the same systems project, each staff unaware that others are engaged in it. Sometimes this situation arises because each staff believes that maintaining a given set of records or exercising a given control is the rightful function of the department it represents. At other times, an interdepartmental systems study that could properly be started at any one of several points is, through lack of communication, begun at more than one of these points at the same time. And in still other instances, a group of projects, which on the surface appear to be of minor importance and of purely departmental interest, become so ramified as they are carried out that each joins the others at departmental boundary lines to become a single major project.

As these duplications of effort become evident, they may lead either to jurisdictional disputes, jealousies, and continued duplication or to constant shifting of responsibility for completing projects from group to group or person to person.

2. Lack of effort.   In contrast to duplication of effort, it is equally probable that many gaps will exist in the joint endeavor. Little or no progress will be made in solving some interdepartmental procedures problems because each staff or department is waiting for some other group to take the initiative. However acute the problem, should it involve explosive organizational issues or an unsuccessful previous effort at solution, it is likely to provoke much talk but little action.

In other instances, the work of one staff may be completed and shelved for a substantial period while awaiting the availability of someone from another staff to undertake and complete his share of the job. For example, the industrial relations procedures staff may develop a new employee interplant transfer procedure, adoption of which must be delayed until the accounting procedures staff—busy pursuing its own program—develops the necessary supplemental procedure for transferring payroll records. Too often, completion of such studies hinges on the passive interest of a group to which the problem is of little concern.

3. Limitations of the departmental viewpoint. The majority of departmental executives cannot reasonably be expected to have the breadth of knowledge and interest necessary to visualize all aspects of an interdepartmental problem and to correlate them properly. In no sense does this statement belittle their talents as managers, for our system of organizational specialization deliberately fosters a proprietary interest in one's own function. The fact remains, however, that under divided responsibility for interdepartmental procedures, a companywide point of view is seldom achieved. Nor can it readily be achieved where many groups with widely varying perspectives and understanding are independently engaged in defining the scope of a common problem and in hacking away at its isolated segments.

A further impediment to applying an overall point of view is the traditional method of conducting interdepartmental systems studies under this divided arrangement. Seldom does any one individual get all the facts. Members of one department ordinarily cannot—or, at least, do not—go into other departments to analyze the records, reports, and routines bearing on the problem. Rather, the most common practice is for the department confronted with a systems problem to develop a proposal based on its own information and to submit it in writing to other departments for comment and approval. Typically that initial step is followed by a long series of consultations, counterproposals, committee meetings, and similar efforts at self-adjustment. Although conflicts and differences of opinion may be resolved by this process, at no time does the process produce the orderly assemblage and analysis of facts necessary to ensure that the decision reached serves the best interests of the organization as a whole.

End results produced by these conditions   Without unity of thought, purpose, and action in the developmental program, there is little warrant for assuming that the systems and procedures adopted will measure up to reasonable standards of simplicity, effectiveness, or economy. Under the departmental approach, the principal quality yardstick applied to systems and procedures proposals is whether they have been approved by the heads of all departments affected. No guarantee of companywide efficiency is implicit in this measurement, for as long as one department head gets what he wants, he will seldom challenge the rights of other department heads to what they want. This attitude is the breeding ground for duplication and complexity. It promotes excessive controls, extra copies of forms, multiplicity of approvals, duplication of records, and a circuitous flow of work.

Where unanimous approval is not spontaneous because of a conflict between two or more department heads, efforts at settlement are generally directed toward reaching a compromise that most nearly satisfies the personal objectives represented. Little if any thought is given to discovering the real needs of the company as a whole. It is not intended to underrate the practical value of compromise in developing workable procedures as distinct from "perfect" procedures. But under divided responsibility for this activity, the process of bargaining, persuasion, and concession too often concentrates attention solely on the viewpoints of the dissenters. The resulting compromise seldom promotes harmony; nor does it, except by accident, produce an efficient system or procedure. If it were possible, at this point, to direct attention to a thoughtful appraisal of how the total organization's needs could best be served, the compromise reached would, in all probability, not only be sounder and more acceptable to the dissenters, but also promote a more unified outlook.

Moreover, the quality of systems and procedures adopted under the divided-responsibility arrangement is not likely to be improved significantly by submitting them to a standing committee of department heads for review. In performing this function, a committee operates under two handicaps. First, its effectiveness is impaired by lack of facts. Clearly it cannot act as a fact-gathering agency itself. And we have already seen that divided responsibility for development work rarely, if ever, results in expert collection and correlation of all the facts. It is hardly necessary to add that sound decisions cannot be reached by applying many minds to incomplete facts. Neither can the deficiency be compensated for by the random impressions, irrelevant discussion, and snap judgments that otherwise characterize committee work.

Second, a committee set up to review and approve systems proposals will probably be little more successful in overcoming the limitations of a departmental viewpoint than its members acting individually. In reality it is only a somewhat more formal substitute for the originator's obtaining the necessary approvals one by one. To be sure, a committee does bring divergent and specialized opinions together and facilitate the interchange of ideas. But that is not enough; the mere assembly of several functional points of view does not of itself produce a well-balanced, companywide point of view. This can only be achieved through a synthesis of thought and unification of objectives that are rarely to be found in a committee of special interests deliberating a general problem.

Thus the first result to be expected under divided responsibility is that the quality of the systems and procedures finally adopted will fall far short of being the best obtainable. A number of other results must also be anticipated. Among them are the following:

1. Progress achieved on specific projects will be painfully, if not dangerously, slow. Lack of coordination will retard developmental effort. And because this effort fails to produce a finished study based on all the facts, the time and energy spent in obtaining approvals, reconciling conflicts, and revising proposals will be even greater. The end result, of course, will be costly delays in developing adequate controls, adjusting procedures to changed requirements, and launching new activities.

2. For want of a planned, integrated approach, attention will be concentrated on procedural problems, not on opportunities. Thus, many of the potential benefits set forth in Chap. 2 will not be realized.

3. Executives and supervisors will find their time seriously absorbed by

systems and procedures matters. Different groups will come to them for advice and information on the same problem. They will be called on to review drafts of conflicting proposals or many redrafts of the same proposal. They will spend time in committee meetings, seeking, by prolonged exchange of opinion, to thresh their way through proposals inadequately supported by fact.

4. Along with costly duplication of effort among systems and procedures staffs, the effectiveness of staff personnel is likely to be impaired by the demoralization that comes from lack of accomplishment. It is unfortunately true that the most competent staff members will be the least willing to continue working in an atmosphere of constant frustration.

5. Installation and effective operation of the systems and procedures finally adopted will be subject to the same delays that lack of coordination produces in the planning and approval stages.

Conclusions on alternative 1   In view of their significance in effective business operation, we may wonder why the development and improvement of inter-departmental systems and procedures are frequently left to so cumbersome an organization arrangement. There appear to be two principal reasons, neither of which is tenable.

First is the not uncommon belief that efficiency of the enterprise as a whole is ensured by obtaining efficiency within each of its individual parts. In the whole span of management practice, there is probably no other more misleading concept than this. It fails to recognize that, while each department may perform its own activities effectively and economically, there can still be widespread duplication and triplication among departments. This condition is particularly common in large organizations where specialization attains its most highly developed form. Here, both lack of communication and failure to develop sympathetic apprehension of each other's functions and needs impose severe limitations on spontaneous coordination. At the same time, the many echelons between related operating groups and a common superior make coordination through that superior equally difficult.

There are few businesses of any size in which overlapping and duplication of activity do not exist. In one company, four departments—storekeeping, mate-rial control, merchandising, and order control—were found to be maintaining records of raw material inventories. In another company, five groups were engaged in the sale of scrap and excess or obsolete materials or equipment. Still another firm kept three separate records of capital equipment—each dupli-cating 80 percent of the information in the other two.

These illustrations reinforce the conclusion that *efficiency of the enterprise as a whole requires a unified performance that is often considerably more than the sum of the organization's many departmental activities.* The great amount of attention that has been given to perfecting skills and techniques in functional areas has created a critical need for specialization in the problems of general management. Perhaps the greatest remaining opportunity to increase the effectiveness and efficiency of today's organization lies in more skillful integration of its pieces.

The second reason why responsibility for interdepartmental systems and procedures is so frequently assigned to all departments is the fear that to do otherwise would not only impair the authority of line executives and supervisors, but also prevent them from applying their knowledge and judgment to procedural matters. The alternative is somehow conceived to be that of handing down each procedure as a dictum that, at best, will be carried out indifferently by those who did not share in its development and, at worst, will fail to meet the test of actual operating conditions.

In response to this point of view, assurance must first be given that the importance of having operating managers participate actively in systems and procedures development is not only conceded, it is vigorously affirmed. Without such participation, neither practicable nor acceptable systems and procedures are likely. But those who use this argument as a justification for the departmental approach to improvement of interdepartmental systems and procedures are missing the very point of their own argument. If the knowledge and experience of operating personnel are of such great moment, it is sophistry to maintain that the application of that knowledge and experience should be left to the chance occurrence of spontaneous coordination. If these assets are to be fully used, they clearly ought not to be subject to the groping, tentative efforts, the confusion and delayn and the lost opportunities that characterize divided responsibility. Nor should they be dissipated in the haggling and crosshauling that lack of leadership breeds. The only protection against such waste is some planned, systematic means for continuously seeking out and utilizing this knowledge and experience. The choice, therefore, is not between organizational democracy and dictatorial rule, but between (1) strong leadership capable of developing effective teamwork, and (2) muddling, delays, loss of effort, and absorption in corporate politics.

Alternative 2: Assigning Responsibility to the Head of a Functional Division

In organizations that have adopted a more positive approach to the analysis and improvement of interdepartmental systems and procedures, responsibility for the activity is generally centered in a single position. In some cases, this position is the head of one of the functional branches, most frequently the controller. The argument advanced to support this organizational alternative is that the controller is the only functional executive having a direct interest in all interdepartmental systems and procedures, for few if any of them fail to involve accounting activity in one way or another. Most forms, records, and primary reports support or feed the financial statements and other executive reports required by top management for control of assets and operations. Consequently many people feel that all such documents should be compiled according to a plan laid down, or at least approved, by the chief accounting and record officer. This premise leads naturally to the conclusion that all clerical activities and procedures should be under either the direct or the functional jurisdiction of the controller. It may also be argued that because the controller ordinarily has the largest group of clerical personnel under his line authority, he is likely to have more experience than any other corporate executive in the efficient use of such employees.

In evaluating this alternative and the arguments supporting it, we should reconsider two points that have already been advanced:

1. Interdepartmental systems and procedures are primarily a means of coordinating the various functional groups that make up an organization.

2. Any staff group engaged in systems and procedures work should report to the person responsible for final results.

Taken together, these two points give rise to a question of organizational principle. Should a controller (or business manager, or administrative vice president) be held responsible for final results in the coordination achieved through the development of interdepartmental systems and procedures?

The answer is properly an affirmative one when all departments that are engaged in paperwork are under the line authority of any one of the foregoing positions. Such an arrangement, however, is uncommon among manufacturing companies. It is more likely to exist, though by no means always, in some types of commercial or financial service businesses.

In the absence of this line authority over all activity involving clerical work, the answer must as certainly be negative. Responsibility for coordination means nothing less than authority to make decisions on such matters. And the authority to make decisions on matters of interdepartmental coordination cannot appropriately be exercised by one of the departments to be coordinated. This authority properly belongs to a general-management executive.

Another question of organizational principle then arises. Is the function of the controller sufficiently related to general-management responsibility to warrant his performing a facilitating role in this matter of coordination? That is, should the controller's relationship to coordination be the same as it is to control as an administrative process? Although control is intrinsically a top-management responsibility, its exercise is made possible through the controller's discharge of his primary duty. It is he who gathers, correlates, and interprets the control data that help top management and divisional executives do a more effective job. Should the controller be charged to the same extent with the tasks of investigation, analysis, and development of recommendations on interdepartmental systems and procedures as a means of facilitating the coordinative process?

There is no categorical answer to this question. Much depends on the controller's position in the organizational structure, the character of his administrative job, and the extent of his personal authority as distinct from his nominal authority.

Where the office of controller is part of a company's treasury or finance division—as it often is—assignment of this responsibility to the controller would be a departure from basic precepts of organization. The vice president for finance or the treasurer performs a financial function, just as a director of research performs an inventive or developmental function; a vice president for manufacturing, a production function; and a general sales manager, a selling function. A treasurer's function is scarcely more similar to the *general* management of his company's business than is that of the sales manager or any other functional executive. Therefore, the treasurer should not be assigned a general-management job any more than the sales manager should.

A second factor to be weighed is the extent of the controller's operating responsibility. Not infrequently the controller supervises many activities not related or essential to his principal function of compiling and interpreting financial and operating results. These additional activities may include sales order processing, credit, billing, stock control, office services, and many others of a routine, clerical nature. The more administrative responsibilities of this character the controller assumes, the weaker becomes the argument that he should perform the systems analysis and development work required for effective coordination. This is true for the following three reasons:

1. As the scope of the controller's operating responsibility grows, his department more and more becomes merely one of the functional groups to be coordinated. In the gathering and analysis of facts and the development of recommendations, any functional group working on a general problem is likely to underrate or neglect those aspects that do not directly concern it and to treat as fundamental those that do.

2. If he is overburdened with administrative or routine activity, the controller has little opportunity to acquire the understanding of company operations necessary to ensure that systems and procedures studies are intelligently conducted and that recommendations are developed in the light of the organization's best interests as a whole.

3. Finally, the greater the volume of operating responsibility discharged by the controller, the more other functional executives will consider his position as coordinate to theirs. They will not regard him as operating in a staff capacity to the chief executive, nor will they consider his work as being any more closely related to the general-management function than their own. And because of this feeling, they are likely to resent having their activities investigated by the controller's staff and having the controller develop recommendations regarding the scope of their operations.

Conclusion on alternative 2   It can be stated as a general principle, therefore, that responsibility for investigating, analyzing, and recommending interdepartmental systems and procedures should not be assigned to a functional executive unless, in actual practice as well as in the minds of other division heads, that executive carries out this responsibility in a "chief of staff" capacity for the top executive or the general-management group.

Alternative 3: Assigning Responsibility to a General-Management Executive

Although consideration of alternatives 1 and 2 has resulted in rejection of one and qualification of the other, it has at the same time clarified the principles on which responsibility for interdepartmental systems and procedures must be based. They can be restated as follows:

1. Interdepartmental systems and procedures are primarily a mechanism of coordination.

2. Responsibility for coordination rests with the leader of the groups to be coordinated. In each division head reposes responsibility for coordinating the departmental activities under him; in the chief executive, responsibility for harmonizing divisional activities and achieving organization-wide coordination.

3. The responsibility for coordination is not divisible. It cannot be delegated to all the groups to be coordinated. Any attempt to do so is not delegation; it is abdication.

4. The position responsible for decision making on matters of coordination should not delegate responsibility for fact-finding, analysis, and recommendation on such matters to one of the functional groups that are to be coordinated. Facilitating work of this nature should be performed only by someone whose point of view and interests are coterminous with those of the principal and who is accepted by others as an agent of the principal.

5. Responsibility for development of interdepartmental systems and procedures cannot be effectively discharged in a passive way. It requires something more of the leader than his performing a conciliatory function if and when conflicts arise. It requires something more of his staff than its acting as a "clearinghouse" if and when procedures proposals are submitted. It requires development and execution of a positive, orderly, and continuous program for uncovering needs and seeking opportunities, for planning and organizing systems and procedures studies, for conducting them according to a set of systematic techniques, and for controlling the progress and results achieved.

6. Active performance of this function by the leader and his staff neither impairs the authority of operating executives in their respective areas nor precludes their participation in the development and improvement of interdepartmental procedures. Quite the contrary. Under strong, capable leadership, "multiple management" should flourish, for it will be continuously encouraged as well as guided into becoming a vital, integrated force. Conversely, without positive direction from the top, much of the potential contribution from down-the-line personnel is seldom realized. As one functional executive stated:

> I decided I should give more attention to developing ideas and suggestions for achieving better overall results, but found there was no one to whom I could turn. I tried one member of the Executive Committee and received encouragement, but another turned me down. In the end, nothing happened; so I concluded that, as a practical matter, I had better just sit back and do my job as best I could.

Conclusions on Where the Staff Should Report

The principles of responsibility and the staff concept developed in the foregoing discussion of three alternatives lead to the following conclusions on organizing the systems and procedures improvement effort in a single-enterprise business.

1. In principle, staff personnel engaged in *inter*departmental procedures studies should report to the chief executive[1] of a relatively self-contained operating unit—for example, either to the president, executive vice president, or general manager of the company; or to the manager of a plant or product division that operates autonomously insofar as organization structure and operating systems are concerned. This is the arrangement that an increasing number of organizations have adopted, especially as the systems effort has shifted from an almost exclusive focus on accounting and finance to more emphasis on all functions of the enterprise—including marketing, personnel, and production, or their equivalents in other types of organizations. In one of the major security brokerage and underwriting firms, for example, the so-called corporate systems staff reports to the executive vice president for services along with such other functions as operations, general services, personnel, and the legal and compliance units. In one of the largest of the all-lines insurance companies, the so-called operations improvement staff reports to one of three executive vice presidents who, together with the chief executive officer, constitute the office of the president.

2. If the chief executive (or other general management executive) delegates his supervision of the staff to a subordinate executive, he should do so only under the conditions already set forth.

LOCATING THE SYSTEMS STAFF IN A MULTIPLE-ENTERPRISE ORGANIZATION

Thus far the discussion has centered on locating systems and procedures staffs within a single-enterprise organization. Hence, the remaining issue to be dealt with is the place of such a staff in the structure of a multiple-enterprise organization—for example, in a make-and-sell company that is organized into several profit-accountable product divisions. As a general principle, the staff's location in this instance should be governed by the extent of the authority that headquarters personnel exercise over the organizational structure and the systems and procedures of the operating units. But this observation serves only to raise the timeless issue of centralization versus decentralization as it applies to the problem of systems control: To what extent *should* the systems function be a headquarters or a local responsibility?

There is, of course, no one answer, just as there is no one answer to the question of centralization versus decentralization in its broader aspect. All this

---

[1]A systems staff does not, of course, need to report directly to the chief executive if it is only one of several staffs serving him. Under this circumstance it might well report to a "chief of staff" (as in one company, where a systems and procedures staff, an organization planning staff, and a strategic planning staff all report to the assistant to the president).

discussion can do, therefore, is suggest some of the criteria that should govern decisions in individual cases.

The two principal considerations to weighed are (1) the similarity of products or processes among the various operating units, and (2) the degree of interdependence among operating units, or between each unit and the central organization.

The more similar the activities of the various profit-accountable units—as, for example, in a chain of retail stores or branch banks—the more exchangeable are the procedures developed at individual points. Here maximum economy of effort is achieved by conducting systems and procedures studies through a central agency.

The more frequent the need for coordinated action between units, the more likely that uniform systems and procedures will be considered essential or desirable. For example, if there is a frequent exchange of personnel or a regular flow of work or components between product divisions, a greater degree of procedures standardization will probably be required than if these conditions do not prevail. Similarly, if the central organization includes a number of operating or service departments—such as sales, purchasing, or general accounting—uniformity is desirable at least in those divisional procedures that are tied into the day-to-day operations of headquarters groups.

If, however, the operating units have little in common—and particularly if they are widely dispersed geographically—local control over most systems and procedures matters is generally favored. Under these circumstances, the alternative of centralized systems development produces little economy of effort and often fails to meet local requirements promptly and intelligently.

However, within the broad limits indicated by these two sets of conditions, there are few companies that do not require some central control over systems and procedures. Even under the extreme form of decentralization, the headquarters organization must usually establish routines governing the flow of control information to the central authority.

Against this background, the general conclusion to be drawn is that a headquarters staff should be responsible for those procedures that are to be standardized throughout the company, and a divisional or subsidiary staff for those that do not need to be standardized. Where staffs exist at both points, it is important that the company policy on systems and procedures standardization be spelled out in enough detail to forestall confusion or disagreement over each group's area of responsibility. One company attempted to resolve this issue by publication of a bulletin stating, "Procedures problems or proposals involving more than one division or a division and the headquarters organization will be referred to the Central Systems Staff for handling." But in practice this statement proved to be little more than an evasion of decision, for it failed to specify the conditions under which a plant procedure *should* be made uniform among all plants. As a result, there were constant jurisdictional disputes and varying unilateral interpretations of the extent of central organization functional control.

Finally, the principles governing the place of systems and procedures staffs in the structure of a single-enterprise organization are equally applicable to the individual autonomous units of a multiple-enterprise organization. That is, *interdepartmental* procedures studies should be conducted by the staff of the autonomous unit's chief operating executive (for example, a divisional vice president or general manager); *intradepartmental* studies should be conducted by the department head or his staff, or by the divisional general staff operating on a service-rendering basis.

DEFINING THE STAFF'S RESPONSIBILITIES

The second major step in organizing for systems improvement is to spell out the staff's responsibilities in detail. One of the chronic causes of ineffectiveness and frustration among systems analysts is vagueness about what the staff is supposed to accomplish and how it is supposed to operate. To avoid this, the executives responsible for creating the staff should reach clear-cut decisions on its basic mission, its authority for initiating studies, and the scope of its coverage. Once made, these decisions should be reduced to writing and distributed throughout the entire managerial group.

Basic mission   This part of the total statement should specify what the staff is charged with accomplishing. Specifically, is its primary responsibility to be that of:

1. Recording existing practices by developing and maintaining a systems and procedures manual?

2. Standardizing and controlling printed or duplicated forms and office equipment?

3. Achieving uniformity of practice among similar units (say retail stores, warehouses, or branch offices)?

4. Straightening out paperwork problems and difficulties as they arise?

5. Increasing the end-result effectiveness of paperwork processes and cutting their cost?

Authority for initiating studies   The second phase of the staff's responsibility that needs clear-cut definition is its authority for getting studies under way. Specifically, a decision must be reached on which of the following two approaches the staff should take:

1. *The service-rendering approach.* Under this method of operation, the staff works primarily on the problems raised or projects requested by operating personnel.

2. *The programmed approach.* Here the staff initiates most of its own studies through some systematic means of identifying the major opportunities for improving the operational and management systems of the enterprise. Even

under this alternative, however, the staff does not ordinarily undertake a project without the agreement of the line executives concerned.

Scope of staff's coverage Finally, to complete the staff's charter, the *scope* of its coverage needs to be spelled out in terms of two dimensions: breadth and depth. First, how broad should be the reach of the staff's work in terms of the organizational units to be covered? Second, in any one study, how deep should be the staff's penetration in terms of the elements of management to be evaluated? Should it deal with all the factors affecting the complexity or cost of the administrative process—including policies, organization structure, and management information requirements—whenever these factors have a bearing on the process being studied? Or should the staff's work be limited to detailed procedural steps and clerical methods?

As an illustration of the job elements just discussed, one company's published statement of the systems and procedures staff's responsibility appears in Fig. 5.

SELECTING STAFF PERSONNEL

The final aspect of organization that needs to be dealt with here is that of recruiting and selecting the persons who will constitute the systems staff. In leading into this aspect of the organizing effort, it is hardly necessary to observe that no matter how vigorously the systems improvement program is launched or how resolutely it is supported by top management, its ultimate success will depend primarily on the quality of the analysts that make up the staff—their problem-solving skills and their ability to work effectively with large numbers of operating people at all levels of the organization.

This need, in turn, underscores the importance of developing guidelines for recruiting and selecting systems staff members. These guides break down into two categories: initial screening criteria to help assemble a pool of possible candidates and qualitative standards to aid in final selection decisions.

1. *Screening criteria.* The systems staff head should make a thoroughgoing (and periodically repeated) effort to identify all or a large group of persons within the organization who fall into the following categories:

   a. *Age 25 to 40.* Persons considered for the staff should be young enough to be adaptable and eager to broaden their exposure to the organization. In other words, they should be individuals who are still gaining experience for a key management job.

   b. *Salary.* As a quantitative measure of achievement, a candidate's compensation should be near the high end of the range for his or her age and experience.

   c. *Experience.* Candidates should have enough experience to establish their personal competence, but not so much as to promote complacency. The most relevant type of experience would be that gained either in a job requiring some analytical ability, or in a sufficiently wide range of jobs to have given the person a diversity of insights.

POSITION: SYSTEMS AND PROCEDURES DIRECTOR

Reports to Executive Vice-President

Responsibilities:

1. Takes the initiative in studying continuously all systems and procedures involving more than one department of the company. Develops, recommends, and assists in installing improvements that will: (a) improve planning, execution, coordination, and control and in other ways enhance operational and management effectiveness; and (b) reduce clerical, paperwork, and other administrative costs.

2. At the request of the department head concerned, studies existing intradepartmental routines and clerical work methods. Makes recommendations to the department head for simplifying or improving these activities.

3. Keeps continuously informed of new developments in EDP and all other types of office equipment. Undertakes feasibility studies and makes recommendations on the acquisition and application of such equipment to the company's operations.

4. Reviews all existing reports and requests for new reports to ensure that each is needed, that it does not duplicate other reports, and that it is prepared in the most economical way.

5. Makes work-measurement studies as the basis for determining the required clerical personnel complement for each department at varying volumes of work.

6. Reviews and makes recommendations regarding requisitions for the hiring of clerical personnel beyond budget limits and for the purchase of office machines.

7. Controls the ordering of printed or duplicated forms. This responsibility includes:

   a. Evaluating the need for proposed new forms and making recommendations for the elimination or combination of forms
   b. Ensuring that forms are so designed that they can be filled out and used with a minimum of clerical effort
   c. Establishing paper and printing specifications that will result in the minimum production costs consistent with the use to which each form is to be put.

8. Develops, obtains necessary approvals of, distributes, and keeps up to date standard practice instructions manuals covering any procedures that need to be reduced to writing in order to be carried out and controlled effectively. This responsibility includes:

   a. Acting as a clearinghouse on all permanent instructions issued to administrative personnel throughout the company
   b. Seeing that such instructions do not conflict with other instructions already issued
   c. Recording and issuing such instructions in standard manual form.

FIG. 5 *How one company defines the responsibilities of its systems and procedures director.*

2. *Qualitative or selection criteria.* Candidates chosen for the staff should possess a substantial measure of the following characteristics:

   a. *Analytical and perceptive mind.* Systems analysts must be able to think clearly, logically, and searchingly. They must know how to isolate and define the real problem, to see beyond symptoms to fundamental causes, to develop relevant facts and draw sound conclusions, and to detect obscure but significant relationships among the data obtained.

   b. *Aggressive curiosity, imagination, and creative intelligence.* Systems staff members must also have a questioning point of view—a willingness to challenge current practices and policies in the search for a better way. And they must be sufficiently resourceful to develop and evaluate a wide range of alternative solutions in arriving at recommendations, and to visualize the probable consequences of each alternative.

   c. *Tough-minded and result-oriented disposition.* In addition, the basic driving force in their makeup must be the desire to produce significant, lasting benefits through the systems studies they carry out, tangible benefits that can be factually supported.

   d. *A sensitive, professional temperament.* Since systems analysts must persuade people to act, they need a keen awareness of the attitudes and values of others. They must also have a professional concern for honoring confidences, respecting privileged information, and offering genuine help to the operating people with whom they work.

   e. *Skill at communications.* Finally, since so much of their effectiveness depends upon their ability to communicate with others, analysts must be able to transmit findings, conclusions, and recommendations—spoken or written—with clarity and impact.

Commitment of the foregoing kinds of people resources is an investment than any organization can well afford to make. Because a systems improvement program of the caliber described in this book can greatly accelerate the career development of talented younger men and women, it can, therefore, add materially to the organization's reservoir of talented individuals available for promotion to key management positions.

FIVE

The Top-Management Approach to Systems Analysis

The preceding chapter dealt with three of the ingredients in a successful systems improvement program: *objectives, policies,* and *organization.* As it emphasized, any such program must be guided by objectives and policies that reflect a general recognition among key executives of the importance of systems analysis as a continuing activity. In addition, the systems staff must be positioned at the proper place within the organization structure if it is to carry out its mission with maximum effectiveness.

Although these three factors are major prerequisites, obviously they cannot by themselves ensure effective action. They can only provide the framework within which action must take place. If the systems improvement effort is to produce the most fruitful results, this framework of objectives, policies, and organization must be buttressed by a broad-gauged, fundamental approach to the conduct of the work. Otherwise systems analysis is likely to take the form of random hacking away at obvious problems while the opportunities for far-reaching improvement are neglected.

In its discussion of the barriers to systems improvement, Chap. 3 spelled out what the basic approach of the systems staff should *not* be. This chapter, in turn, spells out what the staff's approach *should* be. Essentially, it consists of two elements.

1. The viewpoint, the set of values, or the way of thinking that the staff applies to its work
2. The way in which it tackles individual projects

THE TOP-MANAGEMENT VIEWPOINT

The single determinant that most clearly "separates the men from the boys" in systems and procedures work is whether the staff members possess a top-management point of view. The identifying characteristic of such a viewpoint is that it continuously focuses on the objectives of the organization as a whole. To be sure, systems analysts may frequently be concerned with a problem that is— or at least, appears to be—limited to one area of the enterprise. But if they are trained in and dedicated to applying the top-management approach, their minds—consciously or unconsciously—will raise and seek answers to questions such as these:

1. As a general framework for identifying systems problems and improvement opportunities, do we have a sophisticated understanding of the profit economics of our business, the key factors controlling success in this industry, and the impact of each department on overall results? Does our improvement program give top priority to those systems that have a major impact on the quality of the company's performance in each of these key areas?

2. On each project undertaken, are we dealing with real underlying causes or merely with symptoms?

3. If this system were optimal, what conditions would exist or what end-result benefits would be realized? What is the potential significance of these conditions or results to the overall performance of the organization?

4. What is the tie-in with other parts of the business? Specifically, can the problem be solved in this particular area, or is it affected by requirements and conditions in other departments?

5. In the situation under study, what is the relationship among organization, policy, system, and method? What is the relative importance of each to the total solution?

6. Is this activity or proposed solution clearly worth what it costs or is likely to cost?

7. How sure are we that a given solution or course of action will serve the best interests of the organization as a whole instead of merely suboptimizing one of its segments?

From these and similar questions, it can be inferred that the top-management approach is the approach of the broad-gauged business manager rather than of the technician. Someone has said that the trouble with many business organizations today is that there are too many executives and not enough businessmen. In that statement lies the gist of the definition we seek: the top-management approach is the approach of those whose breadth of vision enables them to foresee the impact of business problems and practices on each other, and whose business judgment enables them to evaluate all things in light of their probable effect on overall results. It is the approach that leads a person to apply to every business problem or opportunity and to every alternative course of action the tough-minded question: What will this solution do to promote the health, the competitive vigor, and the long-term profitability of our business?

This compelling point of view is well expressed by Malcolm McNair in his book *The Case Method at the Harvard Business School:*

> William James, the great teacher of philosophy at Harvard during the early years of this century, made a useful distinction between people who are "tough-minded" and people who are "tender-minded." These terms have nothing to do with the levels of ethical conduct; the "toughness" referred to is toughness of the intellectual apparatus . . . not toughness of the heart. Essentially it is the attitude and the qualities and the training that enables one to seize on facts and make those facts a basis for intelligent, courageous action.[1]

THE APPROACH TO INDIVIDUAL STUDIES

The second element in the top-management approach is the method of attack used by the staff in carrying out individual systems and procedures studies. Specifically, the attack should be built around six steps taken in the sequence listed:

1. In planning the study, be sure to *relate its scope to the kinds of benefits* it is aimed at achieving.

2. Determine whether the *basic purpose or objective* of the system is being accomplished effectively.

3. Determine *constraints and ancillary requirements.*

4. Reevaluate *the work that must be done* or that earns a profit.

5. Analyze the *indirect factors affecting the volume of work or complexity* of the system.

6. Finally, determine the *best method* of performing the essential work.

Relate Scope of Project to Benefits Sought

The scope of individual systems studies should differ depending on the kinds of benefits sought. For example, if the project is aimed at improving operational effectiveness, then the *total system* should be studied, not just some of its isolated segments. Alternatively, if the project is targeted at reducing clerical and other office costs, then the *organizational component* responsible for those costs should be studied in its entirety, regardless of the number of systems in which it participates. In other words, the first type of project should follow a flow-of-work approach; the second, an organizational-unit approach.

The total-system approach   The paperwork of any organization, however large and diverse it is, comprises not more than a handful of basic, serialized processes. Each of these processes is an entity regardless of the number of persons participating in its performance or the place of those individuals in the organization structure. Thus, where the objective is to improve the

---

[1]McGraw-Hill, New York, 1954, p. 15.

results obtained from the organization's network of systems, the unit of study should be one of these processes, from beginning to end, through every organizational unit and individual contributing to it. Only when the scope of the study is defined in this way can the spotlight be thrown on the relationships between jobs and organizational segments and on all the other factors affecting the quality of the end results achieved.

This point becomes clear when we consider the series of steps and the organizational groups involved in the typical system for processing sales orders within a manufacturing company. If a company produces its finished goods for stock, this procedure might consist of some or all of the following functions and activities:

1. Order entering and filling
   a. Registering the customer's order and assigning an order number
   b. Editing the order for completeness and accuracy of information and converting it to standard company terms
   c. Establishing the shipping date if the date requested by the customer cannot be met (which in effect amounts to establishing the priority of shipments)
   d. Making adjustments in the quantities, styles, models, or grades ordered
   e. Determining the plant or warehouse from which shipment is to be made
   f. Obtaining approval of the credit risk and terms of sale
   g. Acknowledging the order if shipment is not to be made immediately
   h. Preparing, distributing, and filing copies of
      (1) Shipping orders and notices
      (2) Change orders
      (3) Back orders
      (4) Packing slips
      (5) Shipping labels
      (6) Bills of lading
   i. Posting finished-goods inventory records
   j. Preparing orders to manufacture for stock
2. Order service
   a. Investigating and supplying information in response to customers' inquiries about the status of their orders
   b. Handling customers' emergency requests
   c. Handling adjustments of customers' claims
   d. Keeping the sales force informed of the extent to which shipments are falling behind schedule
3. Invoicing
   a. Pricing and extending shipments and customer returns
   b. Computing discounts, freight rates, etc.
   c. Preparing, checking, and distributing invoices (or statements) and credit memos

4. Record keeping and report preparation
   a. Maintaining files or records of
      (1) Sales contracts
      (2) Open orders and shipments
      (3) Returnable containers
      (4) Accounts receivable
   b. Preparing summaries and analyses of orders and shipments for sales management, market research, and similar purposes
   c. Supplying the manufacturing division with information on unfilled orders for production planning purposes
   d. Preparing commission reports by salesman or broker
   e. Compiling sales tax reports

In companies that manufacture only against specific customers' orders, this list would be expanded by some or all of the following activities:

1. Obtaining and transmitting delivery promise dates to sales representatives and customers
2. Allocating orders to factories
3. Designating the manufacturing sequence of orders
4. Issuing production orders to the production planning and control, engineering, and accounting departments
5. Following up the factory on delivery promises

It is evident that these manifold activities not only involve almost every department of an organization, but also have a high degree of interrelationship. Each flows from one and feeds into another. By approaching them as a unit rather than singly, the systems staff is able to visualize and deal with their relationships more imaginatively—to detect organizational problems, duplication of effort, excessive controls, and similar improvement opportunities that otherwise would be difficult to spot. Most important, staff attention is focused not on improving related pieces of action separately, but on developing a highly integrated process that represents the optimum means for carrying them out as a whole.

The organizational-unit approach   In contrast to the foregoing, when the major —if not exclusive—goal of a study is the reduction of clerical and other administrative costs, the approach taken should be a study in depth of *the work of each position making up a given organizational component.* Only in this way will the analyst be sure of uncovering the many peripheral tasks, nonintegrated activities, and tributaries to the mainstream that the flow-of-work approach does not reveal and that can become a large part of any organization's total office activity and cost.

Determine How Well the System Achieves Its Basic Objective

It has been emphasized elsewhere that a primary purpose of business systems is to increase the effectiveness of the entire organization by improving coordination, facilitating decision making and action, and providing adequate control. Until these end results are satisfactorily accomplished, it is premature and unwise to concentrate on reducing the cost of operating the system itself. This is true for two reasons: (1) However attractive the direct savings in clerical labor may be, they are often secondary to the savings and benefits that result from a better end product of the clerical effort. (2) Unless systems changes aimed solely at clerical cost reduction are evaluated in the light of their effect upon end results, the saving achieved may prove to be wholly illusory. Indiscriminate elimination of office tasks may weaken needed controls or retard the flow of work sufficiently to produce hidden losses greater than the saving.

There is a tendency among systems analysts to become so diverted by the lure of laborsaving opportunities that they lose sight of the fact that systems and procedures are only means to an end. It is hardly necessary to add that the lowest-cost means of achieving an unsatisfactory end result is still an unwarranted extravagance. In developing a top-management approach, therefore, analysts must make a critical appraisal of *how effectively the desired end is achieved* before they give attention to reducing the intrinsic cost of the means.

The various directions that this inquiry might take are indicated by such questions as the following:

1. Does the annual profit planning and budgeting system lead to the imaginative identification of concrete ways in which each manager plans to further improve the performance of his department and cut his unit costs in the coming year?

2. Does the capital budgeting system develop the information and analyses that management needs to make strategically sound tradeoff and resource-allocation decisions?

3. Does the cost accounting system supply management with timely, significant, and highly usable information on the reasons for variances from planned profits and on the extent of each department's responsibility for these variances?

4. Do maintenance and repair accounting routines produce the facts needed to measure performance on recurring jobs, to plan the optimum amount of preventive maintenance, and to establish equipment replacement policies?

5. Does the procedure for developing delivery quotations to customers produce the information required to ensure that delivery promises are realistic?

6. Does the stock control procedure provide the means for preventing or anticipating shortages so that production interruptions, constant rescheduling, unnecessary machinery changes, and wasteful emergency purchases are avoided?

7. Does the procedure for handling customers' inquiries about the status of their orders ensure competitively superior customer service?

8. Do the invoice coding and sales statistical procedures produce the basic internal data required to plan sales strategy, to control selling effort, and to spotlight opportunities for major improvements in the distribution program?

9. Do the clerical activities that support field sales help to increase sales representatives' productivity by maximizing the amount of time that the sales force is able to spend in direct selling?

10. In retail stores, does the procedure for handling customer complaints produce the information necessary to target complaint-prevention efforts?

11. In property and casualty insurance companies, do the premium billing and accounting procedures optimize cash flow for investment?

12. Does the organization maintain personnel records just because it is traditional to do so? Have these records been shaped solely by legal requirements and union contract provisions? Or have they been consciously designed as an effective tool for personnel management? Are their content, arrangement, and accuracy such that they facilitate the analyses required for personnel planning and control? Can they be used for analyzing turnover? For forecasting labor costs or determining the effect of pay increases? For planning the best mix of labor grades to perform a given function? For carrying out a program of promotion from within, along defined career paths?

It is true, of course, that where this type of inquiry reveals profitable opportunities to strengthen end results, development of the desired improvement may have to await more detailed analysis of individual operations. But the systems staff has taken the initial step of achieving the mental orientation essential to a sound solution. Once it has acquired a thorough awareness and appreciation of the system's basic objective, its subsequent work, no matter how detailed, is not likely to spoil its aim or obscure the target.

Identify Constraints and Ancillary Requirements

The next step in a fundamental approach to systems analysis is to determine the constraints that must be observed and the minimal requirements that must be met in the design of a new system or the modification of an existing one.

Such constraints might originate externally or internally and might be imposed by any number of forces: regulatory requirements, competitive practices, company policies, or the skills of the available clerical labor, to name a few. Requirements to be met can be equally varied and might include, for example, the minimum standards of service to customers that must be maintained in the conduct of a project aimed primarily at cutting clerical costs.

Of course, the need for developing a realistic awareness of such limitations and requirements does not mean that they must be accepted blindly. Quite the contrary. One of the responsibilities of the analyst is to evaluate each of these governing factors thoroughly and either revalidate it or invalidate it. And this challenge should not be limited to internal factors such as company policies or operating requirements. The same sort of questioning viewpoint needs to be applied to external constraints and requirements, including even regulatory demands. All too often, these external conditions—whether they be alleged customer requirements, competitive practices, or regulatory restrictions—are either more apparent than real or, although once valid, have since ceased to exist without any corresponding adjustment in a company's response.

An example can be found in the experience of a specialty chemical company which produces a toxic chemical that is shipped in large quantities to many parts of the world. Because of the hazardous nature of the product, it is shipped in high-strength, virtually indestructible, steel drums. Given the cost of these containers (each is worth many times the value of the product inside), the chemical company requires that empty drums be returned from all countries to which they have been shipped. To avoid United States import duty on the incoming shipments, the company, for years, numbered each container serially and, through these numbers, kept detailed records of the movement of thousands of individual containers across the world and back.

The company was convinced that this kind of costly record keeping, with its high incidence of error and high expense of error detection, was required by the U.S. Bureau of Customs. But when this control system was finally challenged, the company discovered that the Bureau of Customs did not, *in fact, demand records by individual numbered drums. It had never required more than simple records of the total number of drums in each shipment out of the country, the total number returned in each incoming shipment, and the total number still in foreign countries.*

Determine What Work Is Worth Doing

The next logical step in the top-management approach is to reevaluate the work that *must* be done in order to achieve the basic objective (within the framework of valid constraints). Here, of course, the analysts' goal should be to eliminate all unnecessary or unprofitable work and all work of marginal value.

This step sounds straightforward and simple, but unfortunately it is not. In fact, it is by far the hardest of the six steps making up the top-management approach. The reason is that the heads of office activities *do not have to show a profit* on their operations. From the supervisor's standpoint this circumstance may seem to be an advantage. But from the company's standpoint, it is a heavy cross to bear, for it does more to inhibit clear, objective thinking than any other condition of business life.

In the face of this built-in obstacle, how do systems analysts go about determining what is unnecessary or unprofitable and what is of marginal value?

Part of the answer is that their whole point of view has to be an imaginative, courageous, venturesome one instead of the typically cautious one. They have to develop the habit of challenging everything that is done instead of accepting it—of persistently asking: Why are we doing this at all? If we once had to do it, do we still have to? And of course, as the other side of the same coin, they must always be ready to take a calculated business risk and to accept something less than perfection.

In addition, to do a good job in evaluating a function, the analyst must exercise unusually sharp judgment because it is easy to fall into the trap of mistaking reasons for justification. The real test of necessity does not lie simply in determining whether a given function serves some useful purpose. Few, if any, activities will fail to pass that limited a test. Just about everything that is done serves some purpose or produces some value. Yet it is so often true that anyone defending a given activity begins with an explanation of the reason it is performed and ends with a recitation of the disasters that would result if it were eliminated.

The real test of necessity is whether the benefit produced by the activity is greater than the cost. Careful probing into this area of value-cost relationship (or cost-effectiveness) often leads to surprising results. It is not uncommon to find that some of the most hallowed business practices least measure up to this standard.

To illustrate: A few years ago, one of the country's largest mail-order houses adopted the practice of destroying all correspondence, shipping papers, and other documents associated with an order as soon as shipment to the customer had been made. Obviously this practice subjects the company to a certain risk that it did not face when it retained these records. Nevertheless, a careful study conducted over a considerable period of time revealed that the cost of filing, storing, and searching through this tremendous volume of papers was greater than any losses that would result from inability to produce positive evidence of shipment.

But outright elimination of an activity is not the only alternative. If systems analysts conclude that a function cannot be entirely eliminated, they should then determine whether the volume of work—that is, the number of transactions or frequency of performance—cannot safely be curtailed.

An example of the potential payout of this line of inquiry can be found in a study of the purchasing department of a large petroleum company. One of the functions performed by the clerical staff of this group was to check vendors' invoices (unit price charged, terms of sale, and arithmetic computations). Many systems analysts would skip quickly over this sort of activity, because the checking of vendors' invoices seems so obviously necessary. Yet, under the discipline of challenging every function, the study team learned that 61 percent of all invoices received by the company were for less than $100 each and that in total these accounted for less than 10 percent of the dollar value of all invoices received. By maintaining error detection records on a sample of 7,500 invoices, they found that the invoices amounting to less than $100 each contained only ten errors. If undetected, those errors would have netted out to a $30.33 gain by the company. Consequently the company stopped checking invoices under $100, eliminated six clerical positions, and thereby saved $55,000 a year.

Another type of improvement that might result from value-cost analysis is a change in the level or quality of service that a company provides. A case in point is the experience of a large consumer-goods manufacturer which ships its products to wholesale grocers and grocery chain stores. To determine the optimum speed of customer service in processing and shipping orders, the company periodically experiments with different speeds for different classes of customers. By this process it determines what is the slowest (and therefore the least expensive) speed of service it can maintain without risking a significant number of customer complaints or lost orders.

One final point needs to be emphasized about determining the real profit-making value of the work being studied: To realize the maximum payout, the analyst should acquire the habit of challenging whole functions or major activities *before* questioning detailed operations. The obvious reason, of course, is that it is pointless to study individual operations if the entire function of which they are a part can be eliminated or sharply curtailed.

Identify the Hidden Causes of Complexity and Volume of Work

Very often the cost of an operational or management system is governed less by the method of carrying it out than by such other factors as company policies, the requirements and practices of other departments, the goals that have been established for speed and accuracy, or the way in which the organization is structured—particularly at down-the-line levels. These indirect factors affect clerical and other administrative costs in three ways:

1. They may produce peaks and valleys in the workload.

2. They may compound the complexity of a routine (*a*) by multiplying the number of units or "pairs of hands" through which a transaction passes, (*b*) by introducing many exceptions to the normal flow of work, or (*c*) by increasing checking or other quality control operations.

3. They may significantly influence the volume of work to be done.

In most systems studies, some attention is generally given to leveling out or reducing the impact of irregularities in the workload. Much less frequently, however, is any investigation made of the factors that determine the complexity of the process or the volume of work. Neglect of these causal factors is as costly as it is pervasive, for analysis of this type can often produce savings that dwarf those resulting from methods improvement.

Causes of complexity   Two of the most common causes of complexity in systems and procedures are lack of well-defined policies and laxness in enforcing policies. Whichever the cause, the inevitable result is a large number of exceptions to the normal routine for carrying on the activity in question—exceptions that result in special handling, extra records, complex controls, more errors, and delays in the processing of work.

When analysts encounter an excessively involved routine, it is particularly important, therefore, that they first examine the *basic requirements or policies* that the routine is intended to satisfy. They will usually find that the path to work simplification lies within those areas, not within the mechanics of the system or procedure itself.

It is obvious, of course, that neither the substance of business policies nor the manner in which they are executed should be governed principally by the attendant clerical cost. But chronic variations in the handling of recurring business matters are seldom the product of considered policy administration.

They usually reflect a failure to define policies or a general lack of understanding of the policies in effect.

An example can be found in an involved billing and accounting procedure that developed in a large service organization from an attempt to keep pace with complex and highly variable sales practices. Few of the policy decisions that had been made were reduced to writing in some permanent form. And on many sales matters, ad hoc decisions were continuously required in the absence of policies in any form. The result was that sales representatives had widely different concepts of their authority to commit the company on prices and discounts, billing and payment arrangements, conversion from one type of service to another, and so on.

Repeated efforts to improve performance within the accounting and billing areas were unsuccessful until the company undertook an intensive program of policy review and development. The outcome of this program was a written manual defining the scope and application of each sales policy. This manual accomplished two purposes: (1) By promoting a clear understanding of what policies had been established and how they should be applied, it gave the company a basis for setting up effective controls to ensure adherence. (2) It provided the uniformity of practice essential to standardization and simplification of procedures.

Causes of work volume   Once the basic causes of complexity have been isolated and eliminated where possible, analysts should turn their attention to the factors determining the volume of work or the overall cost of an activity. Here again the inquiry must be organizationwide in scope, for these causal factors are rarely under the control of the departments responsible for carrying out the system or procedure. The size and diversity of the savings potential in this area are illustrated by the experiences of the following three companies.

The first is a light machinery manufacturing company that undertook a cost-reduction study of all its clerical activities. The analyst assigned to the study found that the company was manufacturing its detail parts in such small lots (or in such short "standard runs," as they were called) that it was frequently necessary to machine the same part every 4 to 6 weeks. Aside from its adverse effect on manufacturing costs, this practice also resulted in substantially higher clerical costs. Investigation of clerical routines in five departments—production control, inventory control, cost accounting, timekeeping, and payroll—showed that the workload of 60 percent of the clerks was determined not by the company's total volume of business, but by the number of shop orders processed and therefore by the average size of the manufacturing lots. These clerks required no more time to process a shop order, material requisitions, and job timecards for 500 of a given part than for 10 of the same part. Thus, when the average size of the company's manufacturing lots was doubled, the workload of these five departments was decreased by 30 percent, with a corresponding reduction in clerical costs.

A second company had centralized files that were maintained by eight file clerks. A study of sorting and filing techniques led to the purchase of sorting racks that saved 4 worker-hours a day or half the time of one clerk. Two years later, when the company's office space became critically short, the same filing activity was restudied. This time the study included an analysis of records retention and destruction policies, and it led to a decision to destroy 40 percent of the material before it ever reached central files. With this decline in volume resulting from a sound change in policy, the work of three clerks was saved.

The final illustration involves a large processing company that carried approximately 25,000 different items in its maintenance, repair, and operating stores. Under the pressure of a companywide program to reduce inventory investment, the storeroom was following a policy of ordering only a 90-day supply of all items but standby parts. A study showed, however, that on a great majority of these items, the acquisition costs of a 90-day inventory were much greater than the carrying costs. (Acquisition costs include the clerical and other costs of requisitioning, purchasing, receiving, and paying for the item.) On the basis of these facts, the company divided its stores inventory into three classes of items and established for each class a different supply level, ranging from a 3-month supply for the highest-value items to a 12-month supply for the lowest. The new system raised inventory investment costs slightly, but it reduced clerical costs sufficiently to produce a sizable saving in the combined total of carrying and acquisition costs.

Not all opportunities the analyst comes across during this phase of the basic approach will be so far-reaching as the three described. But, at the same time, some opportunities of this type can usually be found in every company. For example, significant savings can often be achieved simply by making relatively minor adjustments in internal operating policies, accounting practices, completion deadlines, speed and accuracy requirements, and so on.

Determine the Best Method of Performing Essential Work

Only after the major steps discussed above have been completed should analysts undertake an evaluation of the methods by which the essential work is performed. Even in this final step, however, the approach should still be broad-gauged and fundamental instead of narrow and detailed. That is, before becoming preoccupied with simplifying the methods used by individual workers, systems analysts should explore basically different methods for carrying out the activity as a whole.

In initiating this final step, analysts will find it helpful to begin by describing the *basic nature and purpose* of the function under consideration *in as few words as possible.* This exercise, in effect, amounts to model building—creating a schematic diagram or word picture of the essential characteristics of the total system or procedure. With this picture before them, the analysts' next step should be a wholly theoretical one. That is, without reference to limitations of existing facilities or personnel, they should prepare a list of all the means by which the function might conceivably be performed. The first round in the process of evaluating these alternatives should then lead to eliminating the ones that prove impracticable and to digging more deeply into those that seem promising.

Often analysts will find that they must reserve final decision on the question of basic methods until they have analyzed existing operations in detail and outlined all possible improvements. These steps are necessary, of course, to obtain a fair basis of comparison since, in weighing new methods, maximum operating efficiency is presupposed. Once again, however, initial consideration of basic requirements will reduce the danger of faulty or superficial diagnosis. It will help analysts avoid the common error of patching up an existing method that is fundamentally unsuited to the overall job to be done.

CAUTIONS FOR THE SYSTEMS ANALYST

In developing the concept of a top-management approach to systems and procedures studies, this chapter has emphasized the need for delineating the *areas* to be explored. The basic premise is that the analysis should deal with fundamentals, not frills—that it should probe for causes, not merely treat effects. But penetration in sizing up problems and opportunities is not enough. A well-balanced approach also demands of systems analysts that their recommendations reflect a practical appreciation of the job to be done and of the personnel resources available to do it. Some words of caution should be added, therefore, lest, in their effort to make a searching diagnosis, they lose their touch of realism in rendering treatment or writing the prescription.

Much has already been said in this and the preceding chapters about the control aspects of operating procedures. But, in one's zeal for achieving effective control, there is a danger of assuming that the tighter or more encompassing the control, the better will be the end result. As a practical matter, however, it is as easy to err on the side of developing an unnecessarily precise, and therefore a wastefully expensive, control as it is to develop an inadequate, makeshift one. The problem is nothing more nor less than that of selecting the proper tools for the job to be done. Machinists cannot work to close tolerances without precision instruments. On the other hand, micrometers are not needed in a box factory.

The second caution is that the development effort should be aimed at creating a simple, workable system as distinct from a perfect system. To be sure, the initial effort should be directed toward developing the ideal. But once that point is reached, reappraisal and modification should be made in light of the skill and experience of the work force and the cost of perfection. As one speaker put it, "Hire a few workers to mop the floor so that you don't have to develop a perfect system that will keep 400 people from dropping things on the floor."

This caution is not an invitation to resort to looseness or expediency or to disregard long-range improvement objectives. It is simply an appeal for common sense in applying theoretically sound principles and techniques to practical operating problems.

The final caution is that *top-level* analysis does not mean merely broad-brush analysis. The approach described in this chapter does not excuse analysts, in any way, from "getting their hands dirty" by digging deeply into underlying details to solve any problem. To be sure, it requires that they think imaginatively and conceptually about the broad administrative problems of the organization. But, to the same extent, it demands that they bring to this effort the kind of thoroughgoing, completed staff work that converts basic concepts into profitable operating realities.

SIX

Gaining Acceptance
of the Systems Improvement
Program

The steps outlined in Chaps. 4 and 5 largely complete the drawing-board phase of systems development work. The "machine" has been designed, and the analyst is now confronted with the somewhat more practical job of building it and making it run. At this point, emphasis shifts from problems of design to problems of execution: from planning to doing. Skilled performance in each area is equally essential to achievement but, in each, different factors have to be reckoned with. In a sense, planning deals largely with abstractions—with objectives, structure, and theoretical approaches. Execution is concerned with vital forces—with human relationships, attitudes, and interests, as well as with trained analysts actively applying their skills.

This distinction highlights the nature of the remaining ingredients necessary to effective operation. Although many factors can cause some soundly conceived systems programs to fail and others to succeed spectacularly, most of these factors fall readily into one of the following four categories:

1. The management climate and value system of the organization
2. Relationships between top management and the systems staff
3. The staff's standards of performance
4. Relationships between the staff and operating personnel

THE MANAGEMENT CLIMATE

Once a systems staff has been brought into being, its success will be determined to a large extent by the management climate and values of the organization that the staff exists to serve—more specifically, by the character of the support that the systems staff receives from the executive in whose area of responsibility it is to operate. For a companywide program, this is the support given by the chief executive; for a more restricted program, that given by a divisional or departmental executive.

This is not a new thought. The need for positive and continuing management support of any such undertaking is widely accepted. Its equivalent can be heard wherever papers are read or discussions held about any function that involves several or all departments of an organization. Yet such pronouncements are seldom discussed in helpful detail. The thesis having been advanced, it is usually left to stand as a vague and unadorned generality. In this form, it has limited value as a stimulus to management, and it tempts others to make unwarranted assumptions regarding the nature of the support they have a right to receive. It is worthwhile, therefore, to consider not only how management support should manifest itself, but also what should *not* be expected of it.

What "Management Support" Does Not Mean

Time and again one hears earnest but disheartened systems analysts insist that they cannot put their recommendations into effect because, as a staff, they do not have the complete and unquestioning support of management. Occasionally this explanation is but a convenient catchphrase to defend lack of accomplishment. More often, however, it is the sincere conviction of conscientious staff members who see only that badly needed improvements, which they have worked diligently to develop, are bogged down in the indifference, stubbornness, or inertia of operating managers and supervisors. Confronted with these obstacles and convinced of the soundness of their recommendations, analysts frequently reveal a predisposition to favor what seems the quick and logical solution: They feel that management should give the staff full authority to enforce its recommendations whenever other means of securing their adoption have failed. (The word "enforce" is used intentionally, for it is not uncommon to the vocabulary of systems staff personnel.) A somewhat less extreme position, taken by others, is that top management itself should exercise whatever authority is necessary to prevent worthwhile improvements from being blocked by sheer emotional resistance. This, of course, is only a euphemistic way of saying that when soundly developed systems proposals are not accepted voluntarily, management should "shove them down the throats" of recalcitrant operating personnel.

Fortunately, most managements have been diffident about delegating or exercising their line authority in this manner—and justly so. For it is obviously an anomaly to maintain that the coordination of effort can be promoted through pressure or compulsion. But as long as the systems staff persists in believing that management support is the means of overcoming resistance to specific recommendations, much of its energy will be dissipated in frustration and special pleading.

The only way to prevent this mistaken notion from taking hold is to ensure that staff members clearly understand the nature of their task. Theirs is a twofold job. It is not merely a matter of generating sound ideas through an imaginative, analytical approach to problems. It is just as much a matter of *gaining the acceptance* of those who will put the ideas into practice. No one would seriously contend that management support should be a substitute for technical competence of the staff. While it may be less apparent, it is no less true that management pressure should not be a crutch for the staff's lack of skill in handling the human relations aspect of its work.

We do not gainsay the existence of apathy, lack of cooperation, or resistance on the part of operating personnel. These obstacles are real, and no organization is without them. But they often loom larger in the imagination than they are in reality. What the analyst sometimes mistakes for unwarranted opposition may be a natural disinclination to "buy" a product that has been designed poorly or presented unconvincingly. In a sense, therefore, the resistance of operating personnel—however subjective it may be—has real value, for it challenges the conscientious systems analyst to do such outstanding, fact-based work that its rejection by any of the persons concerned is recognized by the others as pure obstructionism.

In summary, management support should not be construed to mean either the assumption of management's line authority by systems staff personnel or reliance on the top hierarchy for decision making as a substitute for the staff's reaching general agreement at lower levels through its own efforts.

What "Management Support" Does Mean

Management support of a systems program means *belief in and cultivation of an idea*—the idea that continuous adaptation, simplification, and strengthening of the organization's systems and procedures are essential to operating effectiveness, and that the effort exerted toward this end must be carefully planned, thoroughly integrated, and competently directed. Diffusion of this idea throughout the organization cannot be accomplished by passive means. It requires sustained action of a kind that only top management can take and without which systems work must, perforce, be carried on under the most disheartening handicaps.

To create the climate in which this idea can flourish, top management's basic task is to *stimulate among the entire executive and supervisory force an understanding of the problem and the opportunities it holds, and a will for united action.* This task highlights the nature of leadership. It is not just a matter of the chief executive's accepting an idea, but of his communicating and interpreting that idea in a way that gives it power and substance to those who must carry it out. Organizational efficiency and effectiveness begin with top management's determination to have them. They are an attitude of mind that top management must not only consciously adopt but also consistently transmit to all other levels of management.

Development of this attitude is not a one-time job. It must be worked at continually. To be sure, a good start can be made by a wholehearted expression of management interest and encouragement at the time the systems improvement program is launched. But the real need is one of ensuring the durability and intelligent application of the concept by gradually building it into the thinking habits of key personnel. This is a job of managerial and supervisory education. It cannot be achieved by any means other than consistent, demanding, resolute leadership. Without this brand of leadership, there is little possibility of real success.

Following are some of the specific concepts and values that management should seek to instill through this process of education:

1. Top-level executives must develop throughout the organization a full appreciation of the part that operating and management systems play in the running of any large-scale organization.

2. Management must also foster a strong, continuing *improvement* orientation that is built into the very bones and sinews of the organization. This sort of creative, venturesome drive is likely to flourish within an organization only if large numbers of its people have the courage, the imagination, and the intellectual toughness to question everything that is being done—no matter how long established, how unavoidable, or how obviously logical it may seem.

To promote this way of thinking, top management must communicate its determination that the systems improvement program will *not* amount merely to a "raking of leaves" here and there, but that it will be a dynamic force in the management of the enterprise. Far from being just another one-shot improvement campaign, it must have the uncompromising objective of producing a continuing flow of substantial and enduring benefits. Accordingly everyone must understand that the ultimate payout comes not from *making changes* but from consistently *cashing in on them.* As just one example, this *focus on results* means that the organization must discipline itself to convert work eliminations into realized savings through personnel reductions.

One step toward the achievement of a continuing improvement orientation is the following statement, taken from one company's written management philosophy:

> As a management team, we seek the development of an open inquiring state of mind. We are dedicated to bringing the research point of view to bear on all phases of our business. We are never satisfied with things as they are. We assume that everything and anything—whether it be performance results, products, systems, or methods—can be improved.

The chief executive of another company puts it this way:

> Top management must work hard to create an atmosphere of receptivity to change. It is important to make change, or more accurately the consideration of change, fashionable.

3. Management must stress the need to optimize the contribution that each functional segment makes to the achievement of the overall goals of the

enterprise. With respect to this need, it should be emphasized that the sort of strong, continuing improvement orientation just discussed is—by itself—not enough. The reason is that such an orientation could lead the organization to pursue suboptimal objectives—like, for example, improving the efficiency of activities that ought not be performed at all, or in other ways focusing the efforts of competent, dedicated people on all the wrong things. Hence, another critical need that only top management can fill is to ensure that the jobs of all functional executives are defined in terms of the contributions they are expected to make to the attainment of the company's overall economic results. It has been our observation that strong, healthy organizations are managed in a way that fixes everyone's attention on the identification and achievement of companywide goals. In this sort of environment, the focus of each departmental manager is on how he can help build the company, not just on how he can build and strengthen his own function. That is, his objective consists not just of reaching a higher and higher level of technical perfection within his department, but of fixing on the kind of performance and the types of decisions that will enable him to make the maximum contribution to the achievement of the company's end-result objectives.

4. The entire organization must come to understand how widespread the scope and objectives of the systems improvement program are. Here management must take the lead in dispelling the misconception that the program is limited to the office area or to the reduction of clerical costs. The ten types of improvement opportunities or potential benefits discussed in Chap. 2 should serve as a useful checklist in raising the organization's sights on this matter.

5. Management must make the organization aware that significant improvements cannot be gained cheaply. Their achievement almost always requires a great deal of hard, competent, disciplined effort. And, in addition, a real price has to be paid for them in terms of giving things up, taking some calculated business risks, and making some courageous, difficult decisions affecting status symbols and traditional practices.

6. Management must also communicate a clear understanding of the responsibilities of both line and staff for the success of the program. Operating managers and supervisors must see the systems improvement undertaking not as something that is being done *to* them or *for* them but as something for which they have a major responsibility. That is, the basic attitude of line management must reflect not just a willingness to endorse the program or to cooperate with the staff, but a willingness to play a leading role in ensuring that worthwhile results are in fact achieved. At the same time, the line organization must understand that if the systems improvement effort is to be orderly, continuous, and companywide, it must be planned, organized, and spearheaded by staff specialists.

7. Finally, top management must demonstrate that it has a continuing interest in the systems program by giving visible signals that "the name of the game" is never-ending improvement in every aspect of the company's operations. This means, for example, that management must identify and encourage

agents of change within the organization, and must protect people who think differently from others. It must give recognition to those who are willing to stick their necks out by challenging what is now done and by pushing in new directions. But most of all, it must reward the operating managers or supervisors who get results—that is, the ones who do an outstanding job of streamlining their own procedures, cutting their own clerical costs, and cooperating effectively with the systems staff on organization-wide projects. Only through such reinforcing measures as these will top management ensure the *continuing* vitality of the systems improvement program. For regardless of the resolution and enthusiasm with which the program may have been launched initially, down-the-line personnel will judge whether management means what it says by the way it acknowledges and rewards good performance.

RELATIONSHIPS BETWEEN TOP MANAGEMENT AND THE STAFF

In addition to building understanding and commitment among the organization's line executives and supervisors, top management has another responsibility critical to the success of the systems improvement program. This responsibility is one that it shares with the systems staff. We refer to the development of an effective relationship between the two. Development of this relationship is a reciprocal matter; it cannot flourish without a sustained mutual effort.

Top Management's Responsibility for Relationships with the Staff

Top management's responsibility for fostering these relationships consists of two parts: (1) learning how to guide and use the staff, and (2) making certain that the staff gets a fair chance to do an effective job.

Learn how to guide and use the staff This responsibility assumes particular importance when we consider how the systems staff works with the line executive to whom it reports. Other types of planning staffs normally work more closely with their chief. They exist principally to increase his personal effectiveness; they have frequent contact with him and get most of their assignments directly from him. A systems staff, on the other hand, receives many of its requests for service from other sources. As long as it can fit them into its schedule, it seldom discusses them specifically with the chief. The balance of its work is made up mainly of projects that it originates itself as part of a master plan previously approved by the chief. Even on projects that the superior himself initiates, the staff may work on its own for long periods before reaching a stage of completion suitable for discussion with him.

These circumstances tend to break down the communication without which top management's initial interest and understanding can soon be dissipated. Although this consequence may not seem serious in itself, it is quickly reflected in the diminishing enthusiasm of other executives and in the growing suspicion that the staff has lost the endorsement of its superior.

Much of the responsibility for overcoming this hazard rests, of course, with the staff. But again, the responsibility is not a unilateral one. It must be shared by the executive. And because of the circumstances peculiar to systems staff operation, something more than normal effort is required of both to ensure maintenance of a sound working relationship between them. The part that the staff should play is discussed later in the chapter. Following are some of the specific steps that the executive can take in shouldering his share of the responsibility:

1. Review the slate of planned projects in detail with the staff leader to ensure that it has been carefully thought through, that it gives adequate emphasis to the systems aspects of major top-management problems, and that, in general, it provides for applying effort to the areas that need it most or that offer the greatest benefits.

2. Learn to recognize which of the problems growing out of discussions with other executives or revealed by operating reports should be turned over to the systems staff for study.

3. Take the time to follow the progress of the program at regular intervals. Make certain that the staff is not straying from its defined objectives or tackling problems of marginal potential, that the program achieves and maintains momentum, and that it does get tangible results.

4. Learn enough about the techniques of systems analysis to appraise the quality of the staff's work. This requirement does not mean, of course, that the management executive need become a systems expert. It simply means that, with regard to the staff's approach as well as its recommendations, he should be able to apply intelligently a few basic evaluation criteria that will keep him from "buying a pig in a poke."

See that the staff gets a fair chance to do an effective job   This responsibility imposes two requirements on top management: First, management must give the systems staff sufficient *time* to prove itself. The newly established staff is confronted with a major task of gaining the confidence of operating personnel. And if it was recruited from company personnel with good basic qualifications but limited experience in systems analysis and design, it will also be faced with the job of developing a certain amount of technical know-how. There are few shortcuts it can take in achieving either of these goals. Both are essentially matters of education, and education is a slow process. But until both have been attained, the systems program itself cannot be hurried without inviting failure.

At least during the staff's first year of operation, its attention must be concentrated largely on gaining acceptance—on selling its services and winning the voluntary participation of operating executives and supervisors. This means that it must "make haste slowly." While it is sizing up the attitudes of key personnel, it must operate unobtrusively and modestly. It should cut its teeth on noncontroversial projects and avoid meeting major issues or infringing on entrenched proprietary interests at the very outset. In addition, the staff should take time to do its first few jobs with extreme thoroughness, both to avoid early criticism of its approach or recommendations and to provide ample opportunity for developing and testing its techniques.

For these reasons, management must come to understand that if the systems staff is to build soundly, it must be given the chance to build slowly. Its performance during the first year or two should be judged not so much by the number or importance of projects completed as by the progress it has made toward developing sound relations within the organization. During that period, the staff members will doubtless make a number of mistakes. How serious these mistakes are should be of less significance than whether the staff recognizes, admits, and profits by them.

Unless the top-management group is determined to adopt and hold to this long-range view, it would be better never to undertake a planned systems improvement program. To start the program up and abandon it within a year for want of spectacular results would, in all probability, do more harm than good.

The second requirement necessary to give the staff a fair chance to prove itself is that top management must take the steps necessary to *facilitate decision making*. These steps entail (1) working out with divisional and departmental executives a simple but adequate pattern of approvals required on systems recommendations, and (2) "keying up" the organization to reach decisions on such proposals within a reasonable time.

> *In one manufacturing company that lacked this clarification and incentive with respect to decision making, systems staff personnel spent about 40 percent of their time on the creative phase of their work and the remaining 60 percent attempting to obtain final decisions. Considerable uncertainty existed about whose approval had to be secured. Thus, to avoid criticism, staff members sought to obtain the approval of many persons and then to resolve the conflicting views presented. In this environment the review and approval of systems proposals simply opened the door to political maneuvering and self-assertion. Action on important questions was often delayed by minor issues raised by persons whose authority over the matter was assumed rather than granted. These impediments not only reduced the staff's total output but also caused confusion, discouragement, and loss of creative drive among its members. One of them aptly characterized the problem by saying, "No matter how good a systems proposal we develop, when we come to getting decisions and action, it's like trying to tie a rope around a pile of sand."*

The Staff's Responsibility for Relationships With Top Management

So much for the role that top management must play in giving support to the systems staff and its program. If the staff is to merit that support, what specific steps must it take to earn the confidence of its chief?

In large measure, of course, the answer lies in the accomplishments of the staff. We have already seen that success in the field of systems design and improvement is determined not solely by the quality of the ideas developed, but equally by the extent to which those ideas are accepted and put into practice. We can assume, therefore, that the confidence of top management in its systems staff will be materially influenced by the quality of the staff's analytical and systems design work and by the soundness of its relationships with operating personnel. But before discussing those two factors, we may appropriately consider a third—the manner in which the staff conducts its day-to-day relations with its chief.

A crucial point for systems analysts to remember is that they are *assistants* to a line executive; their staff exists to supplement him, not to create more problems for him. However important an organized systems improvement program may be, the staff's responsibility is to make certain that the participation and support required of its chief do not become onerous to him. To bog him down in the details of carrying out the program would force him to neglect his responsibilities for policy formulation and strategic and operational planning which are certainly among the principal sources of corporate growth and strength.

If the staff is to be a blessing rather than a curse to its chief, here are some of the steps it should take:

1. Become familiar with his point of view, his way of thinking, his set of values, and seek to apply them in the development of recommendations as well as in contacts with other executives.

2. Take the initiative in drafting and getting the chief executive's agreement on a statement of the staff's basic mission as well as its scope and method of operation.

3. Achieve a frank understanding with the executive on the barriers to systems improvement work and the parts that both he and the staff will play in overcoming or minimizing them.

4. Prepare material that makes it easy for the chief to indoctrinate other executives in the objectives and importance of the systems improvement program. (As a checklist in developing this material, refer to the discussion earlier in this chapter under the heading "What 'Management Support' *Does* Mean.")

5. Plan the staff's program of projects in such a way that some part of it is always devoted to matters of real interest and importance to the chief executive.

6. Do not undertake, without the chief executive's approval, any major project that is inconsistent with the program previously agreed upon with him.

7. At regular intervals prepare the summary information that top management needs to evaluate staff progress and results achieved.

8. Be sure that the staff's demands on the chief executive's time are necessary and effectively utilized. Avoid running to him with every minor conflict or problem. At the same time, keep him adequately informed on matters in which he is vitally interested. Reach agreement with him on what these matters are. Generally they will include the staff's master program, summary reports of progress, highlights of major projects brought to completion, conclusions and recommendations on projects that the chief has

requested, and important areas of disagreement among operating personnel which the staff has not been able to resolve.

9. Adhere to the standard of *completed* staff work in preparing for meetings with the chief executive. Think through the subjects to be covered and organize the major points so that they can be presented clearly and forcefully in a few minutes. Above all else, never present a problem to him without recommending a carefully thought-out solution.

THE STAFF'S STANDARDS OF PERFORMANCE

The next prerequisite to the success of the systems improvement program is a set of high performance standards to which the staff is dedicated and which it continuously strives to achieve. Taken together, these standards must reflect a desire for excellence that permeates everything the staff does, whether it be

1. Establishment of end-result improvement goals for each project undertaken
2. Fact-finding and analytical work
3. Development of benefit-producing recommendations
4. Efforts at inducing action and effecting meaningful and enduring change.

Part of the means by which the staff meets these kinds of demanding performance standards is by applying the fundamental top-management approach described in Chap. 5. But important as that basic approach is, by itself it is not enough. It must be supplemented by a number of other means of achieving excellence if the staff is to consistently perform in a way that gains respect and builds confidence.

Establishing Project Goals

Much has already been said about the staff's need to develop and maintain a strong benefit-producing orientation. Its members should approach every systems project with the determination to achieve something of significant, demonstrable value as the end product of that effort. This sort of compulsion is the driving force that gives real purpose and meaning to the systems staff's work.

The most effective means of building a goal-oriented approach into the staff's way of thinking and values is to adopt the practice of setting and getting agreement on a concrete improvement objective—expressed in terms of end results—before any systems project is undertaken.

Setting such goals is not difficult, of course, on projects aimed directly at achieving quantifiable improvements—for example, cutting clerical costs in a given area or speeding up shipments to customers. Yet, notwithstanding the relative ease of doing so on projects such as the foregoing, few systems staffs, in our experience, consistently apply the forcing technique of fixing an accomplishment target in advance. And without the challenge of a stretching goal, the project effort seldom realizes the full potential for improvement. As an example, in the conduct of administrative cost reduction studies, specifying in advance the size of the saving that is sought has two constructive values. First, it serves as a stimulus to the project team to maintain a higher level of effort. Second, it helps more than any other step to divert operating managers' attention from *whether* they will have to make changes and savings to *how* they can best achieve the specified goals.

Conceding the relative ease of establishing quantitative improvement goals on some types of projects, we must then ask: How relevant or applicable is this technique on systems studies that are concerned with structure or management processes as opposed to those aimed directly at achieving measurable benefits? The answer is that, in a results-oriented environment, an honest, rigorous effort will always be made to subject every project undertaken to the test of real *commercial significance.* Even though the profit improvement impact of a given systems study cannot be determined precisely, analysts should still force themselves to ask these two questions—at the beginning of such a study and continuously as they proceed. First, in what specific ways will this project on management methods and techniques contribute to better end results for the organization as a whole? And second, even though we cannot measure it, what, in our best judgment, is the order-of-magnitude impact on profits likely to be?

Some management processes can be subjected more readily to the direct-profit-improvement test than the analyst may think. As an example, the systems staff of one major company—in undertaking a broad planning and control study—defined the objectives of the project as follows: "Strengthening the company's profit planning and control system in ways that will directly increase annual after-tax profits by $5 million within two years after completion of the project."

Gathering and Analyzing Facts

Beyond this single overriding characteristic of high achievers in the systems field, the remaining standards that they seek to live by have to do with the way they do their work. The rest of this book is devoted largely to a detailed discussion of how to carry out all aspects of a systems improvement project in ways that produce superior results. It will suffice here, therefore, to give a few guidelines and examples that highlight the kind of demanding standards that the systems staff must adopt.

1. In the fact-finding phase of their work, the analysts' performance can be considered up to or above standard when their whole focus, from the very beginning, is on *solving the problem.* They do not see a systems improvement project as a series of separate, consecutive steps such as fact-finding, followed by analysis, followed by development of recommendations, and so on. And they do not spend several weeks simply gathering all the facts there are to be gathered, and then sit down to analyze them and see what they add up to. Instead, they begin to think fundamentally and creatively about *the problem* at an early point in the study. They either have a natural predisposition to do so,

or they train themselves to develop tentative conclusions as they gather facts. And, from the very beginning, they continuously force themselves to identify alternative solutions and to formulate and test hypotheses.

2. In addition, the staff's analytical work must reflect a tough-minded willingness to challenge the widely held beliefs that could have a significant bearing on the results achieved through the project. This questioning point of view is of particular significance because every organization fosters the growth of a great many shibboleths, myths, and apparent truths that stand as impediments to change. These are the beliefs and contentions that are repeated so often and so widely that they come to be accepted as fact.

For example, the systems staff of a leading life insurance company recently undertook a study aimed at developing more effective control over the policy lapse ratio of its individual sales outlets or agencies. Lapses are a critical problem in the life insurance industry, because the cost of issuing a new policy—including, as it does, agents' commissions, medical reports, and the like—is substantially greater than the first year's premium. Hence, policies that lapse within the first year or two after their issuance are a drain on company surplus and reserves.

Early in the course of this study, a number of the company's marketing or so-called agency people contended that a high rate of growth in sales volume was invariably accompanied by a high lapse rate and that, because the company's new business volume was, in fact, growing rapidly, the company simply had to expect a lapse rate that was higher than the industry average. This belief happens to be widely held throughout many segments of the industry. It grows out of the conviction that the pressure to increase sales produces more and more business of questionable quality.

The systems staff decided to challenge this deeply imbedded conviction by examining the relationship between growth rate and lapse rate for each of the company's branch offices or agencies. In this analysis, members of the study team plotted on a scatter chart each agency's average annual rate of sales growth against the average first-year lapse rate from 1970 through 1972. If the relationship between growth and lapses were positive—that is, if the agencies with the greatest sales growth were also the agencies with the highest lapse rates—the results, of course, would look like those shown in Fig. 6. There, the dots representing individual agencies are concentrated around a line that could be drawn from the lower left-hand corner of the chart to the upper right. In this hypothetical case, each dot represents an individual agency, and the position of the dot shows clearly that the higher an agency's growth rate in sales volume, the higher its lapse rate is likely to be.

But when the study team plotted the actual *results for all the company's agencies, it obtained the picture shown in Fig. 7—a completely random scattering of dots indicating no relationship whatever between the rate of sales growth and the lapse rate. Some agencies had achieved an annual growth in sales ranging from 10 to 20 percent, with average lapse rates in the neighborhood of 20 percent. Others had achieved equally outstanding sales records with lapse rates around 10 percent.*

To demonstrate even more conclusively that this scatter was completely random, the study team superimposed on the chart the broken vertical line showing the company's average *lapse rate and the heavy horizontal line showing its* average *rate of sales*

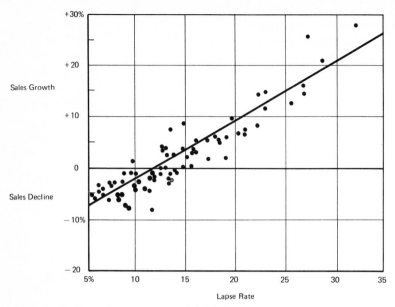

FIG. 6 *Myth: The marketing staff contended that the lapse rate of life insurance rises with a growth in sales volume.*

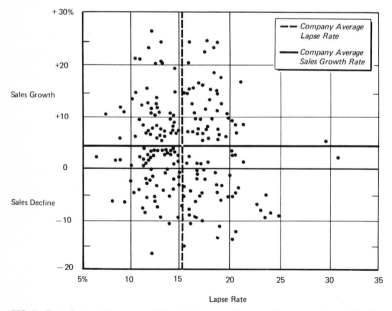

FIG. 7 *Fact: Analysts found no relationship between the rate of sales growth and the lapse rate.*

growth. These two lines, of course, divide the grid into four quadrants, and it is significant to note that the numbers of dots or agencies falling into each quadrant is almost equal—specifically, 23, 28, 27, and 22.

This illustration dramatizes the point that one of the qualities systems analysts must bring to their work is a willingness to challenge and explode the myths, half-truths, and preconceptions that tend to be sources of comfort and barriers to innovation.

Developing Recommendations

In this phase of the systems staff's work, the most important performance standard, by far, is the one already covered—that is, the achievement of significant, demonstrable improvements, either in operating or management effectiveness or in the reduction of overhead costs. But even in this aspect of its work, the staff can apply still other standards to help ensure that the foregoing level of benefits will in fact be achieved. Following are two illustrations of these additional standards:

1. The first has to do with the development of alternatives. On this score, the project team members must ask themselves: How many respectable alternatives have we identified and explored? And how thoroughly have we evaluated each one? On problems that lend themselves to more than one solution, failure to explore all possibilities may well mean that the best solution has not been found. Hence, the truly creative staff will not permit itself to reach a final judgment on the solution to be recommended until it has forced itself to develop and test a whole range of alternatives.

2. The second standard to be met in this phase of a study is that the recommendations the analyst develops not only must have substance, significance, and worth in themselves, but also must be communicated to others in a way that demonstrates their operating relevance and usefulness. One of the identifying characteristics of top-drawer systems analysts is that they have the intellectual toughness and self-discipline to demand of themselves that their recommendations—in addition to being imaginative and fundamental—are, at the same time, hard-packed, well supported, and usable.

An example of what this standard means in actual practice can be found in the contrast between the two pieces of work presented in the appendix at the end of the book. These pieces were prepared by two different analysts working on the same study of marketing management systems in a major life insurance company. In requesting the development of these two pieces of written material, the staff director said:

> The Vice President for Marketing has agreed to adopt our recommendation that he create within the Marketing Department a new position of Director of Financial Analysis and Control. As we all know, this position is radically new—both for the company and the industry. We are all convinced that the position can pay great dividends—but we still have a lot of work ahead of us to make the position come to life. As the first step in this added work, would each of you try your hand at writing

out our concept of this job—the reasons for its existence and the kinds of problems and improvement opportunities the incumbent should be concerned with. This job write-up is of pivotal significance, because it will be the charter for an important position for which there is no precedent or accumulated experience.

The two different write-ups in the appendix are revealing. Notice the vague, superficial, boiler-plate abstractions of example A, and the high standards of thoroughness and operating usefulness of example B.

Getting Action and Effecting Change

The final phase of system analysts' work in which high performance standards must be met is the "payoff" phase, when the staff tries to secure the adoption and implementation of recommendations in a way that converts potential benefits to realized benefits. On this score, staff members must recognize that over the long run their success will be judged not in terms of the number of studies they complete or recommendations they make, but in terms of the number and importance of the *tangible, constructive changes* that result from their work. The staff's image throughout the organization will be shaped not by how hard it works at diagnosing problems and developing solutions, but by the magnitude of the *profitable action* it stimulates.

Effectiveness in this final phase of its work depends largely on the strength of the relationship the staff is able to build with line operating executives and supervisors. For this reason, the standards of performance that the analyst must meet in inducing action and effecting change are covered in the next and final section of this chapter.

RELATIONSHIPS BETWEEN THE SYSTEMS STAFF AND OPERATING PERSONNEL

Notwithstanding the obvious importance of technical competence in fact-finding, analysis, and development of recommendations, it is unfortunately true that the more proficient the staff becomes, the more difficulty it sometimes encounters in gaining the cooperation of operating personnel. Although a mediocre staff may not produce spectacular results, by the same token it is less apt to disturb old habits or probe sensitive spots. But the more penetrating the staff's observation of problems and the more fundamental its diagnoses, the greater is the danger that its efforts will incite suspicion and resistance.

This danger cannot easily be avoided, for it is inherent in the very nature of the work. On the one hand, the task of the systems staff is essentially that of looking for weak spots, of constantly seeking better ways for others to do their jobs. By virtue of their interests and experience, staff members generally possess a highly developed critical faculty. Thinking only of the benefits to be achieved or the credit to be gained through adoption of their plans, they may become impatient and irritated by delayed acceptance or indifferent execution.

On the other hand, operating executives or supervisors are just as prone to feel that investigation of their activities by an outsider may amount to tacit deprecation of their performance. Sensing an implication of criticism, they are quick to point out the staff member's lack of knowledge about the function being administered and slow to recognize the compensating value of analytical skills and specialized knowledge about systems design. And the more fruitful the ideas developed by an analyst as a result of the study, the more resentful a supervisor is likely to feel for not having thought of them himself.

But if the systems staff sees its responsibility as consisting not only of developing ideas but also of stimulating their adoption, it will neither ignore these natural handicaps nor regard them with pessimism. Rather, it will give as careful thought to obtaining the emotional and intellectual acceptance of its recommendations by operating personnel as it does to performing the technical phases of its task. Without such acceptance, the constructive changes made as a result of its work are not likely to be substantial, either in numbers or importance. At best, the staff must suffer passive cooperation or indifference. At worst, it may encounter open antagonism, willful sabotage of its work, or the circulation of harmful rumors about its objectives or competence.

The extent to which operating personnel voluntarily accept and carry out systems recommendations is directly related to the following circumstances: their understanding of the staff's work, their participation in it, the recognition they receive, their confidence in the staff, and the demonstrated results achieved by the staff. Subsequent chapters offer suggestions on practical ways of shaping these circumstances during each of the several steps of a systems or procedures project. The discussion here will therefore be limited to what these terms mean in actual practice and how they can contribute to the staff's effectiveness in stimulating profitable action.

Understanding

Only an exceptional program or plan makes its underlying objectives self-evident to those who have not participated in its formulation. Although the mechanics of a proposed course of action may be clear enough in themselves, the attitudes of those whom that action affects will be determined largely by their understanding of the purpose behind it and the need or justification for it. In short, acceptance depends not so much on an understanding of *what* is to be done as on *why* it is to be done.

The responsibility of top management for spreading this kind of understanding has already been discussed. For the most part, management's educational and promotional effort must be focused on the objectives of the systems improvement program as a whole, the need for it, and the logic of having the activity spearheaded by a staff group. But the real test of acceptance will be found in the attitudes of operating executives, supervisors, and workers toward specific projects affecting them personally. To the staff itself, therefore, falls the responsibility for sharing its knowledge with the operators—for ensuring their understanding of the reasons for each study, the way the study will be conducted, and the help they will gain from it. Instilling this understanding is the staff's starting point in overcoming the uncertainty and suspicion that foster resistance.

Participation

Human nature is such that if we are asked to work on our own idea, we will do so at white heat; if we are asked to work on someone else's idea, we will do so indifferently. There is no greater stimulus to the adoption of systems improvements than creative participation in the work of development by those who will be using the new system. The measure of one's enthusiasm for new operating techniques is generally determined by the extent to which one can recognize them as being partly the product of one's own thinking.

This, in itself, would be reason enough for the systems staff to avoid the temptation of relying exclusively on its own versatility or self-sufficiency. But there are other equally compelling reasons for encouraging participation by the line organization. The first is that operating personnel can make a real contribution to the undertaking—a contribution for which there is no adequate substitute. The analyst's special value lies in his ability to apply systematic techniques to the diagnosis and solution of systems problems. The operating supervisor's special value, on the other hand, lies in his intimate knowledge of existing activities and problems. These two values complement each other and are fully realized only when they are merged. Once that is achieved, the development of improvements becomes the function of both groups. There are no proprietary rights in the generation of ideas.

Thus, quite apart from considerations of strategy in gaining acceptance, the systems staff must seek line participation as a means of building a better product—one that will meet the test of operating realities. Staff members must develop the conviction that their work calls not for highly individualized diagnoses and independent thinking, but for a synthesis of all the intellectual resources that can be brought to bear on a problem. In relative effectiveness, these two approaches are like the difference between a sniper and a task force.

In no sense does this need for interaction relieve the staff of responsibility for final results. It simply means that the staff's work must be conducted in such a way that the cumulative knowledge, experience, and judgment of the operating organization are tapped.

In addition to improving the quality of end results, the participation of line personnel usually speeds up the conduct of systems studies and thus further increases the staff's effectiveness. The fact-finding phase is facilitated by operator knowledge of the sources of information and operator assistance in gathering it. If conclusions and recommendations are reached jointly, elaborate written reports and lengthy discussions are less likely to be required in obtaining final approvals. If operating personnel acquire adequate background throughout the course of the study, they can usually carry out the installation phase of the project with a minimum of staff guidance.

The final inducement can be found in the benefits that accrue to operating groups themselves from participation in systems projects. This experience tends to develop closer teamwork among them and to broaden their points of view. It heightens their interest in systems matters. By being exposed to the use of rational techniques, they learn to think more analytically and to act with greater effectiveness on their own problems.

An illustration of one method of putting these concepts into practice can be found in the experience of a large, multidivision consumer goods company that has one of the most successful systems and procedures programs we have seen.

In this company the systems staff has purposely been kept small (with never more than six or eight analysts), because its members see their function not as one of performing all the systems improvement work themselves, but as one of organizing teams of staff and line personnel to carry out individual projects. Each team ordinarily includes only one systems staff member, but may have two to five operating people assigned to work full-time on the project.

Under this arrangement, the systems staff still performs the functions of planning the company's overall systems improvement program, exploring opportunities and identifying specific projects to be undertaken, organizing study teams, contributing its own ideas, bringing in new ideas and approaches from outside sources, and steering the team to successful completion of the study.

In addition, however, the study team as a group takes several steps to help its line members develop a true sense of participation and responsibility.

1. The team defines the objectives of the study in specific terms. For example, one study of the company's sales order handling system stated the objective as follows: "To develop a system under which, as a matter of regular routine, sales orders can be processed and placed in the hands of the shipping department within 4 hours of their receipt by the company mail room."

Aside from promoting a common understanding of exactly what the team is trying to accomplish, this step has another value in that it marks the beginning point in generating active support. The operating people themselves have, by implication, recognized the opportunities and created the challenge to exploit them.

2. Next the study team typically develops a list of all the ideas to be explored—that is, a list of all the alternative ways by which each function and each operation might conceivably be simplified or speeded up.

3. Then, without trying at this point to judge the merits of any one of the alternatives, the study team prepares a list of all the facts needed or steps to be taken to evaluate each alternative.

From the viewpoint of operating personnel, the second and third steps are valuable mainly because they dispel the shock that line members would otherwise feel were this array of ideas advanced unilaterally by the systems staff. Hence, these steps help to create a predisposition to change before the study itself is scarcely begun.

Lest analysts assume, however, that substantial participation by line personnel is the quick and easy solution to all their problems, some words of caution must be added. This method of operation is not an entirely unmixed blessing. It requires an adroit staff to keep the study from becoming a vehicle for promoting the line supervisor's preconceived ideas.

Three other more specific limitations and dangers must also be recognized. First, line personnel of the right caliber will not always be available to participate in the study. Under these circumstances, the systems staff would do better to carry out the work alone than to accept operating personnel on the project team who do not enjoy the confidence of their line associates. Second, analysts must recognize that by working with operating personnel for long periods, they may tend to lose—or, in the minds of others, appear to lose—their objectivity and impartiality. Third, the use of persons who are unskilled in systems analysis, and who are subjectively involved in the activity being studied, can downgrade the quality of the systems improvement effort.

These cautions, however, are not arguments against an appropriate degree of line participation. They simply emphasize the need for alertness, courage, and skillful leadership on the part of the systems staff.

Recognition

The importance of giving adequate recognition to those whom one wishes to persuade is sufficiently obvious that we mention it only to indicate the forms that such recognition might take. The most common form, of course, is for the staff to acknowledge cooperation received and give credit for ideas contributed. The important thing for the staff to remember, however, is that appreciation expressed privately to the contributor is courtesy; but credit given to the contributor before others is recognition. Systems analysts will do wisely to be neither careless nor reluctant about this matter, for nothing will destroy their standing with operating personnel more quickly than the piracy of ideas— however unintentional it may be. Nor need analysts fear—as some do—that their own reputations will be diminished by the recognition someone else receives from the assignment they have carried out. It is surprising how a readiness to give generous credit to others reflects even greater credit on the donor.

Another form of recognition lies in making use of suggestions received from operating personnel, whether these suggestions concern the manner of conducting the study or the recommendations developed from it. This is not always possible, of course. But the staff should make a sincere effort to adopt each suggestion received or modify it to a usable form. If some cannot be used, the reasons should be carefully explained to the contributor.

The final form of recognition has to do not with the contributions made but with the place occupied in the organization structure by those whom the study affects. This form might be termed the *recognition of position*. It simply means that analysts should avoid making an end run around anyone; they should arrange all their contacts from the top down through each executive and supervisor concerned.

Confidence in the Staff

It is reasonable to assume that the confidence of operating personnel in the systems staff will be substantially influenced by the staff's technical competence—by its reputation for doing each job as well as it can be done. But in its effort to gain acceptance, the staff cannot afford to rely solely on the authority of its ideas, for its relationships with the line organization will be governed to an even greater extent, perhaps, by the personal conduct of its members. The behavior required of them is something more than mere tact or the quality that is loosely referred to as the "ability to get along with people." It is a composite of many factors that add up to faith in the integrity of staff members and to a comfortable feeling toward them and their work.

We do not intend to recite the many personality and character traits generally considered necessary to achieve these ends. Rather, we hope that the practitioner will profit more from a consideration of the type of conduct that begets confidence. In all likelihood, he will gain more practical value from thinking in terms of what he must *do,* rather than what he must *be.*

If the line organization is to have faith in the integrity of the systems staff, the staff members must establish a reputation for intellectual honesty and fair dealing. This is a matter not only of honoring confidences, but also of seeking the correction of unsatisfactory conditions by working out their solution with the supervisor directly concerned, rather than by making the conditions known to his superiors.

A second requirement is that the staff avoid playing corporate politics, or even giving the appearance of it. This is not easy, because politics within most organizations revolve around position, functions, and responsibilities—the very elements with which the staff must deal in its work. Changes in systems and procedures are one of the means by which the manipulator seeks to gain personal advantage.

Two general rules of conduct, however, will help the analyst escape political involvement. (1) Avoid discussing personalities. In particular, do not seek to gain favor with one person by concurring in his adverse opinion of another. (2) Be completely impartial in the formulation of recommendations. Even members of the line organization who are encouraging the recommendation of changes that are advantageous to themselves will respect an analyst for being totally objective and will realize that if he has not favored them he is not likely to favor others.

The third measure of the staff's integrity in the minds of operating people is the action that the staff takes on their suggestions. This point has already been covered, but it is worth emphasizing here that if the staff is not sincere in its intent to incorporate worthwhile suggestions in its recommendations, it would do better not to solicit them at all.

Finally, the staff must be conservative in the promises or claims it makes. This rule applies equally to estimates of savings to be achieved, estimates of time required to complete a study or install a new system, and appraisals of benefits actually realized.

When the staff has earned a reputation for honesty and forthrightness, it will have gone a long way toward bringing about the second condition requisite to gaining confidence: the development of a comfortable feeling among operating personnel toward the staff and its work. But the cultivation of this feeling does not rest on integrity alone. It requires something more than trustworthiness or the assurance that one will be dealt with fairly. Following are some additional dos and don'ts for the staff to observe in its effort to achieve this goal.

1. Establish a reputation for helpfulness, not for masterminding. The systems staff should endeavor to create the feeling among operating personnel that it is not developing a system or procedure to be handed to them as a finished package, but that it is helping them to improve their operations and to think for themselves. In place of making pronouncements about what line operators ought to do, the staff should reason out a course of action with them. It must avoid the temptation to jump to conclusions, and must be wary of the assumption that because it is a group of specialists it must always have a ready answer for every question asked or problem raised. Staff members will find that there is great strength in the honest admission, "I don't know," or "That's an angle I hadn't thought of." In short, the discreet staff is one whose members avoid the implication that they are experts and know all the answers. Instead, they seek to convince the line organization that they know how to investigate and to analyze and that, with this knowledge, they are helping operating personnel reach sound conclusions from established facts.

2. Save face. Much of the line supervisor's resentment of implied criticism can be forestalled if analysts let it be known that some understandable excuse or reasonable explanation exists for the unfavorable conditions they encounter. This does not mean that staff members must manufacture excuses for others, but simply that they must take the time to think about why something went wrong and to show their tolerance and understanding of what they find. And in their zeal for spotting weaknesses and opportunities for improvement, they should not let themselves become casual about the strong points of the systems they study. To be sure, they may be able to do nothing to improve a given system, but it will repay them to let others know that they recognize the system's merits.

3. Avoid the spectacular. Analysts do not necessarily have to engineer a complete reorganization of each acitivity they study to prove that they are on their toes. The ingenuity and imagination required of them are not qualities to be exploited. The staff's function is to develop practical, worthwhile, and acceptable improvements, not to seek spectacular changes for their own sake.

This caveat is not intended to discourage major changes or fundamentally new concepts; it is merely a plea for circumspection in approaching them. The analyst must make sure that his innovations are a better way, not just a different way, and that the measurable benefits they promise to produce will outweigh the cost and confusion of the changeover.

An example of precisely the opposite can be found in a highly automated, computer-based system of policy writing, billing, and record keeping recently installed by a property and casualty insurance company. One of the most advanced systems of its type, it has been referred to as "elegant," "sophisticated," "in the vanguard," and "a systems man's dream." Yet the blunt facts are that this new system costs significantly more than the one it replaced and has not given the company that developed it any semblance of competitive advantage.

4. Think empathetically. In their approach to each study, staff members should concern themselves more with determining what is in the minds of those with whom they work than with expressing their own ideas and convictions. They need to be *people*-oriented as well as *problem*-oriented. They should strive to place themselves in the position of the other person and to think about the problem at hand from that person's point of view. The more often and more skillfully they do this, the more effective they will be in energizing constructive action on their recommendations. By studying the people they work with; by observing their reactions to varying conditions and influences; by learning something of their experiences, prejudices, and ambitions, analysts will know better how to deal with them to gain their confidence. By thinking in terms of the other person's apprehensions or interests, analysts will be able to make their arguments vastly more meaningful and persuasive. By showing a sincere willingness to listen and learn, they will develop a reputation for tolerance and understanding that will gain them far greater acceptance than technical proficiency alone will ever win for them.

Demonstrated Results

The final key to building sound relationships with operating personnel lies in demonstrating, under actual operating conditions, the workability of systems and procedures proposals. Under some circumstances this path may be the only sure one open to the staff—particularly if it has not already gained wide acceptance throughout the line organization, or if its recommendations on the study in question are regarded as radical.

> *Circumstances similar to these were encountered by the systems staff of a large, long-established manufacturing company. The management methods of this company had been slow to change over the years, and its operating executives were generally suspicious of staff work.*
>
> *Operating in this climate, the company's newly organized systems staff came to the conclusion that substantial improvements could be realized from a fundamentally different method of production scheduling and inventory control. But in view of operating management's rigidity, the staff realized that a direct, frontal attack on this problem would surely fail. Accordingly, the analysts got permission to work with one of the company's mill managers to "test out some refinements in the mill scheduling routine." Beginning on a modest basis, the staff gradually awakened the interest of the mill manager in trying more basic changes. By the time the far-reaching character of the staff's innovations became evident to the whole organization, the advantages of the new approach were so readily demonstrable that the other company mills began to clamor for the change.*

SEVEN

Managing the Systems Staff

The final ingredient essential to the success of the systems improvement program is skillful management of the staff's overall effort. How well this requirement is met will be determined, to a large extent, by the quality of staff leadership that is given to the planning, execution, and control of individual systems projects—a subject that is dealt with in the next chapter. But beyond the conduct of individual studies, management of the systems staff encompasses other broader elements that are concerned with the composite performance of the staff. These elements include (1) development of a program of projects to be undertaken, (2) control of the staff's overall performance, and (3) training, development, and evaluation of staff members.

DEVELOPING THE PROJECT PROGRAM

Up to this point, the staff's plan of operation has been considered only in its purely tactical phase—the method of attacking individual systems projects. But however skillful the approach employed on each unit of work, it cannot be fully effective unless it is applied to the areas of greatest need or opportunity. Thus, in addition to providing a sound basic approach to individual studies, the plan of operation should also include a carefully developed program of specific projects to be undertaken.

This does not mean, of course, that all projects carried out by the systems staff can be planned in advance or initiated by its own members. On the contrary, a substantial part of the staff's work will always arise from conditions that cannot be anticipated or will come to the staff from sources outside the staff. Three of the most common origins of this kind of project are

1. Executive or managerial requests for assistance in handling a specific systems or related organizational problem.

2. Federal, state, or local government legislation that creates new procedural requirements.

3. Changes in company policy or in the scope of operations. This category may include launching of new products or services; opening of new plants, warehouses, or offices; development of new channels of distribution; revision of personnel policies; and so on.

Although systems projects originating from these sources cannot be planned in advance, the staff's overall program must be sufficiently flexible to permit prompt handling of them when they do arise. By their very nature, they either require immediate action or represent the special interest of a particular executive. For these reasons, the staff's effectiveness in meeting the needs of the organization is more likely to be judged on its handling of such projects than it is on the studies initiated by its own members.

The percentage of the staff's total time required by projects of this origin will vary widely between periods of change and periods of stability. But in spite of the volume or immediacy of such projects, some part of the staff's effort should always be applied to its own long-range program. This is the phase of its work that, over a period of time, is likely to produce the greatest permanent benefit.

Stated simply, the objective of the planned program should be the orderly and continuous improvement of all systems and procedures falling within the scope of the staff's responsibility. This definition suggests that effective program planning requires (1) sizing up the overall job by listing all systems and procedures encompassed in the staff's area of operation, and (2) developing techniques for spotting problems and opportunities so that priority can be given to projects which offer the greatest improvement possibilities.

Outlining the Scope of the Program

As a guide to the systems staff in outlining the scope of its program, we have listed below a number of typical operating systems and procedures. This illustrative list is not complete for any one type of enterprise, but it does cover a majority of the systems and procedures common to certain types of manufacturing companies and retail stores. Obviously, for other types of businesses additions and deletions are required. Some of the processes included in the list are wholly departmental in scope, but the majority clearly cut across departmental or functional lines.

1. The whole network of sales management paperwork systems, including sales represnetatives' call planning and reporting, lost-sales reports, end use development reports, competitive intelligence, and the like.

2. Development of price quotations and delivery date estimates.

3. Sales order processing, credit approval, shipping, and invoicing. (The comparable procedure in retail stores covers preparation of sales checks, cash control, sales audit, and billing.)

4. Recording of receivables and cash receipts and conduct of the collections function.

5. Receipt, inspection, and pricing of merchandise returns and the authorization and issuance of credit.

6. Handling of customer complaints regarding nondelivery, billing errors, etc.

7. Sales analysis and sales statistical reporting.

8. Purchasing and receiving procedures, including issuance of purchase requisitions, conduct of supplier negotiations and preparation of quotation requests, issuance of purchase orders and purchase order modifications, receipt and inspection of incoming materials, and return of rejected items. (In retail stores this process also includes merchandise marking.)

9. Accounts payable procedures, including verification and approval of invoices, approval and distribution of freight charges, recording and distribution of payables, payment of invoices, and determination of inventory unit prices.

10. Manufacturing order procedures, including preproduction preparation of blueprints, design specifications, bills of material, and operation sheets; issuance of tool job orders, stock requisitions, and factory work orders; movement of work in process; and preparation of inspection reports.

11. Production scheduling, dispatching, expediting, and reporting.

12. Processing and control of engineering changes.

13. Material accounting and control procedures covering product materials, work in process, and salable merchandise; manufacturing, office, and maintenance supplies; tools; packing and packaging materials; customer-furnished materials and parts; and items returned for repair. The procedural elements of this process include maintenance of inventory records; pricing and charging of material receipts and withdrawals; accounting for scrap and rework; costing of sales; control over the size, diversity, and investment in each inventory classification; physical verification of inventories; and detection and disposal of excess or obsolete items.

14. Timekeeping and payroll procedures, including time recording, verification, and pricing; labor cost distribution; payroll preparation and distribution; and maintenance of payroll records.

15. Procedures for processing changes in employment or payroll status, including hirings, transfers, terminations, leaves of absence, rate changes, and reclassifications.

16. Accounting for and control of special job orders, including repairs on returned products, tool and pattern jobs, research and development projects, and building or equipment repair jobs.

17. Record keeping on fixed assets and capital expenditures.

In addition to a tailor-made list of all such *operating* systems existing within the organization, the staff should further delineate the scope of its program by listing all the existing *general-management* systems. In most organizations they include, but are not necessarily limited to, those already named: strategic planning, execution, and control; operational planning, budgeting, and control; budgeting and control of capital expenditures; and the development and motivation of people—especially management personnel.

Spotting Problems and Opportunities

Any such list of systems and procedures delineating the scope of the staff's responsibility is essentially a statement of its long-range objectives. Within that framework, the practical phase of program planning is one of selecting and assigning priorities to projects to be undertaken during the period immediately ahead. This step entails more than randomly selecting some systems for study, or lining up all the systems in some logical sequence and studying them in that order. The elements of the overall program vary too widely in relative importance at different times to justify such methods of selection.

What this job requires instead is ensuring that the part of the staff's effort which is available for the pursuit of its own program is applied, at all times, to the areas where it is most needed or will produce the greatest benefits. Identifying those areas is a research job in itself. Systematic investigation and analysis are needed to find out where the most promising opportunities or most serious problems lie.

There are a number of means by which evidence of systems problems and opportunities can be quickly spotted. Following are six of the most useful.

1. Unquestionably the most fundamental method consists of identifying (and getting agreement on) the *key factors controlling success* in the type of enterprise involved. The approach to defining these success factors is discussed fully in Chap. 13. Hence, to serve our purpose at this point, we need simply say that what the analyst should be seeking to identify are the elements controlling overall performance—the small handful of things the organization must do superbly well to be a leader in its field. Within a private enterprise company, this means developing a keen insight into the *profit economics* of the business, the major profit leverage points, and the basic causes of competitive strength or weakness.

Once these key factors are defined, the next step is to determine the relationship of each of the organization's operating and management systems to them—that is, the extent to which and the ways in which the design of each system can have an impact on overall results.

2. *Informal contacts with executives and supervisors* will ordinarily prove to be another fruitful source of ideas. By keeping informed of the operating problems that are bothering line personnel and by sizing up the difficulties that they would like to have investigated and straightened out, the staff can develop some preliminary measures of urgency and need. Such contacts will usually uncover any significant problems of interdepartmental relationships or delays in the

performance of critical activities. In addition, tapping this particular source can have an important by-product value: It is another useful way in which the staff can demonstrate to line managers a genuine interest in their real-life operating problems.

3. *A brief study of one or two selected performance factors* will often produce valuable supplementary clues. For example, if the procedure is one in which speed of handling is an important requirement, the preliminary appraisal might be confined to a time study of the procedural cycle. In the filling of sales orders from stock, this study would involve determining the range of times and the average time between receipt of the order and shipment of the merchandise. Similar studies of elapsed times might profitably be applied to the processing of requests for quotations, the handling of customer complaints or inquiries, or similar activities in which processing time is competitively significant.

In procedural areas where control problems are greatest, different performance factors, of course, would have to be spotlighted. For example, effectiveness of material control procedures might be measured by the growth (overall and by class) of the dollar investment in material in inventory, by variations from item to item in the number of months' supply on hand, or by the discrepancy between actual and book inventories at the time of taking physical inventory. Effectiveness of production planning and control procedures, in turn, might be determined by such factors as the range of completion times per manufacturing lot, the volume of back orders, and so on.

Still another area worth exploratory probing is any system or any organizational unit in which many people—particularly clerks—participate. The number of personnel by itself is usually a fair measure of savings opportunities. Other indexes that point to possible problems are (*a*) recurring overtime, (*b*) a heavy backlog of uncompleted work, and (*c*) a rate of personnel growth substantially greater than the growth of the company as a whole.

4. Still another source of ideas—and often one of great significance—lies in *comparisons of the results achieved by like organizational units.* This method can be used in any enterprise that has several components (often geographically separated) doing the same kind of work—for example, plants making the same products, warehouses, branch or regional offices, and service or processing centers. Where this situation exists, the staff can develop an inventory of possible operating problems and improvement opportunities by comparing all such units on significant elements of performance—operating expenses, processing time, customer complaints about quality or service, and the like. Almost invariably it will be found that the best-performing unit has developed simplifications or other improvements in its operating routines that have clear transfer value. Simply by identifying the already demonstrated best performance in the enterprise, the staff will have uncovered some high-potential additional projects to add to its study program.

5. *Regular operating and executive reports* are another valuable source of information for spotting systems program improvement targets. Important

among these documents are budget reports showing whether clerical departments are conforming to established expense standards. Production reports showing performance against delivery promises or the amount of idle machine time caused by lack of material may point to deficiencies in material control, production control, or procurement procedures. Narrative reports by the internal auditing staff may also disclose weak spots, particularly if that group makes periodic operating audits in addition to the traditional financial audit. Although the information in these and other reports is generally not conclusive, it nevertheless spotlights problems that will repay further investigation.

6. *Finally, analysis of customers' complaints* by type often points to possible weaknesses in procedures covering the processing of orders or service requests, the receipt and crediting of merchandise returns, the invoicing of shipments, and the handling of customer inquiries.

Admittedly, the problems revealed by these six means will not always fall within the scope of the staff's responsibility. Frequently the causes will lie not in a faulty system or procedure but in poor execution or in policy deficiencies. In such cases, responsibility for taking corrective action rests, of course, with others.

No one of these six sources is sufficient in itself to keep the staff fully informed of problems requiring its attention. But, taken together, they should provide the experienced observer with adequate means for keeping the systems program well geared to the needs of the organization.

CONTROL OF THE STAFF'S OVERALL PERFORMANCE

One of the conditions peculiar to the operation of most staffs is the high degree of variability in their units of work. This is particularly true of the systems staff. No two of its projects are alike in scope, in the conditions encountered, in the time required for completion, or in the character of results achieved. The staff has no regular daily tasks to perform, nor are its members normally subject to close supervision. Their work consists essentially of examination, diagnosis, and deliberation—all mental processes for which no ready standards of measurement exist.

As a result of these conditions, it often happens that little or no effective control, either quantitative or qualitative, is exercised over staff performance. Yet the very circumstances that make the exercise of such control difficult also increase the need for it. Without some systematic means of checking up on the work of the entire staff, management cannot reasonably expect that staff members will continuously put forth their best efforts or that there will be an adequate basis for strengthening staff performance.

The three aspects of the staff's operations that must be controlled are (1) work undertaken, (2) progress made (the quantitative element), and (3) results achieved (the qualitative element).

Work Undertaken

The principal basis for controlling this element is the staff's master program or plan discussed earlier in this chapter. But in addition to operating within these limits, the staff leader should make sure that four other specifications are met before a new project is launched.

The first is that the project about to be undertaken is being tackled in the right sequence in relation to other studies in process or proposed for the future. In other words, the project should be one that (1) is not dependent on the output of other studies now under way or not yet started, and (2) can be pursued now without unreasonable interference with other current projects.

The second specification is that new work should be selected to maintain a sound balance between staff-initiated studies and those requested by operating personnel.

Third, staff-initiated projects should be undertaken in a sequence that reflects the optimum balance between the potential benefits and the probable time and effort required to obtain them. Ordinarily a higher priority should be given to a project that will produce benefits of medium size but produce them quickly than to one that will produce significantly larger benefits but only after 3 to 5 years of effort.

The final control on work undertaken is that the total load or number of in-process projects at any one time needs to be kept at a feasible level. This specification is sometimes difficult for the staff director to apply, particularly when a large proportion of the staff's projects is requested by operating executives and supervisors. Under these circumstances the director may be strongly tempted to try keeping everybody happy. But unless he is careful to prevent the staff from being spread too thin, he is likely to find his analysts buried under a multitude of in-process jobs that they never seem quite able to bring to completion.

Work Progress

The basis for controlling progress on each staff assignment should be an estimate of the worker-days required to complete the study. Projects likely to take more than 2 or 3 weeks should be broken down into their major steps or accomplishment milestones, and a time estimate should be prepared for each. These estimates should be made by the project team leader or analyst assigned to the study and approved by the staff's director before the project is begun.

Such estimates, of course, cannot always be precise. On some studies the team leader or analyst will meet special conditions or unusual delays that could not have been anticipated. On others it may be necessary to redefine the initial objectives or scope as the work progresses.

Notwithstanding these limitations, however, advance estimates of time requirements have many values. They force the staff analyst to plan his day-to-day work more carefully. They keep him alert by providing a bogey against which he can continuously check his own accomplishments. For the staff director, of course, they provide the means for measuring the progress of each assignment. They also enable him to make more realistic promises about the availability of staff members to undertake new studies. Finally, they serve as the basis for the director's reports to the chief executive on the staff's progress and its in-process work load.

The types of records and reports needed to maintain control over this element vary considerably, depending on the size of the staff. For a group of four or five members, a simple card or notebook record of in-process studies, together with periodic oral reports to the staff's principal, should suffice. For large staffs engaged continuously in many projects, the control problem becomes more complex. Here, various visual control devices can be used to great advantage in keeping track of staff operations and in reporting progress to others.

As an example, the systems and procedures director of a large manufacturing company maintains in his office a set of project control boards to facilitate control of his own staff's assignments and to coordinate its work with that of several departmental procedures staffs. The boards show the following information for all projects in process:

1. *Project number*
2. *Project name (or brief description)*
3. *Personnel assigned*
4. *Date started*
5. *Progress against schedule—that is, the scheduled and actual completion dates for the following phases:*
   a. *Initial diagnosis*
   b. *Development of detailed project plan*
   c. *Fact-finding and analysis*
   d. *Development of recommendations*
   e. *Obtaining approvals*
   f. *Development of implementation plan*
   g. *Completion of implementation*
   h. *Measurement of benefits*

To keep progress information up to date, the director checks the status of all projects weekly. At the end of each month he has the boards photographed and sends 10-by-14-inch photographic prints to his superior, the executive vice president, as a summary report on project status.

A less elaborate but equally useful control device is the conventional Gantt chart. Figure 11 in the next chapter shows how this type of chart can be adapted to the scheduling of systems projects and the reporting of progress.

Results Achieved

The ultimate test of a staff's effectiveness is, of course, the character and magnitude of the benefits it produces. This test is not easy to apply since a significant, if not major, part of the staff's effort will be devoted to achieving improvements that cannot be measured in dollars and cents—for example, better management decision making, faster action, or improved control.

Nevertheless, the staff director should keep some record of accomplishments on each completed study. This record should show separately the savings actually realized and those claimed to be obtainable. And even on management process systems, every reasonable effort should be made to develop at least an order-of-magnitude estimate of the value of the improvements realized. Suggestions for making such estimates appear on pages 95 and 302–303.

In using these records, the staff director and management should take care to avoid placing undue emphasis on the measurable savings achieved. In the long run, the intangible benefits of the systems improvement program may well be the more valuable. The staff should not be encouraged to ignore these opportunities because of the pressure for cost reduction.

If this caution is observed, the director and his superior will find these records of real value for an occasional overall appraisal of staff performance. In addition, the very process of assessing and recording results at the conclusion of each study has another value: It helps the director to pinpoint the staff training program by calling attention to opportunities for improved performance on future projects.

Illustrative Reports

These techniques for reporting on the staff's activities can be applied in any number of ways.

The director of one large systems and procedures group submits to his superior a monthly report containing the following information on staff plans, activities, and accomplishments:

1. A list of projects completed during the past month, specifying for each project the savings and other benefits achieved. A sample of this type of report is shown in Fig. 8.

2. A list of projects currently in process and an estimate of the savings and other benefits that each is expected to produce.

3. A brief narrative report summarizing progress to date on each in-process project.

4. A list of requested projects not yet started.

5. A summary of how the staff's effort has been balanced or spread among the various divisions of the company.

6. The performance factors on which the staff's overall accomplishments are based and a summary of how the staff rates on each factor during the current year. Figure 9 is an example of this type of report.

7. A narrative report explaining what the staff needs to do to improve the rating shown on the performance report.

In contrast to this sort of detailed monthly reporting, another systems staff director and his superior schedule a regular semiannual meeting at which the two make a searching review of the staff's work during the past 6 months. This review serves as a

SUMMARY OF PROJECTS COMPLETED THIS MONTH
SYSTEMS AND PROCEDURES STAFF

MONTH ENDING: October 31, 1974

PROJECT NUMBER	PROJECT NAME OR DESCRIPTION	TANGIBLE BENEFITS (P = Potential Benefits; R = Realized)				INTANGIBLE IMPROVEMENTS IN				
		Cost Reductions	Reduction in Investment Requirements	Improvement in Cash Flow	Other	Organization (Structure or Relationships)	Management Decision Making or Control	Customer Service	Employee Relations	Other
37	Study order-filling procedures to identify and capitalize on opportunities to improve customer service and reduce clerical costs	$375,000(P)				Major		Major		
43	Overhaul and strengthen capital appropriation system – decision making and postcompletion evaluation						Major			
45	Identify value analysis techniques within purchasing department and develop systems for capitalizing on them	(Estimated to be very substantial, but not now measurable)				Moderate				
48	Develop computerized system for recording employee information and performance ratings as basis for improved identification of candidates for promotion						Major		Major	
50	Evaluate controls over maintenance, repair, and operating supplies with objective of reducing investment in inventory of these items		$2,750,000 (at 4%, the aftertax return equals $110,000 annually)(P)							

FIG. 8 *Monthly report showing savings and other benefits from completed systems projects.*

117

SYSTEMS AND PROCEDURES DEPARTMENT PERFORMANCE REPORT

MONTH ENDING: January 31, 1975

PERFORMANCE REQUIREMENTS AND STANDARDS	POINT VALUES FOR STANDARD PERFORMANCE	ACTUAL PERFORMANCE (Points attained each month for the year to date)											
		Jan.	Feb.	Mar.	Apr.	May	June	July	Aug.	Sept.	Oct.	Nov.	Dec.
1. 70 percent of completed projects should be "major" in the sense that they result in significant savings or substantial improvements in management planning and control, customer service, etc.	25	20											
2. 80 percent of in-process projects should be "major" as defined in #1 above.	20	18											
3. 100 percent of completed and in-process projects must be "on schedule."	10	9											
4. Tangible dollar savings should exceed 500 percent of the Systems and Procedures Department's budgeted expenditures.	15	25											
5. 100 percent of completed projects should result in excellent relationships with departments served.	15	10											
6. 70 percent of completed and in-process projects should be performed for divisions and departments other than Finance Division.	10	8											
7. 100 percent of unassigned projects should be scheduled for commencement within 90 days.	5	5											
Total	100	95											

Note: *It is more important to reach "standard" on each and every requirement than to excel in "total points." Significant variations from "standard" are discussed in the narrative report that follows.*

FIG. 9 *Monthly report comparing systems staff's performance with performance standards.*

basis for developing up-to-date answers to the following three questions that always make up the agenda:

   1. What significant reductions in overhead costs has the systems staff really helped the company achieve?

   2. What systems has it developed and installed that are actually helping management run the business more successfully?

   3. How competent are staff members in terms of their ability to gain the confidence of operating people?

TRAINING AND DEVELOPING SYSTEMS STAFF MEMBERS

The final element in the systems staff director's responsibilities is that of training the analysts and facilitating their career development. Most of this training and development takes place on the job—that is, in the conduct of individual systems projects. But, in addition, significant developmental value can be gained from formal, off-the-job training activities of various kinds as well as from the periodic performance reviews and counseling that individual staff members are given.

On-the-Job Training

Both the systems staff director and the project team leaders must understand that every study they undertake is an opportunity not only to produce significant benefits for the enterprise they serve, but also to accelerate the development of technical competence and interpersonal skills among the staff members. To meet this second opportunity consistently and well requires that staff leaders do two things:

   1. They must identify the development needs of each project team analyst before the study is launched. Team members' performance weaknesses or improvement opportunities may, of course, cover any aspect of systems work—planning the detailed work, interviewing, analyzing findings and developing recommendations, writing reports, making oral presentations, and so on.

   2. Once the specific development needs of each analyst have been identified, leaders must give adequate time and attention to (*a*) guiding analysts in their approach to planning their work in each area of need, and (*b*) observing and constructively evaluating their performance.

This staff development effort can be exemplified by the approach of one systems and procedures director who sought to upgrade the performance of her staff analysts by inducing them, in all their work, to push harder and faster toward substantive end results. What she said to them was this:

   It's the responsibility of each of you to solve your piece of the total problem. Or, if the project team members are all working together on the total assignment, then each of you is expected to contribute your own share of creative, workable ideas. And to help you approach your work with this responsibility continuously in mind, I urge you to treat me as the key operating executive for whom we're carrying out this assignment. When I get together with you to review progress of the study, don't just give me some incidental thoughts or observations or a random assortment of ideas. And don't make me have to draw out of you, by questioning, what it is I need to know. Instead, make an orderly presentation to me—not so much of what you've been doing, but of what you've been thinking. Give me the tentative conclusions you've reached and the alternative solutions you've identified. Also let me know what problems you've encountered and what you're doing or planning to do about them.

Another important point needs to be made about this on-the-job coaching activity. In his zeal for performing this development task (as well as in exercising adequate quality control over the project), the leader must be careful not to err on the side of becoming too heavily involved in doing work on the project. It is true, of course, that tight control helps to ensure higher productivity, clearer direction, and a margin of safety on quality. But, at the same time, such control is just as likely to frustrate project analysts, suppress their creative thinking, and inhibit their overall development. So the leader must work hard at observing, evaluating, encouraging, and guiding, without disturbing the climate of freedom and individual responsibility for results that is so important to the development and work of systems analysts.

Formal Training

In most instances on-the-job coaching, however effective, can profitably be supplemented by a continuing off-the-job training and educational program. The third section of this book furnishes a guide as well as extensive case material for the development of an in-house program that meets this need.

In addition, systems analysts can draw on a number of other sources to supplement their own experience in both the human relations and the technical phases of their work. As one example, many of the universities in larger cities offer night courses in the development and improvement of systems and procedures. In addition, business periodicals and meetings of professional associations offer increasing opportunities to keep in touch with current thought and practice on systems improvement activities in other organizations.

Performance Evaluation

A final requirement in staff development is to give each analyst prompt feedback on his or her performance. One of the most effective ways for the staff leader to meet this need is by observing the performance of each analyst in a team meeting, an interview, or an oral presentation, and immediately thereafter commenting on the analyst's performance strengths as well as making constructive suggestions for improvement, using real examples.

In addition to these frequent, low-key, informal comments, the leader should fill out a written performance evaluation on each analyst at the completion of every project. After discussing this evaluation with the analyst, the leader

REPORT ON SYSTEMS ANALYST'S PERFORMANCE

Analyst Being Evaluated: _____ Name of Evaluator: _____

Date of This Evaluation:_____ Project Title: _____

Project Duration — From:_____ To: _____ Project Team Leader: _____

Other Team Members: _____

Nature of the Project and This Analyst's Assignment on It: _____

_____

_____

_____

_____

EVALUATION	LEVEL OF COMPETENCE			
In the space provided below each evaluation area, explain your rating or give examples to support it. Such amplification is most important for any rating "Clearly Above Average" or "Below Average."	Clearly Above Average	About Average	Below Average	Not Adequately Tested
A. Basic Problem Solving Skills				
1. Ability to plan the fact gathering effectively — to decide what information is needed and why_____	☐	☐	☐	☐
2. Thoroughness and resourcefulness in digging out relevant facts _____	☐	☐	☐	☐
3. Skill at interviewing — ability to put people at ease and to stimulate from them a flow of useful information and ideas _____	☐	☐	☐	☐
4. Analysis and interpretation: Insight and imaginativeness in drawing conclusions from the findings, seeing the interrelationships among facts, and identifying the key elements of a problem _____	☐	☐	☐	☐

FIG. 10 *A written performance evaluation also identifies training needs.*

	LEVEL OF COMPETENCE			
	Clearly Above Average	About Average	Below Average	Not Adequately Tested

Basic Problem Solving Skills *(cont.)*

5. Ability to develop solutions and recommendations that not only are practical and workable but also produce significant benefits rather than a pedestrian or superficial result _____ ☐ ☐ ☐ ☐

B. **Communication Skills**

1. Ability to express self clearly and convincingly in informal discussions with middle- and top-management executives _____ ☐ ☐ ☐ ☐

2. Ability to organize the structure of written material logically, and to express such material with clarity and impact _____ ☐ ☐ ☐ ☐

3. Skill at making fluent, effective oral presentations _____ ☐ ☐ ☐ ☐

C. **Interpersonal Skills**

1. Sensitivity to others; interest in and effectiveness at gaining an understanding of their viewpoints _____ ☐ ☐ ☐ ☐

2. Apparent ability to earn the confidence and trust of
   a. Front-line personnel _____ ☐ ☐ ☐ ☐

   b. Middle-management personnel _____ ☐ ☐ ☐ ☐

   c. Senior executives _____ ☐ ☐ ☐ ☐

FIG. 10 *Continued.*

	LEVEL OF COMPETENCE			
	Clearly Above Average	About Average	Below Average	Not Adequately Tested

D. Personal Effectiveness

1. Attitude _____ ☐ ☐ ☐ ☐
   _____
   _____

2. Effectiveness in taking the initiative; knowing what to do next _____ ☐ ☐ ☐ ☐
   _____
   _____
   _____

3. Work pace and level of output; performance in setting and meeting deadlines ☐ ☐ ☐ ☐
   _____
   _____
   _____

4. Consistency in doing "completed staff work" _____ ☐ ☐ ☐ ☐
   _____
   _____
   _____

E. Project Management Skills *(To be filled out only if the individual being evaluated had any project management responsibilities)*

1. Effectiveness in planning and controlling progress of the study and in guiding and coordinating the work of other project team members _____ ☐ ☐ ☐ ☐
   _____
   _____
   _____

2. Ability at diagnosing the development needs of team members and effectiveness as an on-the-job trainer _____ ☐ ☐ ☐ ☐
   _____
   _____
   _____

3. Apparent acceptance as a team leader by others _____ ☐ ☐ ☐ ☐
   _____
   _____
   _____

FIG. 10 *Continued.*

SUMMARY

1. Considering this analyst's tenure and experience, how would you rate his/her overall performance?

   ☐ Clearly Above Average   ☐ Above Average   ☐ Below Average

2. What are his/her most important development needs and your suggested action steps for meeting them?

   _____
   _____
   _____
   _____
   _____
   _____

FIG. 10 *Concluded.*

should place it in the personnel file for later use in meeting the need, already discussed, of identifying the analyst's development requirements at the beginning of his or her next assignment.

An example of the performance evaluation form used by one well-managed systems staff appears in Fig. 10 pages 121–124.

PART THREE

Systems Analysis and Improvement Techniques

Planning and Controlling the Individual Systems Project

The results achieved from any given systems project can be heavily influenced by the kind of beginning that is made—by the care that the staff takes to lay the proper groundwork before starting on the study itself. If the preparatory effort is painstaking and imaginative, it will be reflected in the certainty, dispatch, and penetration with which the study is conducted and in the cooperation that is given by operating personnel. If it is casual or superficial, there is likely to be much floundering and misapplied effort by the staff as well as inadequate understanding by those whom the study affects. This may seem obvious. Yet it is surprising how often a staff will plunge into a new assignment without taking time to think it through carefully or to create the conditions favorable to its performance.

The first two steps in a systems study should, therefore, be devoted to planning the project in detail and to paving the way with the operating personnel concerned. Most staffs will profit from an objective analysis of the amount of attention they customarily give to these factors. They will generally find that they are nowhere near the point of diminishing returns—that increased time and thought spent on these initial steps will repay them many times over in faster, more effective performance and in a higher percentage of recommendations acted upon favorably.

But no matter how well the project plans are developed, they will not be an effective instrument unless everyone understands that *they must be taken seriously*. Hence, the final prerequisite to be met in the conduct of an individual systems project is the exercise of control over (1) progress against the agreed-on schedule, and (2) quality of the solutions developed.

PLANNING THE PROJECT

Planning is no more than charting your course before you begin. It is far easier to plan, follow that plan, and modify it, if necessary, than to start by riding off in all directions and to conclude by working under intense pressure long after the study could have been completed smoothly and with the assurance that no important point had been overlooked.

The kind of systems project planning described in this chapter is aimed specifically at helping project team members to

1. Gain common understanding and agreement on the objectives of the study and the approach to be taken

2. Force themselves to think about and plan the project not just in terms of activities (for example, persons to be interviewed) but in terms of key questions to be answered, major issues to be resolved, and end products to be generated

3. Avoid wasting time on excessive or irrelevant fact-finding, on overlap or duplication of effort, or on interesting but marginal side excursions

4. Control their performance more effectively by having definitive yard-sticks (action steps and timetables) for measuring progress

5. Ensure that the approach to be taken is described not in broad, general terms, but in terms that are tailored to the specific requirements to be met—that is, the problem to be solved or the objectives to be achieved

To realize these benefits, the project planning process must be carried out in a way that satisfies two basic prerequisites:

1. *It must cover, fully and explicitly, the two major phases of every systems project.* These are

    a. The problem-solving phase—that is, the approaches to be taken in developing the technical solution that is called for

    b. The acceptance and implementation phase, which covers the team's strategy and tactics for gaining the acceptance of operating personnel and getting them to act on the recommendations developed

2. Next, *the project planning process must be regarded not as a one-time event, but as an ongoing activity.* Ordinarily it should begin when the scope of the study and arrangements for it are being discussed with those who are originating the project or will be directly affected by it. The next—and probably most important—planning session should take place after completion of some preliminary reconnaissance aimed at developing a more detailed project plan than was possible when agreement was first reached to undertake the study. But however exact this second iteration of the project plan may be, it will seldom remain an adequate basis for communication and control throughout the entire study. For this reason, project team members and the staff director should reevaluate the plan at frequent intervals and, if necessary, update it to reflect the latest thinking on definition of the problem or objectives, on alternative solutions to be considered, on any further fact-finding or analyses required, or on the character of the action-induction job that lies ahead.

To build this concept of project replanning into the thinking of his staff members, one systems staff director follows the practice of posing, at every project progress review, these kinds of questions: "What do we know now, and what conclusions—firm or tentative—can we draw? What impact do these findings and conclusions have on the scope, direction, or focus of the work we must still do to complete this project?"

Before we move to the individual elements of the project planning job, one final qualification is needed about the planning process as a whole: While a well-conceived written plan should be prepared on every study, the breadth and depth of the planning required will vary widely from project to project. Following are the principal variables that should determine the amount of planning to be done in any given instance:

1. *Uniqueness of the project.* More formalized planning is usually required for a first-time diagnostic study, where the team is likely to be dealing with an ill-structured problem, than for a type of study that the team has carried out before (for example, developing an improved inventory control system for the third time and for the third product division of the company).

2. *Study size, logistics, and time constraints.* For example, more detailed planning is typically needed when:

    a. The project will be a long one carried out by a comparatively large team.

    b. Team members will be geographically dispersed (say among several plants or regional sales offices) and therefore unable to meet frequently.

    c. The time allowed for project completion is very short.

3. *Experience of team members.* Again, more formalized planning is needed when new staff members or inexperienced line operating personnel are assigned to the study than when a team of old hands is assigned.

Given this background on the aims, prerequisites, and variables to be considered, we are now ready to look at the components of a detailed systems project plan. They are:

1. Definition of the problem or project objective
2. Key questions to be answered and major issues to be resolved
3. Facts to be gathered and analyses to be made
4. The action-induction strategy and tactics
5. Responsibility assignments and timetable

For purposes of describing these elements in some detail, we will consider here the development of a project plan based on the kind of reconnaissance previously mentioned—that is, a brief exploratory survey aimed at acquiring background information about the problem or activity to be studied. Ordinarily this reconnaissance can be made quickly and satisfactorily by conferring with a few key executives and supervisors, including some who are actually engaged in the activity and others directly affected by it. The objectives of these conferences should be threefold (1) to determine the symptoms and possible causes of difficulties or to discover major improvement opportunities, (2) to evaluate the size of the job that lies ahead, and (3) to get a sense of the attitudes of operating personnel toward the study.

Acquiring this background is mainly a matter of asking intelligent questions about the operating experiences and problems of the persons interviewed—their understanding of the specific problem at hand, the nature of the major functions performed, the volume of work and number of employees involved, factors affecting the volume, irregularity, complexity, or quality of the work, and so on. Depending on the scope of the activity to be studied, the time required for this start-up survey might range from a few hours to a few days. But any reasonable amount of time spent in this pursuit will be soundly invested, for without it, development of a concrete and sharply focused project plan is scarcely possible.

Defining the Problem or Objective

This initial step in the project planning job will ordinarily vary somewhat between projects requested by operating personnel and those initiated by the staff itself.

Studies requested by operating personnel When a systems study is undertaken at the request of someone outside the staff—generally because of something that has gone wrong—the first step in planning the project is to think through the real problem. This is a matter of distinguishing between symptoms and causes and of determining whether the conditions that precipitated the study request reflect the whole problem or just a segment of it.

> *The systems staff in a large machinery manufacturing company operating several plants was assigned the task of streamlining all procedures covering personnel hirings and transfers. The complaint of those requesting the study was that the time-consuming paperwork involved in these processes had given rise to two types of problems: (1) serious delays in hiring new employees and (2) grievances arising from failure to pay transferred employees promptly at their new locations. The second problem developed because payroll records were not changed until all the prescribed approving signatures had been obtained on the transfer authorization form. Yet the employee was usually transferred as soon as the form had been filled out by the supervisor requesting the transfer.*
>
> *In light of this initial definition of the problem, the staff's specific assignment was to cut in half the time required to process the pertinent forms by developing recommendations for speeding up their routing, review, and approval. The operating executive who initiated the study made a number of suggestions. One was that extra copies of each form be prepared so that approving signatures could be obtained concurrently rather than consecutively. Another was that "Urgent" stickers be attached to the forms and that no person be allowed to hold one of these forms for more than 4 hours. A third was that physical transfer of an employee be prohibited until all prescribed approvals had been obtained.*
>
> *The staff's exploratory survey disclosed that the company's written instructions*

governing hires and transfers specified as many as four approvals for hiring a new employee and as many as five for a transfer (two in the line of authority from which the employee was to be transferred and three in that to which he was to be transferred). In addition to these officially required approvals, a number of divisional executives had informed their subordinates that they wished to approve all personnel additions within their own line of authority. In large divisions, this requirement often had the effect of adding to the list of approving signatures already prescribed. Thus, in actual practice, the pattern of approvals had been built up by superimposition rather than reduced by delegation.

In digging into the reasons for this cumbersome control mechanism, the staff discovered that although a system of departmental budgets had been in effect for some years, summary reports of actual performance were not being used effectively to control operating results—including the item of manpower expense, which was of immediate concern. Budget reports had fallen into disuse because they were seldom available before the twenty-fifth of the month following that covered by the report. And even when they became available, their use was discouraged by the poor arrangement of data and the sheer volume of information reported. As an alternative, therefore, key executives were seeking to control manpower expense by preapproving individual actions—an approach that, in effect, forced them to labor as their own statisticians.

From this preliminary review, the staff deduced that both the delay in obtaining final decisions on hiring and the large number of grievances arising from employee transfers were merely symptomatic. Nor was the real problem that of finding ways to speed up these activities. Rather, it was (1) to increase the timeliness and utility of operating reports as a control device, and (2) to recognize that preapprovals are but one type of control that must be evaluated in relation to all other types so that a simple but highly integrated control structure could be evolved.

As this illustration shows, the isolation and definition of basic problems are perhaps of greater consequence than all the other thinking that the systems staff must do. They are in themselves creative acts, for they open the way to fundamental and lasting solutions as opposed to the expedient patching up of weak spots.

Staff-initiated studies When the staff itself initiates a study aimed at appraising an existing system and discovering improvement opportunities, it has, of course, no problem to define. Here the initial step in project planning becomes that of defining the *objectives* of the study. This process is essentially a matter of predetermining what is to be accomplished—of thinking through the results to be expected and the form the completed study will take.

If the staff is to find the statement of objectives a useful guide in developing the remaining elements of the project plan, it must take care to ensure that the statement is sufficiently specific. For example, if the activity to be analyzed has an important bearing on the quality of service rendered to customers, merely stating that the objective is to improve customer service is not enough. The analyst must think through exactly *how* customer service is affected and state the objectives in those terms. Thus, in effect, he breaks the general objective into its component parts, each of which points to a specific subactivity requiring detailed study.

This type of breakdown is illustrated in the following statement, which was developed in connection with the planning of major changes in the accounting system of a Midwestern manufacturing company. At one time the activities of this company were almost entirely confined to fabricating job lots of detailed parts and components for manufacturers engaged in the assembly of finished products. A few years ago, the company developed its own line of finished products, which have since become an increasingly large part of its business. The statement below represents one of several projects undertaken during this period to adapt its systems and procedures to the requirements of a different type of manufacture.

> The general objective of this study is to provide the basis for more effective control of manufacturing progress and costs by developing procedures for collecting accurate cost, time, and quantity data applicable to each manufacturing operation at all plants.
>
> Specifically, the project is designed to accomplish all or as many of the following subsidiary objectives as prove to be worthwhile.
> 1. With reference to the Manufacturing Division:
>   *a.* To supply data for
>   (1) Reporting machine activity on productive operations and rework as well as nonproductive operations and toolmaking
>   (2) Developing allowed times
>   (3) Adjusting machine-load schedules
>   (4) Reporting production progress
>   *b.* To facilitate the analysis of scrap and rework
> 2. With reference to Accounting:
>   *a.* To develop operation cost data for estimating purposes
>   *b.* To provide the basis for checking attendance hours against labor-cost-card hours
>   *c.* To facilitate the accounting distribution of labor
>   *d.* To supply data for
>   (1) Determining actual cost of scrap and rework
>   (2) Achieving tighter control over work-in-process inventories

Unless the objectives of each undertaking are defined with precision such as this, the study is likely to be characterized by confusion, waste motion, and random pursuit of whatever fancied opportunities happen to capture the analyst's attention.

Besides giving specific direction to the study, a clear statement of what the staff is trying to accomplish will generally prevent the launching of projects that should never be undertaken at all. For example, if the study is aimed at achieving measurable gains, the range of probable savings can usually be estimated with a fair degree of accuracy in advance. Where this is possible, a tough-minded comparison of the savings estimate with the probable cost of carrying out the project will forestall unprofitable ventures, will make possible the assignment of priorities to others and, finally, will provide a rough standard for measuring performance on projects completed.

Identifying Key Issues and Questions

Once the problem or project objective has been defined, the next planning step required of team members is to begin thinking imaginatively and systematically about the method of attack. This does not mean deciding what facts to gather or what people to interview. Far more, this step, like problem definition, needs to be a creative process aimed at charting a course that clearly will lead to an optimum solution. It requires identification of the key issues to be addressed— that is, the relatively few major questions (usually five to ten) that, when answered, will yield the solution the team is seeking. These are questions that perform such critical functions as (1) putting the problem into perspective, (2) pinpointing the key areas of improvement opportunity, (3) identifying alternative solutions, (4) defining major constraints, and so on.

For example, in planning the project described in Chap. 1 to significantly improve the quality of its company's service to independent agents, the systems staff of the property and casualty insurance company identified the following key questions:

1. What is the true competitive significance of improved service to independent agents—in the aggregate and for each component of service (for example, speed of response, accuracy, creativity in underwriting special risks)?

2. What standards of service would be optimal—that is, would achieve the best balance between agent satisfaction and the incremental operating cost that would result?

3. What is our company's already demonstrated best performance, how does it compare with the optimal standards, and how much does it cost to achieve? That is, among our sixty-one branch offices, what is the consistently best quality of service we have rendered to agents during the past 3 years?

4. If the already achieved performance of our best two or three branch offices is still not up to the optimal standard of service, what are the alternative means through which the desired standard can be met?

5. What are the major constraints that we must take into account—for example, operating cost and its effect on the company's expense ratio, state insurance regulations, competitive practices?

6. What are the criteria to be used in evaluating each of the alternative solutions identified?

Determining Facts To Be Gathered and Analyses To Be Made

Where the project to be undertaken is aimed at improving an existing system, a common predisposition is to think of the fact-finding phase as one of describing in detail the processing steps that make up present practice. But the fact-finding requirements—and also the kinds of analyses to be made—need to be planned with far greater imagination than any such descriptive exercise calls for. Specifically, the fact-finding plan should include all the opinion evidence, "hard" (i.e., quantitative) data, hypothetical exhibits, comparative analyses, and similar diagnostic efforts that will lead naturally to the identification of improvement needs and the evaluation of solutions. These facts above and beyond a mere description of the present system might, for example, include frequency distributions of processing times (both overall and for each key step in the system), the range in unit operating costs among similar units or over a period of years within the same unit, the error volume by type of error, the quantity and trend of customer complaints, and employee turnover, skills required, and training time for each job classification.

This illustrative listing underscores the point that the analysts' fact-gathering and analytical plan should be the end result of their thinking deeply about the overall objectives of the project and about the key questions to be answered. Take the first of these needs—that is, thinking about the overall objectives. If, for example, the project to be undertaken is aimed at cutting clerical costs, the fact-finding and analytical plan should call for:

1. Identifying and ranking all tasks according to total worker-hours expended.

2. Applying to each task the mental forcing device implicit in the question: To what extent does or could this task consume excessive time and effort because of

    a. Unnecessary or marginal steps?

    b. Inefficient methods?

    c. Idle time due to poor work flow?

    d. Poorly trained operators?

Next, in thinking about the list of key questions to be answered, the team can further shape the fact-finding and analytical phase of the project by formulating—even at this beginning point—some working hypotheses on what the answer or answers to each question might be. This sort of imaginative conjecturing will do much to stimulate early creative thinking and, therefore, make fact-finding more purposeful.

Finally, the analyst should decide how each set of facts is to be obtained—whether by interviewing managerial personnel and workers, by relying on existing operating statistics, by making an actual audit of the work as it is being performed, by requesting operating personnel to accumulate certain data, or by some other means.

Applying this kind of rigorous thinking to the fact-finding and analytical phase of the project plan will be repaid many times over. For even with a clear statement of the problem to be solved or the objectives to be achieved, systems studies have a tendency to spread out and cover more ground than necessary, to drill a great many dry holes, or to result in analytical overkill unless their scope and limitations are first delineated in terms such as those just set forth.

Planning the Action-Induction Strategy

The next part of a comprehensive systems project plan should be a detailed statement of the action steps and timetable for gaining acceptance of the study and implementing the final recommendations. The important point to be emphasized here is that this phase of planning should not be pushed off toward the end of the study and thought about only after final recommendations begin to take shape. Systems analysts often do just that, partly because they tend to be oriented more toward the technical than the human relations aspects of their work, and partly because they reason that meaningful action-induction plans cannot be developed until the character of the recommendations is known.

In contrast, however, the more experienced and skillful analysts recognize that the means by which they gain acceptance, build confidence, promote understanding among operating people, and persuade them to act are so varied and involve so many people that—from the very beginning of a study—they need to give as much attention to imaginatively planning such steps as they do to planning the technical or problem-solving aspects of the project.

Following are the key steps the analyst should take in shaping this part of the project plan:

1. Identify the ultimate decision makers—that is, everyone who is likely to have a voice in the acceptance of the ultimate recommendations or to play an active role in their implementation.

2. Determine what attitudes among any of these people might represent roadblocks to action. Such roadblocks could take the form of (a) a strong preconception that significant improvement is not really possible, (b) a conviction that the activity to be studied involves such highly specialized technical knowledge that outside analysts cannot readily comprehend it to say nothing of improving upon it, (c) a concern that significant improvement may reflect unfavorably on the person now responsible for the operation to be studied, or (d) the assumption that the action required to effect improvement threatens the status or security of key employees.

3. Develop at least an initial statement of how much factual and opinional evidence will be required to gain conviction. Some executives will make decisions based on very little evidence, providing it is basic; others of a more cautious nature will not move until a mass of evidence has been assembled and placed before them. In view of these variations, more time spent at the beginning in determining *who needs to be convinced about what* will not only target the fact-finding more effectively, but in all probability reduce the total amount needed. All too often analysts gather surplus and useless facts because they have not thought carefully about how the facts will be used in convincing people to take specific courses of action.

4. In light of the foregoing steps, specify—as part of the written project plan—the completion points at which the project leader or team members will

    a. Make progress reports to, or otherwise communicate with, the persons who will influence the results of the study.

    b. Seek to gain a commitment for action. For example, upon the development of recommendations, should the project team ask for a commitment to pilot-test them, or for a commitment to immediate nationwide implementation?

Fixing Responsibilities and Timetable

The final two closely related phases of systems project planning are (1) to spell out the responsibilities of each team member, and (2) to set completion dates for each activity or milestone already recorded as part of the study plan. The staff director should make sure that *all* responsibilities of each person working on the study are listed, whether these responsibilities cover gathering or analyzing facts, preparing end products, preparing for and conducting progress reviews, maintaining relationships with operating personnel, or coaching and training other staff members. This sort of formal approach to the allocation of team members' responsibilities gives the director a basis for logically dividing up the work to be done, for evaluating each person's performance on the project, and for making individual assignments that meet each analyst's specific training or development needs.

    Once such assignments of responsibility have been made, each team member should develop, weekly or monthly, his or her own personal work plan and timetable, identifying (1) the major phases of work to be done and the number of days to complete them, (2) areas requiring coordination with fellow team members, and (3) specific interviews requiring advance scheduling.

Using Project Planning Forms

In carrying out the foregoing steps, systems staffs will generally find it useful to adopt some kind of standard form for drawing up project plans. By providing a framework of the information required, such a device tends to ensure greater uniformity and completeness in the plans developed. In addition, the requirement that plans be put in writing forces staff members to be more diligent and circumspect about project planning than they might otherwise be. Finally, the value of a written commitment as an incentive to the analyst and a basis for control by the leader has already been mentioned.

    The form used by one systems staff for this purpose is reproduced in Fig. 11. Its four sheets have been filled out with a hypothetical project plan to show how they can be adapted to the type of definition that has been recommended. Of particular interest is the third sheet showing the use of a Gantt chart for scheduling the various steps in the plan. As the project is carried out, red lines indicating actual progress are drawn on the sheet above the black schedule lines appearing in the chart. These data are not only the staff leader's principal control tool, but they also serve as a monthly progress report to the leader's superior. For this report, the staff director draws a vertical line on the schedule

Project General Survey of Order Processing and
title Invoicing System

Project No. _____
(To be assigned when approved)

Requested or
initiated by M. A. Hansen

Date: March 3, 1975

PROJECT PERSONNEL

A. M. Fuller — project leader
R. G. McBride
L.R. Carroll — central order department

PROJECT OBJECTIVES

1. **Cost Reduction.**
   a. To reduce time required of sales personnel in preparing order forms.
   b. To reduce clerical expense of order processing and invoicing.
   c. To minimize idle time in stock and shipping rooms resulting from irregular flow of orders.

2. **Improvement in Customer Service.**
   a. To speed up the filling of sales orders by accelerating the procedure cycle and by facilitating stock picking.
   b. To increase the speed, accuracy, and completeness of order-status information supplied to customers.
   c. To increase the accuracy of stock picking.

3. **Improvement in Controls.** To increase the timeliness and completeness of order-summary data supplied to various operating departments.

ANTICIPATED RESULTS

The present clerical force engaged in order processing and invoicing at the branch sales offices, main office, and plants totals 73 persons. It is estimated that a 15 percent reduction in personnel requirements, or an annual payroll saving of about $100,000, can be achieved through procedures simplification and methods improvements.

In addition, a substantial reduction appears possible in the 1,700 hours of idle time that occurred in the stock and shipping rooms during the past six months because of the irregular flow of orders.

No attempt has been made to evaluate possible intangible benefits in the form of improved customer service or more useful order-summary data. A preliminary survey indicates, however, that a number of worthwhile improvement opportunities exist in these two areas.

SCOPE OF PROJECT

1. **Accelerating Procedure Cycle and Reducing Clerical Expense.** Flow of a sales order from the time it is prepared by field sales representative until shipment has been made and invoiced.

FIG. 11 *Sample project planning form for a hypothetical project.*

sheet through the point indicating the date of the report. In this way, the chief executive has a convenient visual representation of progress achieved against the original plan. Where necessary, the staff leader also includes a brief narrative report on the last sheet of the form under the caption "progress comments."

    a. Areas to be covered:
        (1) Field sales representatives (three working out of New York branch).
        (2) Branch sales office (New York).
        (3) Order department.
        (4) Credit department.
        (5) Stock room (Ashton plant).
        (6) Shipping department (Ashton plant).
        (7) Billing department.
    b. Elements to be covered:
        (1) Sales-order policies.
        (2) Clerical operations, methods, and equipment.
        (3) Forms used, records maintained, reports prepared.
        (4) Volume and irregularities in flow of work.
        (5) Length of procedure cycle.
        (6) Communication costs in order handling.
        (7) Interdepartmental operating problems.

2. Simplifying Order Preparation by Field Sales Representatives.

    a. Discuss order-preparation problems with New York branch manager and three sales persons.

    b. Explore possibility of printing product names on order form, and of equipping sales representatives with small portable dictating machines.

3. Improving Order-Information Service to Customers.

    a. Organization considerations:
        (1) Determine extent to which order department and plant stock rooms have direct contact with customers regarding orders and shipments.
        (2) Determine extent to which sales representatives communicate directly with plant stock rooms to expedite shipments.
        (3) Determine relationships between order department and sales division in establishing priority of shipments.

    b. Analyze sample of 200 customer order-status inquiries to determine time taken to reply and completeness of information given.

    c. Determine extent to which unsolicited information regarding shipment delays is supplied to customers.

4. Facilitating Stock Picking and Reducing Idle Time.

    a. Determine extent to which arrangement of commodities on shipping-order form varies from arrangement of items on stock shelves.

    b. Analyze flow of orders to stock room by one-hour intervals for period of one week.

    c. Determine reasons for irregularities in order flow.

5. Improving Order-Summary Control Data.

    a. Ascertain order-summary information requirements of sales, production, and merchandising departments.

    b. Determine present availability of required data, both as to completeness and timeliness.

FIG. 11 *Continued.*

Testing the Project Plan

As an added means of ensuring the soundness and completeness of each project plan, the systems staff director should follow the practice of consistently applying the following checklist of questions to every plan developed by the staff:

    1. Does the plan meet all the tests set forth in the top-management approach to systems analysis outlined in Chap. 5?

PROJECT PROGRAM

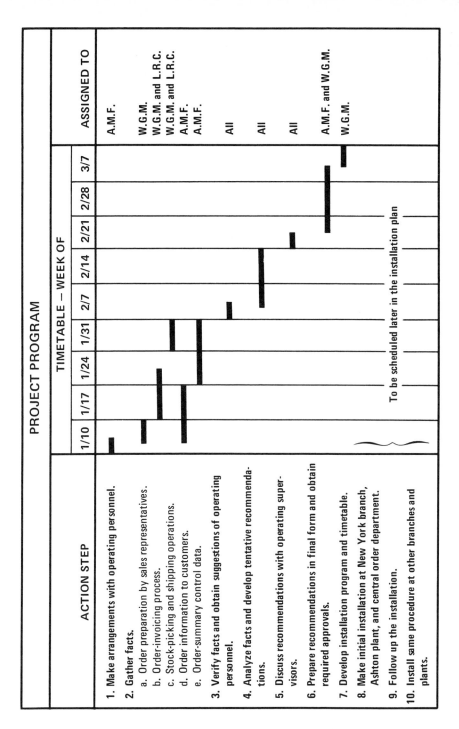

ACTION STEP	TIMETABLE — WEEK OF									ASSIGNED TO
	1/10	1/17	1/24	1/31	2/7	2/14	2/21	2/28	3/7	
1. Make arrangements with operating personnel.	▮									A.M.F.
2. Gather facts.										
a. Order preparation by sales representatives.	▮	▮								W.G.M.
b. Order-invoicing process.			▮							W.G.M. and L.R.C.
c. Stock-picking and shipping operations.		▮	▮							W.G.M. and L.R.C.
d. Order information to customers.				▮						A.M.F.
e. Order-summary control data.				▮						A.M.F.
3. Verify facts and obtain suggestions of operating personnel.					▮					All
4. Analyze facts and develop tentative recommendations.					▮	▮				All
5. Discuss recommendations with operating supervisors.							▮			All
6. Prepare recommendations in final form and obtain required approvals.						▮	▮			A.M.F. and W.G.M.
7. Develop installation program and timetable.								▮	▮	W.G.M.
8. Make initial installation at New York branch, Ashton plant, and central order department.	To be scheduled later in the installation plan									
9. Follow up the installation.										
10. Install same procedure at other branches and plants.										

FIG. 11 *Continued.*

PROJECT APPROVAL

Project approved by   Date

PROGRESS COMMENTS

COMPLETION REPORT

Date project	Date trial	Date final
recommendations	installation	installation
approved	completed	completed

Summary of benefits realized

FIG. 11 *Concluded.*

2. Is the plan aimed specifically at achieving the defined objectives of the study, or does it appear to contain a good many diversionary side excursions?

3. Are the fact-finding and analytical parts of the plan built around a listing of the key issues to be resolved and major questions to be answered?

4. Does the plan contain a complete, well-thought-out series of steps for gaining understanding and acceptance and inducing action?

5. Does the plan make concrete provision for using operating personnel in the most effective way?

6. Finally, does the plan provide for bringing to bear all the relevant accumulated experience gained on previous projects that are similar, in one way or another, to the one now being undertaken?

PAVING THE WAY

The steps required in laying the proper groundwork with operating personnel will vary, of course, depending on the origin and importance of the project. At one extreme in the range of variation are projects that operating executives request the staff to undertake. Here the initial requirements for developing confidence are generally satisfied by the staff's reaching agreement with the executive on the nature of the basic problem and obtaining his approval of the plan for tackling it. At the other extreme are the projects of major significance which the staff itself initiates. On these the task of kindling interest and support is more demanding. For that reason, we have selected the latter type of project to illustrate some of the steps to be taken in the process of paving the way. Details on other types of projects would have to be modified, of course, to fit individual situations.

Securing Executives' Support

The staff should begin by discussing the project openly with the head of each major organizational unit affected. Although these division and department heads will not ordinarily take a direct part in the study, their attitudes toward it will be reflected by those of their subordinates with whom the staff works. Here are some suggestions for the staff to follow in carrying on these discussions:

1. *Give each executive the full background of the study.* Explain its purpose, the reasons behind it, and the results that are expected. Emphasize the benefits most likely to interest the executive. If the anticipated benefits will not affect that person's operations directly, stress the interdepartmental implications of the project so that the executive can see how the cooperation of his or her unit is required to solve a problem that makes itself evident in another area.

2. *Forestall a defensive attitude.* If the executive is likely to feel even remotely responsible for the unsatisfactory conditions that the project is aimed at correcting, provide a reasonable explanation for those conditions early enough in the discussion to forestall any implication of criticism.

Under circumstances similar to these, one systems staff director offered the following explanation to an operating executive, with completely disarming effect:

During the past few years everyone has been so busy worrying about more pressing and immediate problems that there hasn't been much time left to keep an eye on our operating methods and procedures. The result has been that a good many of them have become complicated and unwieldy, and a few have broken down altogether.

It's easy to understand how this came about. Take, for example, the system we've been discussing. When it was originally designed, it was quite simple and exactly suited to the requirements. But, over a period of years, new operating personnel and new conditions have brought a good many changes and additions to the original structure. Most of these additions were not important enough individually to stir up any great excitement at the time or to give rise to much thought about how they fitted into the overall plan. But taken together, they have had a substantial, cumulative effect on the scope of this activity, on its importance to the internal operation of the business, on the number of people it involves, and so on.

And so, just as the company takes a periodic physical inventory, we would like to work jointly with you and the other department heads concerned to take stock of this operating system—to ensure that it's the simplest and most efficient way of meeting present requirements.

Your subordinates probably have a good many worthwhile ideas in the backs of their minds on ways to short-cut and improve existing operations. They've probably had them for some time but just haven't had a chance to do anything about them. As we see it, one of our principal functions in working with these people is to crystallize their ideas and see that the best use is made of them.

We assure you that we don't want to move too fast or abandon any necessary controls. And, as the study progresses, we'd like to discuss it with you and get your approval of any ideas affecting your operations before we discuss them with others.

3. Review the project plan with each executive. By reaching advance agreement with key personnel on the steps to be taken in the study, the analyst avoids the risk of having the results of the project later discredited through criticism of the method by which it was carried out. And here again, the staff member has an opportunity to establish rapport by being completely open and forthright with others. The very act of laying a carefully developed plan of action in front of them builds assurance that the analyst clearly knows how to get at the job. It is also true that most executives like the feeling of being consulted—of being asked for their ideas and their approval of the project. Finally, the executive can often save the staff considerable time by offering helpful suggestions on sources of desired information.

4. Request the executive's assistance. If the project is comprehensive in nature, the staff will want to formalize operator participation by asking the executive to designate a subordinate to work actively with the staff during the survey and design stages. Where less formal participation is required, the staff should get the agreement of department heads on any specific data to be accumulated by their groups in connection with the study. It should also inform executives to what extent the assistance asked of their personnel will affect the department's regular work. Finally, the staff should seek the help of the executive in laying the groundwork for the project with front-line supervisors and workers. Together they should decide whether any formal explanation of the reasons for the study need be given to the persons with whom the analyst will work and, if so, what the explanation should be and who should make it. Where operators are called together as a group for this purpose, it is preferable that the executive make the explanation in order to erase any doubt about his or her support of the project. In later contacts with individual supervisors and workers, the analyst will have adequate opportunity to amplify the executive's earlier explanation in whatever way seems necessary.

Securing the Support of Supervisors and Workers

If by these steps the project team members succeed in gaining the initial support of the executives whose operations are to be studied, they must not make the mistake of assuming that the attitudes of supervisors and workers are any less important. In short, the staff must do the same sort of job to arouse interest and support at this level as it did at the higher level. Winning support down the line often requires even more careful planning and skillful execution than were necessary at the executive level since the interests and habits of the workers and their immediate superior are so much more in jeopardy.

In all probability, the factor that exerts the greatest influence on operator's attitudes is their understanding of the purpose behind the study. Granted, it may not always seem wise to reveal all the objectives sought, and the reluctance to do so is understandable when the principal goal is to reduce operating costs. But the analyst must remember that the real purpose of the undertaking cannot long be concealed from most workers, and a frank acknowledgment of intent may, therefore, be infinitely less harmful to morale than rumor, suspicion, and uncertainty.

One company that undertook a cost reduction study of all its overhead activities assured its employees that any resulting reductions in personnel requirements would be translated into actual savings through normal turnover and transfers rather than through layoffs. The departments that were affected employed some 650 people, and 90 jobs were eliminated by the study. The full amount of this saving was realized within 8 months after completion of the study without a single layoff and with incalculable benefit to morale throughout the organization.

Such assurances cannot always be given, of course, but they can be given much more often than they are even thought of.

In addition, during his initial contact with supervisors and workers, the analyst must begin to make them aware of his desire to solve problems or develop improvements *with* them, not *for* them. He should make it clear that their knowledge and ideas are essential to achieving the most profitable results and that they will receive full credit for their contributions.

CONTROLLING PROJECT
PERFORMANCE

Quite obviously, the project plan and timetable are made to be followed. But they will not be followed unless the project team leader or systems staff director uses them regularly as a basis for checking progress and accomplishments.

To be fully effective, the monitoring process should have four purposes:

1. Identify any slippage against the agreed-upon schedule and determine its cause and significance. Sometimes this sort of review will reveal that the team has been drawn off into side issues or has pushed fact-finding or analysis well beyond the limit of need. But even if the failure to keep on schedule is fully justified by developments not originally foreseeable, the director may conclude that the delay is causing or will cause problems in the future—for example, with the operating personnel involved. In this event, the monitoring process can help point up the need for adding or reassigning project team members, or for taking some other kind of corrective action. But even though such corrective action may seldom be required, regular progress reviews serve as a useful controlling device to keep the project team purposefully directed and on its toes.

2. Evaluate the adequacy or the quality of the analytical work that is being done. This sort of qualitative evaluation is perhaps the most important part of the project control process. It can take a number of different forms, ranging from the staff director's or project leader's review of each analyst's work to a team session in which team members review and evaluate each other's work.

3. Facilitate project replanning. Another value of formal project progress reviews is that they help to sharpen team members' thinking about major issues, stimulate new ideas on alternative solutions, and identify additional facts needed to evaluate those alternatives. The review may disclose that project priorities need to be changed, that certain issues or analyses should be eliminated, and that others should be added. Such disclosures do not necessarily mean, of course, that the original planning was ineffective. In working on fundamental systems problems, the staff should expect that as it digs deeply into them, uncovers new facts, and does more penetrating analyses, the best approach to the problem will change—sometimes quite markedly. In this event, the project plan and schedule *should* be revised, and the revised plan then becomes the new basis for checking progress.

4. Keep operating personnel informed of progress and changes in plans. As a final by-product, the project progress review provides the team leader or staff director with another appropriate opportunity to communicate with the key operating people concerned with the study, to test tentative thinking with them, to obtain their agreement to changes in the project plan or schedule, and generally to maintain their understanding of and interest in the whole undertaking.

NINE

Gathering the Facts

A top merchandising executive of a large fashion-goods manufacturer summed up one of her company's basic operating systems in these words: "We've been trying to straighten out and simplify this process for years without much success. Our biggest stumbling block is that the system cuts across so many departments and has become so complicated that probably not more than two or three people in the whole organization understand it completely."

Investigation showed that low as her estimate was, it still overstated the facts. Some front-line supervisors knew what actually took place within a major segment of the system, but no single individual understood it wholly. In addition, the information that did exist was tainted by a good deal of supposition and misconception. Parts of the routine were being performed quite differently from the way supervisors thought they were being done or from the way written procedural instructions said they should be done. Erroneous ideas on the causes of difficulty and delay had been voiced so freely that the whole clerical and administrative force had tacitly come to accept them. In short, no one had all the *facts* essential to the intelligent analysis or improvement of the system.

These circumstances highlight the importance of rigorous fact-finding as the first step in the execution of a systems improvement project. This is the same fundamental research process that underlies the orderly solution of any business problem. Its significance never varies, regardless of the type of project undertaken. Development of an entirely new system requires knowledge of the objectives to be met and the resources available to meet them. Improvement of an existing system requires thorough understanding of its purpose, its detailed operations, and the conditions affecting its performance. Both demand a great deal of hard work and resourcefulness to ensure that the information gathered is complete, and diligence to ensure that it is accurate.

The purposes of this chapter are to:

1. Provide a checklist of the types of information to be gathered in systems and procedures studies, and point out some common pitfalls that lie in the way of getting complete and accurate information

2. Show how the essential facts can be recorded in a way that will reveal significant relationships and facilitate analysis

3. Suggest some useful pointers on how to ensure good survey results

WHAT INFORMATION TO GATHER

The previous chapter stated that the factual and opinional evidence to be obtained on any project should be governed by the defined objectives of the study. An obvious corollary is that not all the types of information outlined in this chapter are applicable to every project. For example, a study aimed at strengthening operating controls would ordinarily call for the gathering of facts somewhat different from those required for accelerating a procedure cycle. The same may be said of solving a specific, well-defined systems problem in contrast to making an overall diagnosis of a given process for the purpose of uncovering improvement opportunities of all types.

Granted that specific information requirements are a function of the scope of an individual project, systems and procedures studies as a group require information on the following points:

1. Objectives and requirements to be met
2. Organization and personnel
3. Policies
4. Details of the existing system
5. Operating costs
6. Effectiveness of the system
7. External factors

Before considering each of these classes of information, we should point out that frequently some part of the data required will already be available and will therefore not need to be gathered during the formal survey phase. For example, the staff director or project leader may have gathered facts during the initial exploratory survey that forms a part of the project planning step. Other essential information may be a matter of general knowledge, or may have been acquired during previous studies of related procedures.

Objectives and Requirements

If the analyst is to determine how effectively the objective of a system or procedure is being achieved, he must first find out what that objective is. He must thoroughly understand the basic requirements that the routine exists to satisfy. There are few types of projects on which this knowledge is not required. Even on those concerned solely with procedural mechanics, the analyst cannot make changes intelligently until he has evaluated their possible effect on end results.

If information on the objectives of a procedure is to be useful, it must be clear, detailed, and complete—not merely an obvious generalization inferred from the name or nature of the activity under study. It is not enough, for example, to observe that the purpose of sales statistical procedures is to provide the information required for planning and controlling the company's marketing operations. Rather, the analyst should think through the objective of that particular activity and express it in some such terms as the following.

The purpose of the sales statistical procedures to be covered by this study is to produce the internal sales data needed for effective planning and control of the following types:
1. Sales planning
   a. Establishing sales volume objectives by territory, industry, and channel of distribution
   b. Developing the advertising budget and planning the allocation of funds for local advertising
   c. Evaluating the character and size of the product line
   d. Developing additional outlets and new channels of distribution
   e. Determining the need for and location of additional warehouses
2. Measurement and control of sales results by:
   a. Salesman
   b. Geographic area
   c. Distribution channel
   d. Type of industry
   e. Product class

Obviously systems analysts should investigate the requirements or objectives of statistical or reporting procedures more thoroughly than those of procedures designed principally to carry out the routine transactions of the business. But, even on the latter, they will find that careful digging into all the purposes of the activity will supply additional points of reference needed to discover basic weakness.

Organization and Personnel

Step two in the survey should ordinarily be to gather organizational facts. These consist of three types of information:
1. The structure of organization, which includes the position titles and job classifications of the persons participating in the system and the lines of authority among them
2. Names of people who now staff the organization
3. Principal functions of each unit

The first two types of information can usually be obtained quite readily, although the analyst must guard against accepting formal organization charts without making sure that they are up to date and that they accurately represent the present situation rather than some proposed or idealized arrangement. As part of this fact-finding step, the analyst should also seek to learn about any changes in organizational structure that are being planned and the reasons behind them.

Determining the major functions of each unit will often prove more difficult. In the absence of any written definitions of responsibility, the analyst is likely to encounter considerable uncertainty or disagreement about functions, particularly where the activities under study include planning or control rather than routine operations. But even where some formal delineation of responsibility exists, there may be significant variations between the plan and the realities. Here again, therefore, the analyst must exercise care to ensure that he discovers the functions *actually* being performed and not merely those set forth in some written statement.

This part of the fact-finding process assumes its greatest importance whenever a large number of organizational units participate in the system being surveyed. Here the job is essentially one of determining the exact working relationships between departments on matters of common interest that may represent areas of overlap or potential conflict. The specific organizational facts to be obtained in this area will vary, of course, with each system or procedure studied. As an illustration, however, here are some of the questions regarding division of functions that might be raised in a study of the order processing, shipping, and invoicing procedures of a manufacturing company:

1. Are separate organizational units responsible for order processing and invoicing? If so, what is the line of demarcation between their respective functions?

2. What is the relationship between field sales offices and the headquarters credit department on establishing lines of credit and approving individual orders?

3. To what extent do sales headquarters, order, production control, and manufacturing departments have direct contact with customers regarding orders or shipments?

4. To what extent do the field sales offices deal directly with the manufacturing departments?

5. What are the relationships between the order department and the sales force in determining the priority of shipments?

6. What are the relationships among the order, production control, and factory departments in establishing shipping dates, determining the manufacturing sequence of orders, or specifying the plant from which the order will be shipped?

7. Is a clear and logical distinction made among the duties of the sales, order, and production departments for the control and replenishment of finished-goods inventories?

8. If the company manufactures several products at several plants, are the functions of the order department organized or divided among the personnel by product, by plant, or by clerical activity?

This discussion of organizational facts may prompt the reader to question whether organizational problems properly fall within the scope of a systems study. The answer is clearly yes: If anything approaching a fundamental diagnosis is to be made, the effect of organizational structure and division of responsibility on performance of an operating or management system must be thoroughly explored. To be sure, a systems improvement project should not be used as an excuse for launching broad organizational studies as such. The analyst's concern with organization should be limited to its impact on the system itself. But it is worth emphasis that organization structure has a greater influence on system costs and effectiveness than is generally supposed. It will often be found, for example, that a highly complex system merely reflects a faulty or cumbersome organizational structure. Similarly, poor organization may impede the flow of work or promote duplication and overlapping.

A second reason for digging out the organizational facts is that existing routines, records, reports, and other paperwork cannot be intelligently evaluated unless the analyst thoroughly understands the responsibilities to be discharged. The organization chart in Fig. 12 is useful in illustrating this point. The sales organization shown is divided into two major parts—a field selling group responsible to a general sales manager, and a headquarters staff reporting to a chief product manager. The headquarters group is responsible for various sales-planning, policy formulation, and control functions. The regional managers, on the other hand, are responsible for execution of the company's marketing plans, policies, and programs through their direction and control of the branch offices.

Obviously no far-reaching program of systems improvement could be carried out in the area covered by Fig. 12 unless the analyst first learned in detail the functions of the two groups involved, their relationships to each other, and the information they need to perform their jobs effectively. Any analyst who doubts this should try to develop a written procedure or structure of management control information involving a number of departments without having an organization pattern into which procedural responsibilities can be fitted. Such tasks are hopeless and invariably result in shelving the procedural phases of the study until related organization lines are clarified and jurisdictional disputes settled.

Finally, the organizational facts are of value to the analyst in planning the details of a study. They provide an overall picture of the scope of the project, the division of work, the number of persons and classes of labor employed, and the diversity of functions performed. This information will guide the staff member in determining how much interviewing needs to be done, who should be interviewed, and in what sequence.

Policies

The third element of background information consists of the policies that the system or procedure is designed to carry out. If the analyst is studying a billing procedure, he will want to know the company's pricing and discount policies; an order processing procedure, its sales and credit policies; a complaint and

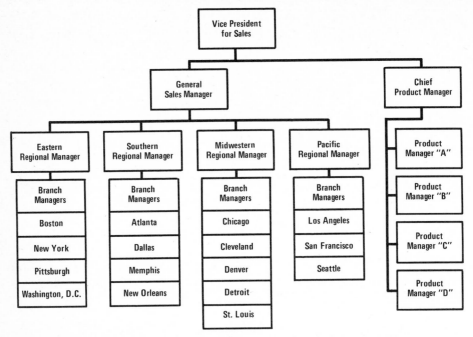

FIG. 12 *Organization chart of a sales division shows how responsibilities are divided.*

adjustment procedure, its return and refund policies; a payroll procedure, its compensation policies; a filing system, its records retention and disposal policies.

In addition to discovering what policies are in effect, the analyst should find the answers to such questions as these: Are existing policies reduced to writing, and are they clearly understood by those who must execute them? Are they uniformly interpreted and administered? If not, what are the nature and frequency of the exceptions? Are there any important recurring matters on which policies have not been established?

Often analysts have to defer this part of the fact-finding until they have studied the details of the existing system. But eventually these questions must be dealt with, since they have an important bearing on the substance and effectiveness of the system itself.

Details of the Existing System

Once project team members thoroughly understand the objectives, organization, and policies related to the system under study, they are ready to dig into the details of the system itself. Following is a checklist of facts to be obtained in this part of the survey.

1. Functions and operations
   a. *What* work is performed?
   b. *When*—that is, in what sequence—are the operations performed?

      c. Who performs them?
      d. How?
      e. Why?
  2. Quantitative data
      a. Volume of work
      b. Worker-hours required
  3. Flow of work
      a. Degree of stability
      b. Frequency of movement
      c. Direction
      d. Distance
  4. Qualitative data

Functions and operations Dissection of an existing system or procedure is essentially a matter of breaking it down first into its major elements or functions, and then into the detailed action steps associated with each. The survey should begin with the operation or document that initiates the routine and should follow some logical sequence until all activities within the defined scope of the study have been covered. The analyst should record each separate operation and inspection, and each movement or delay of work, in the order of its occurrence and in sufficient detail to permit its evaluation. In other words, he should determine not only the work that is done but also the position title or job classification of the person doing it, the manual or mechanical method by which it is done, and the reason for it.

    Obviously, the more detailed this breakdown is, the more opportunities for improvement will be uncovered. Even minute elements of operating practice, which analysts are inclined to ignore, may hold clues to basic problems or needs. For example, the method of filing documents and the arrangement of each file may not seem worth attention. Yet in one study of an order handling procedure, information on this point brought to light a serious deficiency in the company's order follow-up practices. The study revealed that the order department's file of unshipped orders was arranged by serial numbers. Since this sequence bore no relationship to the promised shipping dates appearing on the orders, the analyst asked what basis the order department used for following up the factory to ensure fulfillment of delivery promises or to notify customers of probable delays. The answer given was that the entire open-order file was audited occasionally for overdue shipments but that no procedures existed for regular factory follow-up or customer notification.

    Irregular or unusual work is another element that is often overlooked in systems and procedures surveys. Analysts tend to follow the normal flow of work to the exclusion of other phases that often have an important bearing on the problem. These exceptions may include special methods for handling nonstandard work, contributory or facilitating activities, or recurring tasks performed at irregular intervals. Questions should always be asked to bring such exceptions to light.

In breaking down the system into its operational elements, analysts eventually reach a point of diminishing returns beyond which further probing does not pay. It is better, however, for inexperienced analysts to err on the side of gathering too much rather than too little information. In this way they will be less likely to overlook important clues while gaining the experience that will enable them in time to distinguish quickly between significant and immaterial details.

The survey of clerical work *methods* is largely a matter of determining what types of office machines, equipment, and manual work aids are in use and what part they play in performance of the system. Where forms or reports are prepared on a special type of office machine, the analyst should determine to what extent the nature of the document is governed by the characteristics of the particular machine.

In addition to the specific uses of each office machine, analysts need certain supplementary information if they are to evaluate the application of the machine intelligently. They should, for example, inquire into machine speeds (average output per operator, not maximum machine capacity), skills required to operate the machine, and breakdown and repair experience.

Finally, they should determine the degree of utilization for all types of machines and equipment, whether owned or leased. The importance of such information on rental machines is readily evident. But even on equipment owned by the company, unused capacity may point the way to additional applications, or may forestall the purchase of similar equipment by another department whose requirements have increased.

Determining the *purpose* of each function, document, or operation is such a fundamental part of systems analysis that its importance hardly needs emphasis. For most systems and procedures, this type of inquiry invariably points the way to some curtailment or outright elimination of work. Yet this part of the fact-finding job is often the most poorly done, because the real use of a report or record is frequently quite different from its apparent or intended purpose. Inexperienced analysts, therefore, may be easily misled by their own assumptions or by the unintentional misinformation of others.

To avoid such pitfalls, systems analysts should remember two important rules:

1. They must not be dissuaded from their search by the assurances of operating personnel that a given activity, record, or report is necessary. They should persist in determining its exact end use so that they can satisfy themselves about its value or necessity.

2. Wherever possible, they should seek to verify, by test checks or actual observation, the statements that operating personnel make about the purpose or end use of any clerical product or process.

The reason for many detailed clerical operations is, of course, readily evident and therefore need not be a matter for special inquiry. For example, the analyst does not have to find out why invoices are extended. But aside from such exceptions, the investigation should be thorough and complete. Particular attention should be given to forms, records, and reports and to checking or inspection operations. The analyst should trace each copy of a report or form to the people who receive it, and should question them about the use they actually make of it as well as the purpose it is supposed to serve. Similarly, each record should be studied to determine the source, nature, and frequency of references to it or requests for information it contains. For each checking and inspection operation, the analyst should find out the type and volume of errors detected and the risks that would be incurred if the operation were eliminated.

The following case history illustrates the importance of thoroughness and persistence in digging into the end use of the clerical product.

The headquarters sales organization of a nationwide company maintained an expensive record-keeping system to accumulate operating and maintenance costs on each passenger car in a fleet of 3,000 cars assigned to company sales representatives. The analyst studying this activity was told that the purpose of these records was to reach decisions on when to trade in any given car, and to develop the summary information needed to determine which make of car was most economical to own and operate. But on checking more deeply at other points in the organization, the analyst found that (1) the company followed a uniform practice of trading in each passenger car at 65,000 miles or 3 years of age, whichever was reached sooner; (2) for reciprocal-trade reasons, the company consistently divided its new-car purchases among the "low-priced three" according to predetermined percentages.

Hence, the costly record-keeping system had no bearing whatever on the decisions for which it was allegedly being maintained.

Quantitative data In addition to understanding fully the detailed activities and the purposes of the present system, the analyst generally needs to know how much work is performed and how much time is required per unit of work. These quantitative data are important in determining the cost of an operation, in appraising the suitability of present work methods, in comparing them with possible alternatives, in measuring the effect of work elimination on personnel requirements, in spotting bottleneck operations, and so on.

Where a number of clerical operations are performed on a given document, obtaining volume figures on each is not necessary since these can be established by a count of the documents themselves. For example, the preparation of an invoice might consist of pricing, extending, typing, proofreading alphabetic information, reextending, separating, mailing customers' copies, and distributing internal copies. Determining the number of invoices processed per day, week, or other period will automatically produce volume figures for each of these operations.

The analyst can usually develop figures on the quantity of work performed by reference to supervisors' records of daily production, by the use of serial-numbered forms, or by an actual count of postings, orders, inquiries, forms, letters, and so on. Since estimates of work volume are frequently misleading, they should not be relied on except when other means of measurement are unavailable.

Unlike volume statistics, performance times should generally be obtained for each individual operation or task. This measurement is necessary because performance time can vary widely from operation to operation and because individual operations are more subject to change than the system or procedure as a whole. Whenever an operator spends all or a measurable percentage of the workday on a given task, realistic unit times can be established by dividing by the known volume. Under other circumstances, satisfactory estimates can usually be made through close observation of the work without having to resort to stopwatch measurement. As a closely related fact-gathering requirement, whenever speed of processing time is among the elements being studied, analysts will want to determine not only the time required to perform a unit of work at each work station, but also the average time that a piece of work waits for processing at each station.

In addition to these data, analysts should be alert to the need for other quantitative information that has a bearing on the choice of methods by which the procedure can best be performed. For example, in the study of an accounts payable routine they will want to know what percentage of customers pay from monthly statements and what percentage pay from invoices covering individual shipments. In analyzing an inventory control system, they will want to know the quantity of stores withdrawal requisitions for each value category (for example, 0 to 25 cents, 25 to 50 cents, 50 cents to $1). In surveying an order-processing procedure, they should find out the number of items in the company's product line; the average number of items per order; the volume of back ordering or partial shipments; and the volume of other changes—that is, reduction or increase in quantities, substitution, deletion, or addition of items, and so on.

Flow of work   Closely related to work volume and unit times are the degree of stability in the flow of work and the frequency, direction, and distance of its physical movement. Since these factors have an important bearing on clerical costs and processing time, they should be carefully analyzed whenever labor savings or acceleration of the system or procedure cycle is among the project objectives.

Evenness of work flow. The analyst's investigation of the rate at which work flows through the system should be built around three questions:

1. What are the size and frequency of peaks and valleys in the workload?
2. What effect do these irregularities have on the system's costs or time?
3. What causes them?

The answer to the first question can be found by analyzing the daily volume of work flowing into a given department over a representative period of time. Although day-to-day variations in volume are generally the most frequent and controllable, systems analysts should also be on the alert for abnormal fluctuations from week to week, month to month, or season to season. Finally, within each department participating in the procedure, they should look for unevenness in the flow of work between major operations.

The effects of irregularities in volume usually take the form of backlogs and delays, overtime work, and idle time beyond the workers' control. Some of these factors can be measured precisely; others can only be estimated. Nevertheless, a reasonable effort should be made to determine the extent of each in order to create an awareness of the full impact of volume fluctuations. In addition, any staggered work schedule, installed for purposes of smoothing the flow of work, should be carefully studied to determine how the various operations are merged.

The causes of an uneven work flow may be internal or external, controllable or noncontrollable. When analysts find that a fairly uniform quantity of work is always available at the starting point of the procedure, they know that the irregularities are created within the procedure itself. To isolate the cause of the irregularities, they should ask such questions as these: Is any one department or position accumulating in-process work and dumping it in large batches onto the next department or position? Is special, irregular, or nonstandard work permitted to interrupt the normal routine? Do the numbers of workers assigned to various operations accurately reflect differences in time required to perform the operations? Or is one part of the work force overloaded and another underloaded? Have the supervisors whose units participate in the process planned and scheduled the flow of work carefully? Is the schedule arranged to keep work from reaching critical stages toward the end of the working day? How much worker time is lost in waiting for facilities or supplies or for the supervisor to open mail, distribute work, and issue instructions? What delays are incurred in getting approval signatures on forms before they can be released for action?

Where variations in volume are caused by departments not directly involved in the procedure, the needs of these units should be carefully evaluated and compared with the cost of meeting them. Once the facts are known, a minor adjustment in activities or requirements at these external points will often make possible a substantial leveling of the workload in the area under study.

If the irregularities in the flow of work are found to be noncontrollable—as will frequently be the case when their cause lies outside the company—the analyst must then seek ways to minimize their effect since he cannot remove them. Suggestions for doing so are given in Chap. 11.

Physical movement. Step two in surveying the flow of work is to determine the frequency, direction, and distance of the movement of personnel and documents from desk to desk and department to department. Although these facts will generally be recorded during the survey of individual operations, they may be made the subject of an auxiliary study. Where the team members discover excessive movement or backtracking, they should find out whether the cause lies in the system or procedure (for example, in the sequence of operational steps) or in the physical arrangement of the office. If poor layout appears to be a contributing factor, they should prepare space and layout diagrams showing the location of personnel, equipment, and records and the flow of work among them. An illustration of this type of charting appears in Fig. 18 later in this chapter.

Qualitative data   The final area to be explored in the survey of existing procedures is *how well* the individual functions and operations are performed. Here the analysts should determine the general condition of records and files, the legibility and completeness of forms, the volume and type of errors detected at various checking points, and so forth. They should also talk with operating personnel at various points in the procedure to find out what difficulties or problems they have experienced because of errors made at earlier points in the process. Where unsatisfactory conditions are found, an effort should be made to discover whether they are caused by inadequate skills or training, by weakness or complexity in the procedure, by backlog pressures, or by still other factors.

Report, form, and record checklist   Since forms, records, and reports represent the end product of most clerical effort, they provide convenient points of reference for summarizing much of the foregoing material. Accordingly, this section on details of the existing system concludes with a checklist of the information to be gathered for each of these three classes of documents.

Systems analysts should make a practice of accumulating and including among their working papers completed copies of all forms, records, and reports used in each system or procedure they study. This practice will keep the notes they must make during the survey to a minimum and ensure that all essential information is in front of them when the analysis is carried to completion.

1. Reports.
    a. Information reported. (If this is not clearly indicated by the captions on the report form, enter in the appropriate columns or sections a full description of the information required.)
    b. Period covered by the report.
    c. Frequency of preparation.
    d. Age of information reported.
    e. Source of each part of the information.
    f. Method of compiling data.
    g. Method of preparing or reproducing the report.
    h. Verification or checking procedures.
    i. Responsibility for preparation (organization unit and position title).
    j. Worker-hours required.
    k. Number of copies.
    l. Complete routing of each copy.
    m. Purpose of report.
    n. Use actually made of each copy by recipients.
    o. Effectiveness of report. Does it meet the requirements?
2. Forms. Most of the points of information appearing in the report checklist are equally applicable to business forms. In addition, however, the following supplementary facts should be gathered whenever forms are studied:

    a. Volume of use.

    b. Form cost.

    c. Information entered at time of origination.

    d. Information added subsequently.

    e. Volume and significance of errors.

    f. Signatures of approval required.

    g. Use made of each piece of information on the form.

    h. Information transcribed to other forms, records, or reports.

    i. Ultimate disposition of each copy.

3. Records.

    a. Information recorded in each column or space.

    b. Source of each entry.

    c. Volume of posting.

    d. Frequency of posting.

    e. Method of posting.

    f. Responsibility for maintenance.

    g. Worker-hours required.

    h. Method of verifying posted data.

    i. Method and frequency of summarizing posted data.

    j. Equipment in which record is filed.

    k. Arrangement or sequence of filing, type and frequency of visual indexing, etc.

    l. Purpose of record; nature and frequency of references to it or inquiries for the information it contains; types of reports, if any, prepared from it.

Since *files* of forms, correspondence, or other documents are essentially a specialized type of record, many of the points immediately above will also apply to them. In addition, when studying files, the analyst should obtain such further information as the period of time that documents remain in file before being destroyed, the legal or company requirements governing their retention, and so on.

Operating Costs

Having completed the survey of individual functions and operations, the analyst is now ready to conclude the fact-finding process with an investigation of three broad areas applicable to the system or procedure as a whole. These are: How much does it cost? How effective is it in accomplishing its basic objective? And what external conditions or practices not already studied contribute to its cost or complexity?

The facts regarding system costs should include the following elements: the salary range and average salary for each job classification, supervisory salaries, depreciation and maintenance charges on owned equipment, rental charges on leased equipment, forms and supply expense, and a pro rata share of space costs. In addition to such normal elements, the analyst should look for unusual expense items that are peculiar to the system and that may lead to the development of savings ideas. These items might include heavy communication expense, the cost of services rendered by an outside agency, and so on.

Armed with all relevant cost figures plus the worker-hour and machine-hour data previously accumulated, the analyst is in a position to determine the cost of individual operations, of methods used in performing various parts of the work, or of the system or procedure as a whole. Under some circumstances, he may also find it worth while to develop certain unit cost figures, such as the overall cost per account, per invoice or purchase order, or per direct-labor dollar. Information on unit costs is particularly useful when some meaningful standard of comparison is available—as it would be, for example, among similar operating units within the same company.

Effectiveness of the System

The effectiveness of a system or procedure in accomplishing its basic objectives is sometimes susceptible of fairly precise measurement and is equally often a matter of judgment. Occasionally analysts are able to develop factual information that, by itself, will provide the answer. In other cases they will have to rely largely on the opinion of operating personnel. Most often, however, their appraisal will be based on a mixture of fact and opinion.

The following illustrations will serve at least to indicate the possible range of variation in type of inquiry made and evidence gathered during this phase of the survey.

Example 1: Factual evidence   In evaluating the effectiveness of a shipment scheduling procedure, the analyst would presumably study an adequate sample of shipped orders to determine the relationship between shipping promise dates and actual shipping dates. One such study produced the statistics in the accompanying table.

Shipping pattern	Number of orders	Percent of total
Orders shipped as promised	186	37.2%
Orders shipped late		
1 day	71	14.2
2 days	49	9.8
3 days	35	7.0
4 days	38	7.6
5 days	28	5.6
6 days	14	2.8
7 days	13	2.6
8 days	10	2.0
9 days	8	1.6
10 days to 2 weeks	17	3.4
2–3 weeks	15	3.0
3–4 weeks	10	2.0
4–5 weeks	6	1.2
Total	500	100.0%

This simple classification of data not only served as a measurement of system effectiveness but also indicated the seriousness of the problem that had to be dealt with.

Example 2: Opinional evidence The analysts will usually find that they must depend more on opinion than on fact when they are evaluating systems and procedures designed to provide control information. Assume, for example, that a study is to be made of the clerical routines performed for or by the regional sales offices shown back in Fig. 12. Here are some questions that must be asked: How effectively does the present structure of reports and records assist regional managers in controlling the sales results and selling expense of the branches? Do any reports fail to meet requirements fully? Are some control requirements not covered by existing reports? This is the point at which analysts must obtain the views of the regional managers on the adequacy, timeliness, and utility of the existing information. The study team will, of course, weigh these opinions and supplement them with its own analysis, but in either case the evaluation is a matter of judgment, not of fact.

Example 3: Factual and opinional evidence   In the hypothetical systems project plan appearing in Fig. 11, Chap. 8, two of the survey steps listed are

    1. Analyze sample of 200 customer order status inquiries to determine time taken to reply and completeness of information given.

    2. Determine extent to which unsolicited information regarding shipment delays is supplied to customers.

Although these facts would be of real value in appraising system effectiveness, they would not by themselves be conclusive. Opinions would still have to be gathered to determine what constitutes unsatisfactory customer service in terms of the time taken to reply to inquiries, the conditions under which unsolicited information should be furnished, and the like.

Additional illustrations of the type of inquiry involved in this phase of the survey can be found in Chap. 5, in the section "Achievement of the Objective," and in Chap. 7, in the section "Spotting Problems and Opportunities."

External Factors

The final area to be investigated encompasses all the factors outside the framework of the system itself that have a bearing on system cost or complexity. Many of these factors—including organization structure, policies, and externally caused peaks and valleys—will already have been studied in earlier parts of the survey. The principal remaining consideration is the basic factor that determines the volume of work to be done. For example, the workload in maintaining a continuous inventory record depends largely on the number of withdrawals from stores. The clerical load of the purchasing department is geared to the number of purchase requisitions to be processed.

In examining this factor, analysts should try to find out whether the practices of outside departments are such that the volume of transactions is higher than necessary. Thus, in studying an inventory record procedure, the team will want to determine whether departments are requisitioning stores items so frequently and in such small quantities that an excessive number of requisitions is being processed.

HOW TO RECORD THE FACTS

To compile this vast assortment of data in the most useful form, the systems project team must "take a dose of its own medicine." That is, in recording the information gathered during the survey, analysts should adhere to the same standards of orderliness and simplicity and the same principles of systemization that they apply to the work of others. Unless they do, their subsequent analysis of the facts is likely to be haphazard, incomplete, and unnecessarily prolonged.

The optimum method of recording the facts varies widely depending on the objectives of the project as well as its magnitude and complexity. For a fairly simple study of a few weeks' duration, an orderly set of interview notes—together with filled-out copies of forms, records, and reports—is usually adequate. But for studies of longer duration and broader scope, particularly those involving several project team members, more rigorous documentation techniques are desirable if, indeed, not essential. Such techniques include various methods of charting or diagramming much of the information described in the preceding section. These forms of documentation can provide a number of advantages over narrative descriptions:

1. Charts are a simpler, more compact way of presenting a complex process. They make it possible for the analyst to visualize the whole activity more readily and to detect significant relationships among its detailed parts.

2. The use of standard charting techniques can speed up the survey by lending precision and direction to the fact-finding process, by removing uncertainty about what information should be gathered next, and by reducing the need for descriptive material and cross-reference notations.

3. Graphic presentations provide greater assurance that all operations in the existing routine have been accounted for.

4. They also facilitate analysis, since complexity, duplication, backtracking, and other weaknesses stand out more vividly in graphic form.

5. Charts or diagrams can be used to advantage in comparing present and proposed procedures and in dramatizing the effect of recommended changes.

Systems analysts have devised a number of charting techniques, many of them merely variants of each other. Five of the most common types are discussed and illustrated in this chapter. These five are sufficiently flexible to cover most requirements. Although they have met the test of long use, unquestionably many other formats can be equally satisfactory. The important consideration is *not* that a staff adopt any specific set of charts, but that those which are adopted meet the following specifications:

1. They should be simple, concise, and easy to understand.

2. They should be sufficiently adaptable to portray complex as well as simple situations with equal clarity and completeness.

3. They should be used uniformly by the entire staff, so that analysts who

have not participated in a particular survey can quickly assimilate the essential facts and assist in the analysis and development phases of the study. This means that everyone on the staff should be trained to select the same type of chart for a given situation, to use the same symbols, to furnish the same kinds of information, and to record it in the same detail.

The five kinds of documentation described below are functional organization charts, task lists, work-distribution charts, system flowcharts, and layout flow-charts.

Functional Organization Charts

The organization chart is normally thought of as a device that pictures the *structure* of organization—that is, the individual jobs or positions making up a unit and the lines of authority and responsibility between them. Systems analysts will find that this conventional type of chart also provides a convenient means for recording such related information as the names of personnel, the salary range of each job or position, and the major functions performed. Figure 13 shows one method of arranging this information.

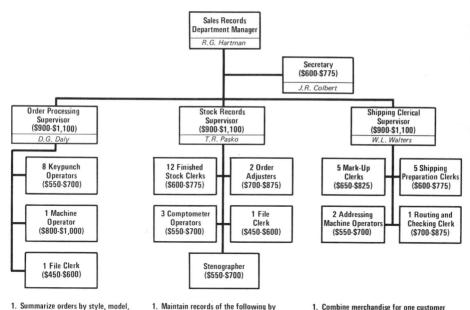

1. Summarize orders by style, model, and size as basis for
   (a) Limiting sales to piece goods purchase commitments and to factory capacity on various types of work
   (b) Supplying sales requirements information to production
   (c) Supplying to merchandising the information required for placing reorders
2. Process order changes and adjustments
3. Prepare shipping orders for stock room

1. Maintain records of the following by style, model, and size
   (a) Sales orders
   (b) Cuttings
   (c) Seconds
   (d) Finished-goods inventories
2. Apply finished stock to orders
3. Postpone or adjust quantities on orders

1. Combine merchandise for one customer
2. Determine when sufficient merchandise is available for one customer to constitute an economical shipment
3. Prepare shipping labels
4. Route shipments
5. Maintain records of shipments
6. Review orders and call past-due items to attention of stock records section

FIG. 13 *A functional organization chart gives the analyst a good picture of basic relationships and responsibilities.*

An organization chart, however, should *not* be used to record the individual tasks making up the work of the unit. The list of functions should be limited to broad areas of activity so that the analyst can quickly visualize basic responsibilities and relationships before becoming immersed in the details of a routine.

Task Lists

As Chap. 5 indicated, whenever the primary objective of a project is to reduce clerical and other administrative costs, the approach taken should be an in-depth study of the work of each position in the organizational component covered by the project. For this purpose, the form illustrated in Fig. 14 is a useful device for recording all the tasks performed by the incumbents in each position, as well as the time they spend on each task. Generally these forms should be filled out—at least as first drafts—by the workers themselves. Then the analyst should review the forms with them to clarify, amplify, and standardize the task descriptions where necessary, and to obtain samples of completed work (that is, records, forms, or reports).

Work-Distribution Charts

A companion form of documentation to the task list is the work-distribution chart illustrated in Fig. 15. It is a useful means of grouping task-list activities according to the major functions of the organizational unit being studied.

Properly prepared, and used in conjunction with task lists, the work-distribution chart can be one of the most practical and versatile tools employed by systems analysts. It assists them in determining where their effort should be applied. By focusing their attention on the more important parts of the routine, it helps them avoid the mistake of spending too much time trying to improve a minor activity. It is equally valuable in bringing to light many contributory or auxiliary activities and special or irregular jobs that workers might otherwise forget to mention when flowcharts are being drafted. In this way it serves as a guide to the analyst in preparing process charts (if they are needed) and as a further means of checking their completeness. Finally, the work-distribution chart is an important tool in the analysis phase of the improvement project. It highlights weaknesses in the division of labor and helps in the evaluation of savings resulting from systems and procedures improvements. With this record of the time spent on each task, an analyst is less likely to be drawn into controversy over the effect of work elimination on personnel requirements.

Systems Flowcharts

Among systems analysts, process charts or flowcharts are probably the most widely used type of graphic presentation. Essentially they are nothing more than a descritpion of individual operations connected by flow lines to indicate sequence and movement. Within the flowchart, symbols are commonly used to designate the nature of each operation. Such symbols not only permit the abbreviation of descriptions but also facilitate the classification and analysis of operations.

TASK LIST

Employee's Name: _____ Date: _____

Job Title: _____

Department: _____

1. Briefly describe tasks usually performed in an average week, listing them in order of amount of time spent.

2. Estimate amount of time per week each task requires.

3. For monthly or quarterly tasks, indicate by (M) or (Q) and show time on a weekly basis.

4. Total hours should equal hours worked per week and should not include lunch time.

5. Time spent on tasks taking less than one-half hour per week each may be shown as a single sum.

TASKS	ESTIMATED TIME PER WEEK (To the nearest ½-hour)
Total Time	*37 hours*

FIG. 14 *A task list describes work performed and time spent.*

WORK DISTRIBUTION – PLANT SHIPPING DEPARTMENT

FUNCTIONS	TOTAL HOURS PER WEEK	POSITIONS									
		Supervisor W.K. Forman	Hours per Week	Typists (3) J.R. Lang, D.L. Hilton, R.S. McCauley	Hours per Week	Senior Clerk G.L. Lindquist	Hours per Week	Junior Clerk J.M. Martin	Hours per Week	Clerk-Messenger D.F. Jacoby	Hours per Week
Recording and Controlling Orders	17	Checking orders against accompanying manifest	2			Checking scheduled shipping dates of orders in open file for release to factory	3	Date-stamping and registering orders / Posting shipped orders to register / Filing completed orders	5 / 3 / 4		
Routing Shipments	5	Obtaining routing instructions by telephone from Traffic Department	5								
Preparing and Dispatching Carloading Instructions	50	Proofreading loading tickets	3	Typing three-copy loading tickets (Form 1318A)	30	Proofreading loading tickets	4			Separating and distributing loading tickets to factory departments and picking up shipped tickets / Filing copy of loading tickets	10 / 3
Preparing Shipping Notices	51			Typing five-copy shipping notices (Form 863C) / Typing daily list of shipments (Form 2182)	30 / 10	Proofreading shipping notices and daily list of shipments	8			Separating copies of shipping notices and delivering them to mail room	3

164

FIG. 15   A work-distribution chart that groups tasks by functions helps analysts pinpoint their efforts.

Function	Hrs	Worker 1	Worker 2	Worker 3	Worker 4	Worker 5
Preparing Bills of Lading and Switching Orders	79	Checking computations of product and dunnage weights — 10	Typing four-copy bills of lading; Typing two-copy yard switching order (Form 1146) — 25	Proofreading bills of lading and switching orders — 6	Computing product and dunning weights — 11; Filing copies of bills of lading and switch-orders — 7	Separating copies of bills of lading and delivering to office of carrier for signature — 12
Preparing Order and Shipment Statistics	17	Reviewing summaries of open orders and shipments — 2		Preparing semiweekly summary of open orders for factory departments; Drafting weekly shipping and finished stock reports — 4	Preparing daily summary of shipments by product and size — 8	
Following Up on Shipped Orders	16	Answering Order Department inquiries on status of orders — 7		Answering Order Department inquiries; Preparing weekly list of pastdue orders — 2		Obtaining information from factory departments on status of orders released for shipments — 5
Processing Order Changes	6				Attaching order-change notices to file copy of order — 2	Distributing order-change notices to factory departments — 4
Supervision	17	Maintaining attendance and production records — 3; Spot-checking work of department — 5; Answering workers' questions — 6; Miscellaneous personnel matters — 3				
Miscellaneous	22	Conferences and contacts with other departments — 4	Typing correspondence and reports — 15			General filing — 3
	280	40	120	40	40	40

165

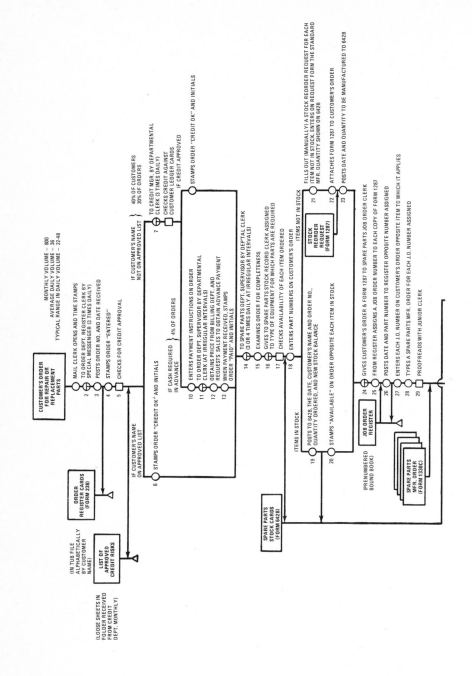

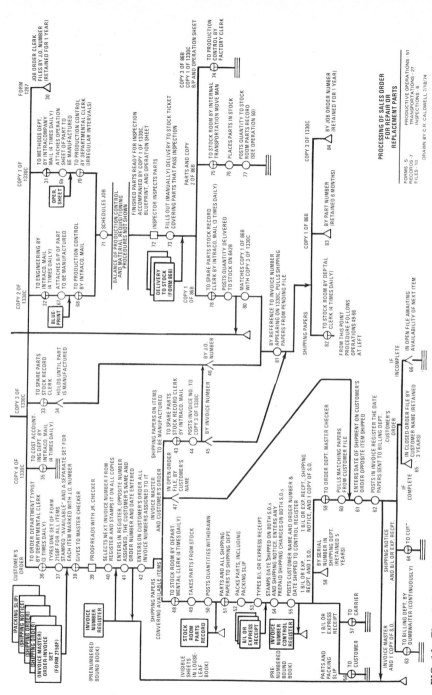

FIG. 16 *The skeleton flowchart is a useful device for recording complex operations.*

One of the most useful devices of this type is the "skeleton" or vertical flowchart illustrated in Fig. 16. Although it is equally adaptable to the recording of simple and complex routines, it is particularly recommended when the analyst needs a minutely detailed record of existing operations and when the process to be analyzed has some or all of the following features:

1. Many tied-in supplementary routines or tributaries to the main flow of work

2. Many records, each of which is used at several different points in the procedural cycle

3. Considerable backtracking

4. Alternative methods of handling work under different circumstances

5. Division of work into various classifications, each of which passes through a somewhat different routine

The structure of a skeleton flowchart is made up of the following symbols and lines.

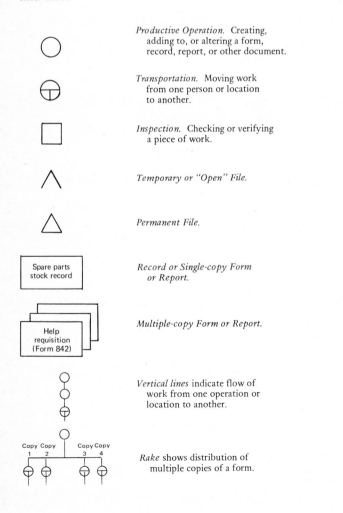

Productive Operation. Creating, adding to, or altering a form, record, report, or other document.

Transportation. Moving work from one person or location to another.

Inspection. Checking or verifying a piece of work.

Temporary or "Open" File.

Permanent File.

Record or Single-copy Form or Report.

Multiple-copy Form or Report.

Vertical lines indicate flow of work from one operation or location to another.

Rake shows distribution of multiple copies of a form.

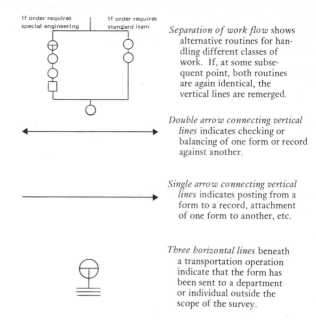

If order requires special engineering   If order requires standard item

Separation of work flow shows alternative routines for handling different classes of work. If, at some subsequent point, both routines are again identical, the vertical lines are remerged.

Double arrow connecting vertical lines indicates checking or balancing of one form or record against another.

Single arrow connecting vertical lines indicates posting from a form to a record, attachment of one form to another, etc.

Three horizontal lines beneath a transportation operation indicate that the form has been sent to a department or individual outside the scope of the survey.

In addition to showing what work is done and in what sequence, skeleton flowcharts are convenient for recording most of the other significant facts about existing systems or procedures—for example, the method by which each step is performed, the class of personnel performing it, the volume of work, and the frequency of occurrence.

The following suggestions help simplify the construction of skeleton flowcharts and increase their usefulness:

1. Describe the operations appearing opposite each symbol in brief but clear and definitive terms. Avoid the use of general terms such as "handles" or "processes." Do not simply write "checks order" if the operation consists of checking the order for completeness and for adequacy of total weight.

2. Do not use words to describe the operation implicit in a symbol. For example, do not enter "files" opposite a temporary-file symbol or "sends" opposite a transportation symbol.

3. Do not enter on the chart the specific information recorded on each form unless that information is peculiar to the operation itself. For example, if a given operator completes the heading of a form, it is sufficient to state, "Fills in heading of Customer Claim Record (Form 1318B)." The details of the operation can then be obtained, if necessary, by reference to a copy of the form, which will be among the analyst's working papers. However, if the operation is limited to the addition of one or a few specific entries, the information recorded should be described in terms of the captions appearing on the form.

4. In specifying the means by which operations are performed, include information on the types of office machines, equipment, and work aids used; the manner in which work is transported; and so on. For example:

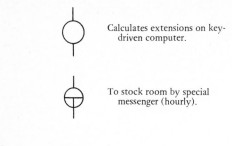

Calculates extensions on key-
driven computer.

To stock room by special
messenger (hourly).

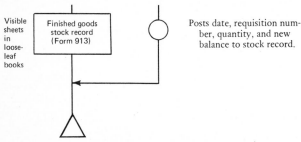

Visible
sheets
in
loose-
leaf
books

Finished goods
stock record
(Form 913)

Posts date, requisition number, quantity, and new
balance to stock record.

5. Where productive operations are not performed daily, indicate their frequency. Simplify this information by entering one of the following abbreviations to the right of the operation description:

W—Weekly
PP—Pay period
M—Monthly
Q—Quarterly
S—Semiannually
A—Annually

6. Where acceleration of the procedural cycle is among the project objectives, be sure to indicate the frequency of movement in the flow of documents. This information should appear as part of the description of each transportation operation. For example:

To stock room by pneumatic tube (continuously).

To engineering department by departmental clerk (two or
three times daily at irregular intervals).

To credit department by intracompany mail (four times daily).

7. Make certain that every vertical line on the chart ends either with a permanent-file symbol or with three horizontal lines indicating that the limits of the survey have been reached. Use of these symbols is important, since open lines immediately suggest that part of the routine has not been completely recorded.

8. To facilitate reference to the chart, number the operations in the sequence of their performance. If different parts of the routine are performed concurrently, assign consecutive numbers to the operations in a given flow line until a logical breaking point occurs. Then number those in the parallel flow line.

9. Identify the chart by:
      a. A title that indicates clearly the process covered
      b. The name of the person who prepared it
      c. The date prepared

10. At the bottom of the chart, record the totals of each symbol. For example:

Summary

Forms	7
Records	3
Files	6
Reports	2
Productive operations	59
Transportations	21
Inspections	3

A second useful technique for recording the details of a paperwork process is the columnar or horizontal flowchart illustrated in Fig. 17. The distinguishing feature of this chart is the arrangement of data in columns headed by the names of the organization units or positions participating in the process. When the chart covers an *inter*departmental procedure (such as the one illustrated), the names of the various departments appear as column heads. When it covers an *intra*departmental procedure, the names of sections, subsections, or individual jobs within the department are used as column heads.

This type of chart has two advantages over the skeleton flowchart just described: (1) It shows more clearly the total operations of a given job or organization unit as well as the relationships between jobs or units. The reason is that all the tasks of a given position, section, or department appear in a single column, with each horizontal line marking the movement of work or point of contact between positions or units. (2) Since it can be unfolded and read from left to right rather than from top to bottom, as is necessary with a vertical flowchart, the columnar format is somewhat easier to handle.

Notwithstanding these advantages, columnar charts have definite limitations. If the process involves considerable backtracking, frequent movement of work, or many multiple-copy forms, columnar charting will produce a maze of crossing flow lines that confuse the picture rather than clarify it. In addition, columnar charts do not readily lend themselves to processes involving split operations or alternative treatments. In general, therefore, the skeleton flow-

RECEIVING AND INCOMING

RECEIVING

1. Receives copy of each purchase order (Form 534) and files numerically in tub file

2. Upon receipt of incoming shipment, pulls purchase order from file

3. Counts or weighs items received and compares count with quantities shown on purchase order and supplier's packing list. (If first count disagrees with packing list, a different receiving clerk makes recount)

4. Fills out seven-copy receiving report (Form 832)

   a. If quantity received differs from packing list, enters on receiving report: amount of discrepancy, names of both checkers, and how quantity was measured

   b. If any items are damaged or broken, enters on receiving report: quantity and condition of these items and condition of package in which received

5. Records quantity received on purchase order. If this completes order, files it in closed file; if not, refiles it in tub file

6. Files copy 7 of receiving report numerically and distributes other copies as follows

   1
   2 } With undamaged items
   3
   ·4 With packing list
   5
   6

7. Holds damaged or broken items and files claim against carrier

INSPECTION

8. Inspects items and enters on copies 1, 2, and 3: quantity accepted, quantity rejected, and reasons for rejections

9. For items that fail to pass inspection, fills out rejected materials identification tag (Form 1312), attaches it to rejected items, and sends them to shipping room for return to supplier

10. Distributes copies of receiving report as follows

    1
    2
    3 With accepted material

FIG. 17 *The horizontal flowchart lends itself best to simple processes.*

INSPECTION PROCEDURE

ACCOUNTS PAYABLE

11. Files copy 4 alphabetically by supplier name until receipt of supplier's invoice

12. If necessary to protect against loss of discount, follows up Incoming Inspection Department for copy 1 containing results of inspection

16. Matches copy 1 with supplier's invoice, copy 4 of receiving report, and copy of purchase order

17. Checks (a) invoice quantity against quantity accepted on copy 1 of receiving report, (b) invoice price against price shown on purchase order, and (c) invoice computations

    a. If prices differ, refers to Purchasing Department for approval or renegotiation with supplier

    b. If quantity received differs from quantity billed, or if invoice contains arithmetic error, corrects invoice and fills out two-copy invoice correction notice (Form 913)

18. Prepares (a) voucher register and account distribution and (b) remittance check to supplier

22. Mails check to supplier together with

    a. Copy of invoice correction form, if any prepared

    b. Copy of debit notice covering any rejected items

23. Attaches copies of all supporting documents to copy of remittance check and files by vendor name

PURCHASING

13. Posts quantity received to follow up copy of purchase order and destroys copy 5 of receiving report

14. If this receipt completes the order, files it in closed file; otherwise refiles it in supplier follow-up file

21. Fills out three-copy debit notice (Form 771) for any rejected items. Attaches one copy to department's file copy of purchase order and sends first two copies to

MISCELLANEOUS

Issuer of Purchase Requisition

15. Copy 6 serves as advance notice of receipt of items requisitioned

Store room (or other material delivery point)

19. Posts receipt to inventory record

20. Files copy 3 of receiving report numerically

Description of Operations
For Figure 18

1. Customer's letter of complaint delivered to distribution clerk, who attaches tracing slip (Form 240A) and assigns complaint to an adjuster.
2. Passes complaint to correspondence file clerk.
3. Correspondence file clerk obtains any previous correspondence from or to the customer and attaches it to the complaint.
4. Passes the complaint to the adjuster designated on tracing slip.
5. If complaint covers a charge purchase, adjuster checks customer ledger card in accounts department for record of the purchase.
6. If ledger card contains a record of the purchase or if complaint covers a cash-payment purchase, adjuster gives documents to delivery slip clerk for further checking.
7. Delivery slip clerk checks files for evidence of delivery of the merchandise and enters findings on tracing slip.
8. If no evidence of delivery is found, complaint is handed to duplicate sales check clerk.
9. Duplicate sales check clerk searches files for evidence of purchase and enters finding on tracing slip.
10. If a cash-payment purchase, passes complaint to refund clerk.
11. Refund clerk searches files to determine whether an adjustment check has already been drawn and sent to customer. Enters findings on tracing slip and returns complaint to adjuster.
12. Adjuster examines findings, writes adjustment instructions on tracing slip, and gives complaint to typist.
13. Typist prepares letter to customer, attaches to complaint, and gives to supervisor.
14. Supervisor reviews all documents, signs original of letter, and gives to distribution clerk.
15. Distribution clerk mails letter.
16. Supervisor gives carbon copy of letter plus all other complaint documents to correspondence file clerk.
17. Correspondence file clerk files documents.

chart should be used when any of these complexities exist, and the columnar chart should be reserved for simpler processes.

Layout Flowcharts

Analysts will find that they can more readily determine the effect of office layout on the flow of work if they prepare a conventional layout diagram with flow lines to indicate the movement of work from desk to desk, as in Fig. 18. The area and equipment should be drawn to scale so that actual distances of movement can be determined and compared with distances on alternative layouts. The manner in which backtracking and excessive movement can be eliminated by changing the layout is illustrated in the "after" chart of this same area shown in Fig. 25, page 240.

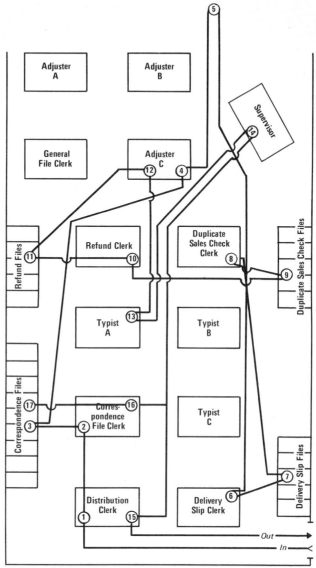

FIG. 18 *A layout flowchart shows movement of work from desk to desk.*

Other Working Papers

Not all the information developed during the survey will be recorded in chart form. The project working papers will also include statistical data, narrative information, results of tests or experiments, samples of forms and reports, and still other types of material. Analysts should train themselves to prepare these papers in an orderly way at the first writing, to identify them clearly, and to index or organize them so that subsequent reference will be facilitated. Untidy or badly organized notes not only encourage haphazard thinking but also cause fumbling and waste of time.

HOW TO ENSURE GOOD SURVEY RESULTS

The preceding sections of this chapter have been concerned with the *end product* of a systems survey. The first covers its substance, or the kinds of information to be gathered; the second, its form, or the manner of recording the facts about the existing system. This final section contains additional pointers that will help the analyst get better survey results as well as create the basis for successful performance of the remaining project steps.

Tapping Sources of Information

There are four principal means of digging out the information previously discussed: (1) by interviewing operating personnel, (2) by observing actual performance of the work and studying the product created, (3) by reviewing written material on structure and process (for example, formalized organization charts, definitions of responsibility, job descriptions, and written operating instructions); and (4) by analyzing existing operating records or reports (say, on inventory levels, machine downtime, customer complaints), or by making special arrangements to gather such statistics for purposes of the study. No one of these means should be relied on exclusively. Each has its place in the survey process, but each also has its limitations.

Assuming that the way has been properly cleared with line executives, the initial point of contact with each organization unit should, of course, be with the manager or supervisor in charge. The specific information to be gathered at this point will vary widely depending on the nature of the study and the amount of background information needed by the study team. Unless its members are already thoroughly familiar with all the conditions surrounding the system to be studied, they should discuss with the supervisor the nature of the work performed and its relation to other departments or functions. If the system closely follows or controls some manufacturing, materials movement, or storekeeping process, they should also observe the physical operations sufficiently to visualize the tie-in of those activities with the related paperwork.

Having thus oriented themselves, the analysts should obtain from the supervisor most of the other background information necessary to understand the actual performance of the procedure. This information includes statements of policy, organization structure and personnel, major functions performed, and the external conditions or requirements affecting the unit's operations. Finally, the analysts should find out enough about the procedure itself to guide them in making the detailed survey. They should ask the supervisor to outline the major steps in the process chronologically and to suggest the names of key employees who should be interviewed because of their job knowledge or importance to the routine.

But the study team should not as a rule rely on the work supervisor to supply the full details of the procedure, for too often there is a vast difference between what is actually being done and what the supervisor thinks is being done. The team, therefore, must *see for itself*. It should follow the routine step by step, sitting down with individual workers to find out exactly what they do and how they do it. And because even the workers may neglect to mention some significant points, team members should watch the actual operations, examine the finished product, and look into all relevant statistics on operating performance.

Interviewing Operating Personnel

Operators' attitudes toward any systems study and the recommendations it produces are largely conditioned by the analyst's fact-finding interviews. Chapter 6 outlined the principles of conduct that analysts should observe in all their contacts with the line organization. Following are some additional suggestions applying specifically to the conduct of systems survey interviews:

1. Before each meeting, find out something about the interests, background, and relevant experience of the person to be interviewed. Then carefully think through the information that is needed. In most instances, these information requirements should be reduced to writing in the form of a list of subjects to be covered or specific questions to be asked.

2. Be considerate of the interviewee's time. Remember that he or she has a regular job to do. Make a definite appointment that fits conveniently into that person's working schedule.

3. Try at the outset to overcome suspicion by taking the mystery out of the project. Build a background and receptivity that will make the person want to throw light on the problem under study. Even though some general announcement about the project and its objectives may already have been made, take time to explain its purpose simply but thoroughly. Raise the employee's sights by describing how the study is being conducted and how his or her job fits into the overall system or procedure.

4. Conduct the interview on a friendly, informal basis. Guard against being intense or overly absorbed in the subject. A light touch, skillfully employed, can often help to relieve tension at the beginning. A casual manner plus a genuine interest in the operator's work and problems will help greatly to maintain a comfortable atmosphere throughout the interview.

5. Keep from dominating the conversation. Be a good listener. Let the operator do most of the talking. In this way, even though the conversation gets off the track occasionally, the analyst is likely to find more clues to operating problems than by trying to channel the interview too tightly. But also remember that being a good listener is an *active*, not a passive, role. It requires that the analyst think searchingly about what is being said and, in doing so, ask the sorts

of questions that will encourage the operator to make the most revealing responses.

6. Use care in phrasing questions. In trying to uncover the purpose of an operation, be sure that the questions asked do not contain an implication of criticism, When seeking opinions, keep from asking leading questions.

7. Encourage operating personnel to talk freely about the difficulties they have experienced or problems they perceive. Ask for their ideas on possible improvements. Make clear to them that the project is a joint undertaking and that everyone who participates in it will share in the credit.

8. Explain to them why the information requested is needed and how it will be used.

Verifying the Facts

As information is gathered—whether it is factual data or opinions—the analyst should seek to validate it before accepting and incorporating it into the findings on which conclusions are based. For example, where opinion evidence is relevant, the analyst will want to obtain the judgments of several people. Where quantitative data are gathered and given to the project team by others, team members should always test the figures for reasonableness—a process that might range from "order of magnitude" judgment checks to sample or spot reviews.

At the conclusion of the survey, the project team should ask the people with whom it has worked to help it check the completeness and accuracy of the project task lists, work-distribution charts, flowcharts, and other working papers. This process of verification has several important values: (1) It assures the team that the data on which it will base its conclusions are trustworthy. It enables team members to fill in the gaps and clear up the misinterpretations that are bound to develop during the initial survey. (2) It gives the team another opportunity to obtain suggestions on improvements from the operating personnel. Some of the best ideas are usually contributed at this point, because the work supervisor gets a better perspective of his or her own operations in relation to the whole. (3) In anticipating this check on the thoroughness and accuracy of its fact-finding, the team is likely to dig more persistently, to organize material more logically, and to depict facts more clearly. (4) By being given a chance to examine the survey results, the operators again feel that they have been taken into the team's confidence. By seeing what has been recorded, they can satisfy themselves that their work has been properly represented and that none of the facts on which conclusions will be based have been withheld from them.

Developing Improvements During the Survey

As discussed in Chap. 6, skillful systems analysts begin to think fundamentally and imaginatively about *the problem* from the very beginning of the study. As they gather the facts, they are continuously seeking to identify and test ideas for improvements and to think about alternative solutions. As operating personnel explain each function or routine, possible shortcuts or improved methods will occur to the analysts. They should not simply jot down these ideas for future use but should try, as the alternatives arise, to think of the evidence that will ultimately be needed to evaluate them. In this way their fact-finding will be more purposeful, and they will save a great deal of callback time during later stages of the project.

Moreover, by thinking creatively as they gather the facts, analysts are often able to suggest improvements that can be put into effect immediately. However minor these changes may be, they create a predisposition early in the study to act on the recommendations that the project team develops.

Modifying the Project

As Chap. 8 pointed out, planning a systems improvement project should be a continuing process, not a one-time event. Throughout the survey, project analysts should be alert to the need for revising the scope or objectives of the study or the approach taken to it. Meeting this requirement does not mean that they should chase blindly down every avenue that the survey opens up, but rather that, when new evidence alters their conception of underlying conditions, problems, or opportunities, they should be quick to adjust the project plan to the revised needs or to abandon nonproductive parts of the work.

TEN

Improving the Performance of Line Operating Functions

The preceding chapters have emphasized that the basic objectives of systems design or improvement projects vary widely from one to the other. The full extent of this variation is reflected in the next five chapters which, taken together, cover most if not all of the different kinds of benefits that the systems analyst might pursue. In their broadest sense, these potential benefits—as we have already seen—range from increasing operational and general-management effectiveness, to cutting clerical and other administrative costs, to overhauling the structure of management information, to optimizing the use of the computer.

Although each of the next five chapters focuses on a different one of these broad classes of benefits, we do not mean to suggest that studies aimed at capturing such benefits are mutually exclusive. To be sure, somewhat different analytical approaches need to be followed in the pursuit of different objectives. For example, the information gathered and analyses made in the design of a management system (say, strategic planning) naturally differ from those in a clerical cost reduction study. But the point to be emphasized is that individual systems projects are often aimed at achieving more than one objective—for example, speeding up service to customers as well as cutting paperwork costs, or improving information on account profitability as well as optimizing the use of field sales representatives' time.

Given this overview, the remainder of this chapter is concerned with those systems that directly support the day-to-day execution, coordination, and control of the major operating functions of an enterprise. In a make-and-sell business, for example, these are the sales, manufacturing, engineering, purchasing, and distribution functions. Within these operating areas, the specific benefits that the systems analyst may work at achieving are as varied as the types of enterprises (public and private) and the multiplicity of functions with which this book is concerned. The following examples, drawn from the writer's experience, illustrate the great diversity of operational improvements that can result from skillful systems design and application:

1. Allocation of manufacturing orders for primary metal products among the company's several mills in a way that minimizes the sum of variable manufacturing and distribution costs for each major product group

2. Significant improvement in the accuracy and speed of an automobile company's defect-recall system

3. Improvement in the selection and "rating" (pricing) of property and casualty insurance risks through development of a more effective system for analyzing previous loss experience over a wide range of risk *subclasses* within existing risk categories

4. Reduced shutdown or turnaround time of major production facilities (for example, a catalytic cracker) because of better maintenance work scheduling

5. Sharp reduction in the loss of investment-services clients through the virtual elimination of clerical errors in customers' accounts

6. For a major railroad, a 35 percent increase in freight car utilization (in terms of loaded car-miles per day), a reduction in the number of empty cars on hand per customer from 7 to 4, and an improvement in service performance (the percentage of time that shippers receive empty cars on the date requested) from 85 to 95 percent

7. For a commercial bank, a significant increase in profits through an improved system for making decisions on major controllable sources and uses of funds—for example, money market supplies and short-term investments

8. Improved techniques for measuring and forecasting workload input, processing, and output of a hospital's professional services departments as the basis for more effective personnel planning, staffing, and utilization

Within this wide range of operational activities and the systems through which they are performed, the following discussion will focus on four areas to illustrate various methods of systems analysis and design that should yield optimum benefits. These areas are (1) service to customers, (2) inventory management, (3) cash management, and (4) operations of a public sector organization.

AREA 1—CUSTOMER SERVICE

All businesses exist to ship their products or render services to customers. Their speed and accuracy in doing so are important counters in the play-for-keeps game of competition. The more standardized the industry's product line and price structure, the larger these counters become.

An organization's effectiveness in serving its customers depends to a great extent on the speed and quality of its paperwork. Streamlined systems and procedures, efficient methods, and tighter controls accelerate both the clerical and the manufacturing cycles. They cut the time required to edit sales orders, to check credit, to transmit order filling instructions to all points of action. They speed up the material procurement process. They protect against lost time in factory operations. They facilitate traffic clearance of shipments.

And the contribution of improved systems to better customer service is not limited to speeding up shipments. Better records make possible faster, more realistic replies to customers' inquiries on the status of their orders. Simpler procedures and improved clerical methods reduce packaging and billing errors and ensure accurate identification of items shipped. If a company sells a number of different items to each customer, tighter material and production controls keep output geared to composite customer demand. Thus, sales orders can be filled in a few large shipments instead of many small ones. From the customer's point of view, this means that each shipment contains a better balance of merchandise.

Against this background, here is a series of guidelines for systems analysts to follow in evaluating and improving the quality of service that their organization provides for its customers.

Information To Be Gathered[1]

Although the details of the factual and opinional evidence to be gathered will vary somewhat between service-rendering organizations (public and private) and manufacturing organizations, the following is a representative list of information needed in most studies undertaken to improve customer service.

1. The organization's present policies and standards governing the quality of service to its customers.

2. The significance of service.

   a. What are customers' minimum service requirements and expectations? The answer might be found by analyzing customer complaints and order inquiries, or by talking with a sample of customers.

   b. What quality of service are the organization's principal competitors providing?

3. Present speed of service. To develop this information the analyst should prepare a flowchart of all steps in the order filling cycle and record such other quantitative data as

   a. The time that any given order spends at each work station in the system, broken down between processing time and waiting time.

   b. A frequency distribution of transactions by number of days to complete them.

---

[1]The preceding chapter set forth a generalized list of the kinds of facts to be gathered in a typical systems study. This chapter supplements that list by outlining the special kinds of information needed in the four areas of systems analysis chosen to illustrate the approach to improving operating performance.

    c. Trend in the average time and range of times required to process an order; trend in back-order volume.

    d. Comparison of the speed of service rendered by similar units (say, regional offices and plants or warehouses shipping the same products).

4. Present quality of service, expressed in terms of errors made either in paperwork or in the physical aspects of order filling (for example, wrong item shipped).

    a. Volume of errors by type.

    b. Point at which errors are caught (by the company or by the customer).

    c. Significance of errors. For example, clerical errors in a commercial bank's check-processing operation can cause not only confusion and annoyance but also personal embarrassment to a customer.

5. Handling of customer order status inquiries.

    a. Steps involved in developing delivery promises to customers.

    b. Trend in the number of such promises that are broken.

    c. Number and types of customer order status inquiries and delivery complaints; time taken to reply.

    d. Volume of unsolicited information given to customers (or field sales representatives) regarding order status and delivery expectations.

Skillful collection and portrayal of these facts will, almost by themselves, point to the sorts of improvements needed and even suggest some of the principal ways of achieving them. As an example, in seeking to improve the speed and quality of the service it renders to its independent agents, the property and casualty insurance company referred to in Chap. 1 conducted a systems study that produced the following facts:

1. For the company as a whole, a significant volume of transactions, both for commercial customers and for personal customers, is issued several days later than the elapsed time that represents agents' minimum service requirements. For example, 31 percent of commercial new-business policies are issued after 8 days and 15 percent after 12 to 24 days (Fig. 19).

2. Transaction processing times vary widely among the company's branch offices, with some offices performing satisfactorily. For example, the five best branch offices issue 93 percent of their commercial endorsements (changes in existing policies) within 14 days (the agents' minimum speed-of-service requirement on endorsements); the five poorest performers issue only 36 percent on time (Fig. 20).

3. Actual *processing* time (the time that all persons spend in the aggregate working on a specific case) is an extremely small part of the total time that elapses before the transaction is completed and sent to the agent (Fig. 21). Hence, waiting time consumes most of the total issue time for every type of transaction.

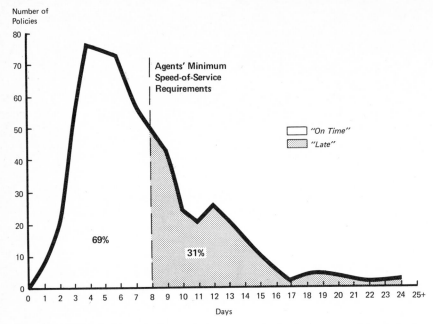

FIG. 19 *This systems study showed that almost one-third of commercial new-business policies are issued late.*

4. Waiting time can be broken down into three components (Fig. 22):

   a. Time that a transaction spends at each work station while waiting for all other transactions in the batch to be processed.

   b. Time that is spent waiting for the backlog of other batches to be processed.

   c. Time that is spent delivering batches of transactions between work stations and departments.

5. All three kinds of waiting time are incurred at each of the four principal work stations (Fig. 23).

6. A minor reduction in total waiting time will speed service more than a major reduction in processing time. For example:

   a. Reducing the average *waiting* time for commercial new-business transactions by 20 percent will speed issuance by 1½ days.

   b. But reducing the average *processing* time for the same transactions by 50 percent will speed issuance by only 30 minutes.

7. The major cause of the long delays is clerical error. When errors are detected, the incorrect transaction is moved back to an earlier point in the processing cycle, where it goes to the bottom of the pile at that work station.

If the analysts were gathering information on customer service within a make-and-sell business rather than within a service organization, as in the case above, they would of course have to look beyond the paperwork processing activities to the relevant physical operations. They would also consider, for example, inventory policies, factory scheduling and order assembly practices,

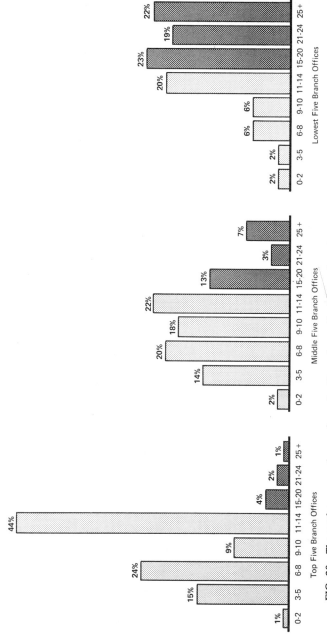

FIG. 20 *The performance of some branch offices shows that agents' minimum speed-of-service requirements can be met.*

185

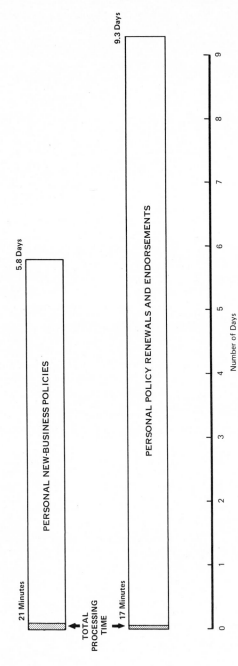

FIG. 21 *Average processing time is only a minute part of average issue time.*

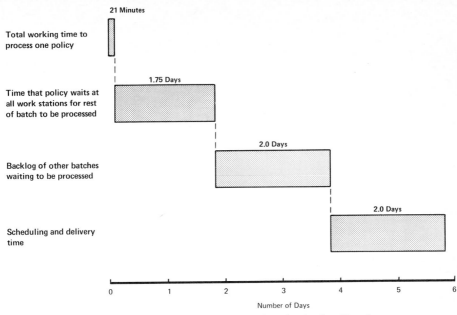

FIG. 22 *A personal new-business policy encounters three kinds of waiting time.*

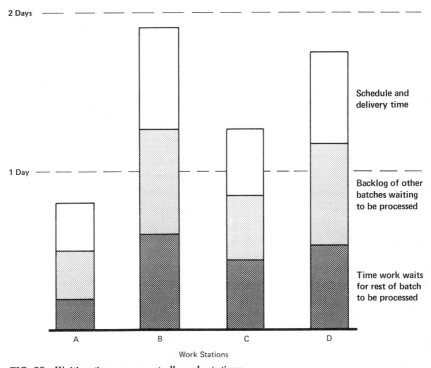

FIG. 23 *Waiting time occurs at all work stations.*

alternative routes and methods of shipment, and the like. These and related issues are covered in the following section on inventory management.

Typical Improvement Opportunities

In searching for ways to improve the systems that control the quality and speed of service to customers, analysts will invariably find dozens of opportunities flowing from the facts they have gathered. The following illustrations are typical:[2]

1. Batch units of work (that is, incoming transactions) by the date they are due out, rather than on a first-in, first-out basis.

2. Mount a one-time effort to significantly reduce the size of the work backlog and maintain it thereafter at the reduced level.

3. As a further means of reducing waiting time, cut the size of work batches so that they move more often from station to station. Where processing speed is of unusual significance, consider carrying this step to its ultimate form by setting up a continuous "assembly line" type of operation.

4. To overcome the unfavorable effects of offsetting peaks and valleys in the workload of various stations:

   a. Cross-train some workers so that they can move from one work station to another with the shifts in work peaks.

   b. Backlog some of the less urgent orders or delay some of the postponable operations such as filing, issuing credit memorandums, replenishing tub·files, etc.

   c. Build up and maintain a pool of part-time workers (often former full-time employees) who will work either regularly scheduled part-time hours per week or month, or irregularly scheduled hours whenever they are called in to meet unexpected peaks.

   d. Use workers from one of the agencies that specialize in providing temporary help.

   e. Stagger work hours during the day—for example, from 7:30 A.M. to 6 P.M.

5. Arrange to have some operations performed concurrently instead of consecutively. For instance, if the system under study produces a multiple-copy form, two or more copies might be worked on simultaneously. As an example, shipments might be speeded up by sending a copy of each order to the shipping department before the merchandise arrives from the stock room. If the shipping crew knows in advance what to expect, it can usually schedule its work better and prepare for any special packing or crating requirements.

6. Allow some operations to be done after the fact—that is, after the critical customer service deadline has been met. As an example, the property and casualty insurance company referred to earlier in this section adopted the

---

[2]These improvements are also applicable, in most instances, to any other system or procedure in which speed of processing is paramount—for example, in preparing payrolls, in getting month-end reports into the hands of management, or in developing cost estimates for a bid or price quotation.

practice of underwriting after the fact on certain prescribed cases; that is, it made the risk evaluation and selection decision after having issued a new policy to the agent. In the rare instance where this after-the-fact review resulted in rejection of the risk, the company was protected by its prerogative to cancel the policy.

    7. Reduce the volume and severity of errors through

        a. Better worker training.

        b. Improved work aids.

        c. Simplification of the work steps in which most of the errors are occurring.

        d. More effective checking routines.

        e. Reassignment of the more complex transactions to specially trained workers.

    8. Streamline the transaction processing operations through work elimination and simplification. A wide range of suggestions for achieving this kind of improvement is given in the next chapter.

    9. Establish customer service standards, and regularly measure and report actual performance against them.

To keep the matter of improving customer service in perspective, remember that the critical test for all proposed improvements is *whether they will be cost-effective*—that is, whether they represent the best balance between the measurable benefits to be gained and the cost of achieving them. This consideration can be expecially important in a customer service study for two reasons:

First, changes in customer service standards can sometimes be very expensive because production, inventory, and distribution costs rise almost geometrically as high levels of service are approached. But second, improvements in service do not always automatically produce this cost increment, and the analyst must guard against any such mistaken notion. Particularly in service organizations, where transaction processing is largely a paperwork activity (as it is in the property and casualty insurance company previously cited), improvement in service often results from the sort of process streamling that, at the same time, actually cuts operating costs. Even in make-and-sell businesses, systems analysts will find that, by being imaginative and resourceful, they can help their organization achieve competitively superior customer service with the kinds of improvements that also result in optimum manufacturing costs, better balanced finished-goods inventories, and efficient order assembly.

AREA 2—INVENTORY MANAGEMENT[3]

Perhaps better than any other business function, the field of inventory management illustrates the diverse ways in which improved systems design can

---

[3]This section draws extensively from Richard F. Neuschel and Alan C. Fuller, "Inventory Control," in H. B. Maynard (ed.). *Industrial Engineering Handbook,* McGraw-Hill, New York, 1963, sec. 7, chap. 4. Copyright © 1956, 1963, 1971 by McGraw-Hill, Inc. Used with permission of McGraw-Hill Book Company.

strengthen an organization's operating effectiveness. This is true because the control of inventories is one of the most complex and far-reaching of all business activities. It is the focal point of many seemingly conflicting interests and considerations, both short- and long-range. Its planning and execution require participation by most of the functional segments of a business: sales, production, purchasing, finance, and accounting. And since the end result directly affects the quality of service to customers, production costs, and soundness of working-capital position, it has a major bearing on the company's financial strength and competitive position.

Two facts reinforce this estimate of the importance of inventory planning and control in modern business operation.

1. In most manufacturing companies, the management of finished-goods inventories is the heart of the day-to-day problem of coordinating sales and production.

2. In most businesses, inventories represent a substantial investment. A random sample of twenty-five industrial companies reveals that inventories range from 17 to 41 percent of total assets. In this light, it is not surprising that inventory losses have been a primary contributor to business failures, and have been widespread in most cyclical business declines and recessions.

Benefits To Be Gained

Although a well-planned, competently administered system of inventory management benefits a company in many ways, the following advantages stand out.

1. *Improved customer relations.* This gain is achieved through faster, more reliable delivery service, which in turn results from
   a. Maintenance of a better balance among the quantities of finished items on hand
   b. Better geographic deployment of inventories among field warehouses

2. *Improved labor and community relations.* This benefit results from a greater leveling of production peaks and valleys and the consequent increase in stability of employment.

3. *Reduced manufacturing costs.* A well-balanced system of inventory management can reduce manufacturing costs in the following ways.
   a. By increasing the utilization of labor, supervision, and facilities through elimination of idle time caused by shortages of raw materials and component parts
   b. By making possible more economic manufacturing runs in place of the small lots, constant rescheduling, and expensive setup changes that are needed to compensate for unbalanced or hand-to-mouth inventories
   c. By minimizing machine downtime that is caused by the unavailability of critical spare parts

4. *Reduced purchased-material costs.* This reduction can be achieved because the need for emergency purchasing seldom arises. Thus, the company is spared

the payment of overtime or special setup charges and similar premiums to suppliers.

5. *Reduced inventory costs and inventory losses.* Good inventory management makes these savings possible by

  a. Relating inventory level and mix more responsively to current and projected demand, thereby reducing the total quantity of finished goods needed to provide competitive service to customers
  b. Maintaining the most economical balance between inventory-carrying costs and inventory-acquisition costs on purchased items
  c. Preventing the buildup of work-in-process inventories which comes from unanticipated shortages of raw materials and purchased parts
  d. Simplifying and standardizing the company's lines of component parts, raw materials, and maintenance and supply items
  e. Minimizing inventory losses that result from
    (1) Declines in the market value of raw materials or finished goods
    (2) Spoilage or deterioration
    (3) Product obsolescence caused, for example, by shifts in market demand, competitive pressures, raw material shortages, or failure of the company to take existing inventory levels into consideration in planning the timing of design, style, or package changes
    (4) Failure to physically verify the quantity or condition of inventories, or to provide adequate safeguards against pilferage or excess waste

6. *Strengthened financial position.* Since investment in inventories is held to the lowest point consistent with the production and sales requirements of the enterprise, the company has more liquid working capital at its disposal. Its capital position is further strengthened by the reduction of capital requirements for field warehouses and other storage space, and for manufacturing facilities to meet peak production demands.

7. *Reduced clerical and other office costs.* This gain can be achieved through a reduction in

  a. The costs of purchasing, follow-up, and expediting activities
  b. The costs of physical inventory verification
  c. The time spent by office personnel in administering complex back-order routines and in order tracing, rescheduling, and similar activities essential to answering order status inquiries and to expediting emergency orders of key customers

Symptoms of Poor Inventory Management

Weaknesses in the planning and control of inventories are usually indicated by, or expressed in the form of, complaints about specific symptoms rather than about the inventory system as a whole. In launching a study of their organization's inventory management performance, systems analysts should ordinarily begin by looking for and, wherever possible, quantifying the following kinds of symptomatic evidence.

1. Wide fluctuations in the ratio of inventory investment to sales volume or production volume

2. Continuously growing inventory quantities in the face of a fairly constant or increasing order backlog

3. Widely varying rates of inventory loss or turnover among the company's plants or branch warehouses; or widely varying rates of turnover among major inventory items

4. Large inventories of slow-moving items and consistently large inventory write-downs because of price declines, distress sales, or the disposal of obsolete or slow-moving stocks

5. Consistently large write-downs at the time of physical inventory taking

6. Frequent or severe back orders—that is, a large and growing volume of partial shipments to customers

7. Lengthening delivery times or delivery times that appear excessive in relation to those of competitors

8. For make-to-order businesses, failure to meet delivery promise dates

9. High rate of customer delivery complaints or order cancellations

10. Excessive effort by sales department personnel to expedite shipments to customers

11. Excessive machine downtime and sizable manufacturing cost variances because of tool, raw material, or parts shortages

12. Sharp peaks and valleys in production volume, with resulting variations in overtime, layoffs, and hirings

13. Frequent need for uneconomical production runs or split lots to meet sales requirements

14. Cannibalizing of parts from in-process assemblies that are in an early stage to complete assemblies that are in a later stage

15. Heavy expenditure of time for parts expediting and order chasing

Additional Information To Be Gathered

Beyond these symptoms of poor inventory management, the analyst will ordinarily want to obtain the following information in the fact-finding phase of an inventory management study.

1. What is the 5-year trend in the size of the investment in·each class of inventory and the turnover rate of each class, including raw materials, component parts and assemblies, work in process, finished goods (at plants and field warehouses), packing and packaging materials, and maintenance and operating supplies? Also, what percentage of the total investment in each inventory class is accounted for by the 10 percent of items having the highest value? By the 20 percent having the highest value?

2. At how many different locations (plants and field warehouses) are inventories carried? What are the comparative inventory turnover rates among those locations (by inventory classes) and among representative major items?

Does every field warehouse contain the company's entire product line, or only part of it?

3. What policies has the company established affecting the overall size or composition of inventories? For example:

   a. What is the company's goal on deliveries to customers? Specifically, what is the maximum time that should elapse between receipt and shipment of an order?

   b. To what extent will inventories be fluctuated to level out the peaks and valleys in sales demand and make possible greater stability of production?

   c. To what extent, if any, is forward buying beyond normal requirements permitted in anticipation of price rises or material shortages?

4. What organizational unit is responsible for

   a. Determining what items are to be carried in inventory?

   b. Determining when and by what quantities to replenish inventories?

   c. Verifying inventory quantities and condition by physical count and inspection?

   d. Identifying and disposing of surplus, slow-moving, obsolete, or damaged inventories?

   e. Furnishing summary information on inventory position for control purposes?

What procedures does each unit follow in carrying out the foregoing responsibilities?

5. What is the company's present approach to planning and controlling the overall investment in inventories? For example:

   a. Does the organization regularly prepare a balance sheet budget that—among other things—specifies the amount of funds available month by month.for investment in inventories? By what method is this budgeted investment developed? Is it supported by a breakdown among major inventory classes? Are actual results monitored against the budgeted amount? How is that control exercised?

   b. Have standard turnover rates been established for each major inventory class? How were they developed? How is control exercised?

Typical Improvement Opportunities

Once again, the very process of digging out and portraying the facts about inventory management will bring to the surface the specific improvement opportunities that the project team should pursue. Among the broad categories of benefits that the team should look for in the management of inventories are those reflected in the actions discussed below.

Eliminate items from inventory Every inventory category—from raw materials through finished goods—should be analyzed to determine what items or classes of items can be eliminated either because they perform no cost-effective function or because the function can be performed better elsewhere.

1. Initiate a program to eliminate unprofitable or marginally profitable items from the company's product line. Typically these will be items for which the demand is low and erratic.

2. If a slow-moving item cannot be weeded out of the product line, evaluate the economics of lengthening the delivery time on that item and making it to order, rather than shipping it from stock.

3. On maintenance, repair, and operating supplies, conduct the sort of item elimination analysis carried out by a large chemical company and described on page 19. This program led to the outright elimination of a large number of inventory items either because they were slow-moving or because they represented unimportant or unnecessary variations in type, form, grade, size, shape, or material composition.

4. Maintain inventories of component parts, assemblies, and finished items only at stages that represent the best balance between customer delivery time requirements and overall investment in product inventory. For example, if a company uses a given assembly in several different products, it usually can achieve substantial inventory savings by stocking that assembly rather than each of the finished products.

5. Stock only high-turnover items at each location, and consolidate inventories of special and slow-moving items at a few central locations.

6. Persuade vendors to stock larger quantities of major items supplied to the company and to shorten their delivery time for those items.

7. Free warehouse space by shipping extra stocks to customers on consignment, or by arranging for wholesalers or distributors to carry larger or more diverse stocks.

8. Inaugurate an aggressive program to identify and sell obsolete materials and parts. Also, to minimize such obsolescence in the future, develop procedures for using up existing stocks before putting planned engineering changes into effect. (The potential gain from this step can be approximated by estimating the value of material that annually becomes obsolete.)

Once analysts have eliminated all the items they can, they should apply the two following steps in the improvement process on a *selective* basis. That is, they should focus attention and control effort on the relatively small percentage (say, 10 to 20 percent) of items in each inventory category which account for the greatest part of the inventory investment.

Shorten inventory replenishment time   The investment in most if not all classes of inventory is directly affected by the time required to replenish stocks, whether through purchase or manufacture. This is true, of course, because the shorter the replenishment time on a given item, the lower the inventory on hand can be permitted to fall before replenishment ordering, and the lower will be the average inventory carried on that item.

To identify opportunities for reducing inventory replenishment time, analysts should study each step in the replenishment process in much the same way that they analyzed the factors affecting speed of service to customers (pages 181 to 189). Here are some of the opportunities that appear most often:

1. Reduce production time by improving scheduling, dispatching, and follow-up procedures.

2. Reduce replenishment time for purchased items by working with major vendors to accelerate their handling of the company's orders.

3. Further reduce replenishment time for both manufactured and purchased items by cutting the time for order initiation and processing.

4. Reduce average processing time by improving control over the physical handling of material in receiving or shipping departments, in stock rooms, and on the factory floor.

5. Reduce transit time to company warehouses by using alternative shipping methods (if economically feasible), by coordinating carriers' schedules more closely, and by scheduling shipments to take advantage of weekend travel.

Establish economic order quantities (EOQs) The third step in seeking to improve inventory management is to determine the optimum quantity of each item to be ordered when replenishing inventory. Setting order sizes requires achievement of the most economical balance between a number of conflicting factors:

1. The costs of carrying inventory

2. The clerical and other overhead costs of ordering—that is, costs of requisitioning, purchasing, inspecting, receiving, putting the item into stock, and paying for it

3. The cost penalties of ordering in small quantities (higher makeready or setup costs for manufactured items, higher unit costs for purchased items that are subject to quantity discounts)

4. Risk of loss through obsolescence, deterioration, or price decline

5. Costs of additional storage space if present warehouse capacity is exceeded[4]

A final and critical point is that EOQs, once calculated, do not remain fixed, simply because the basic elements used in calculating them are continuously changing. For that reason, EOQs need to be recalculated periodically—at least on the large-volume, high-cost inventory items—as changes take place in

1. The forecast rate of demand

2. Replenishment cycle times (on purchased or manufactured items)

3. Utilization of production capacity within the company

4. The cost of money (invested in inventories)

5. Setup or other makeready costs

6. Customer delivery service requirements

Cut inventory clerical costs The most common opportunities to cut the clerical costs associated with inventory management are described in detail in the next chapter. In summary, they consist of these shortcuts:

1. Eliminate duplication of inventory records.

2. Reduce the number of items for which records are maintained.

---

[4]Many formulas are in use for calculating economic order quantities. For a detailed treatment, see "How Much To Order," *Industrial Engineering Handbook,* pp. 7–58 to 7–59.

3. Charge off some items at time of acquisition—charge off small-value materials, parts, and supplies when received rather than when withdrawn from stores.

4. Simplify inventory record posting.

5. Simplify replenishment requisitioning.

AREA 3—CASH MANAGEMENT

In illustrating how systems can strengthen an organization's operating effectiveness, this chapter has looked, first, at the marketing function and the achievement of competitively superior customer services; second, at physical operations in the form of improved inventory management. This third example illustrates opportunities to increase the effectiveness of the financial function through aggressive management of cash resources.

The character and magnitude of the opportunities in this area—opportunities that remain largely untapped in many organizations throughout both the public and the private sectors—are highlighted by two pieces of evidence.

First, the amount of non-income-generating cash held by financially strong companies in a given industry may vary, as a percentage of sales, from a low of 2 percent to a high of 10 percent. This fact alone suggests that many companies have much idle or surplus cash that could safely be put to profitable use—at the very least in the form of short-term, liquid investments. As a concrete example, on December 31, 1974, the 100 largest industrial companies in the United States held a total of approximately $9.2 billion in cash. Merely a 10 percent reduction in these cash holdings (mostly compensating balances or demand deposits) would have resulted in a before-tax profit contribution of approximately $90 million among these companies.

Second and more important, most organizations have an indeterminate amount of so-called hidden cash—cash that is in transit or in "float." If identified and made available more promptly, this cash could add large sums to an organization's short-term income generating portfolio. To illustrate the potential, a *single* day's reduction in the accounts receivable cash-in-transit time of a large organization could produce an annually recurring profit contribution of several hundred thousand dollars.

It is in this second area that systems analysts can make the greatest contribution to improved cash management. By overhauling an organization's cash gathering and disbursing systems, they can identify the various forms of hidden cash and free them for profitable use.

Information To Be Gathered

The end-result objectives of a cash management study dictate the information to be gathered. The first objective is to increase the volume of investments in earning assest by accelerating the flow of hidden cash and by drawing on available cash that is now in non-income-generating forms. The second objective is to increase the level of earnings from short-term investments such as Treasury bills, time deposits, and commercial paper. To achieve those objectives, the analyst will need information on the following factors.

1. *Size of cash assets*
   a. The 5-to-7-year trend in the company's cash and "cash equivalent" as a percentage of sales, compared with the same trends among other companies in the industry.
   b. Stability of cash receipts and disbursements over the months of the year.
   c. Amount of short-term borrowing to level out cash fluctuations.
   d. The nature of the company's daily cash forecasting system. How is it prepared and used, and how accurate has it been?
   e. History of liquidity ratios (for example, the trend in the ratio of current assets to current liabilities).
2. *Bank accounts maintained and business relationships* with each bank used by the company
   a. The purpose of each account.
   b. The average size of, and fluctuations in, the balance for that account as shown on the company's books as well as on the bank's ledger.
   c. The activity of the account (the number of withdrawals and deposits per month).
   d. The bank services the company is using, such as lockboxes, depository transfer checks, or wire transfers.
   e. Borrowing arrangements and the average outstanding loan from that bank.
   f. Time required for transactions to clear. Is the bank located in the Federal Reserve city that serves the collection area? Does the bank use the available wire system for transferring funds? What percentage of checks sent to the bank for the company's account are received 1 day after they are mailed?
   g. Importance of the connection. Is maintenance of good relationships with this particular bank vital for some special reason? What is the approximate cost to the company, as well as the profitability to the bank, of each such banking relationship?
3. *Invoicing and accounts receivable practices*
   a. Trend in the amount of the company's accounts receivable, as a percentage of sales, over the past 5 to 7 years compared with the same trends among other companies in the industry.
   b. The timing of invoice issuance (range in number of days and average number of days after shipment that invoices are prepared and sent to the company's customers).
   c. The number and value, per month, of invoices that customers pay after the due date, broken down by number of days they are late.
4. *Characteristics of inward cash flow*
   a. The average length of time between customers' mailing of their remittances and receipt of those remittances at each stage of the collection process—up to and including the clearance and availability of the funds for use at a disbursing bank.

      b. The geographical distribution of receipts by source (in a petroleum products company, for example, remittances by credit card customers, independent distributors, and independent service stations); for each such source, the number of customers and the average size of individual payments.

      c. The points at which money is first received by the company.

      d. The average amount of money handled and number of transactions taking place at each receiving point per calendar day.

      e. The various paths that money takes to reach disbursing banks.

      f. The devices used to handle and control the movement of money between stages in the various cash gathering processes (lockboxes, wire transfers, depository transfer checks).

5. *Accounts payable and outward cash flow*

      a. Trend in the amount of the company's accounts payable, as a percentage of sales, over the past 5 to 7 years compared with the same trends among other companies in the industry.

      b. The actual versus the required time of payment on accounts payable.

      c. The points from which funds are disbursed; the geographic relationship between those points and the location of payees; the amounts disbursed to payees.

      d. The methods and timing used to transfer money into disbursement accounts.

      e. The average size, and the range in size, of the so-called "float" in each disbursing account. (This refers to the difference between the amount of money actually available in a given bank account and the account balance shown on the company's books—a difference accounted for by the time that elapses between the writing of a check and its ultimate collection by the payee.)

6. *Short-term investments*

      a. Policies governing size and character of the short-term portfolio.

      b. Mix of types and maturity dates of securities actually held in the portfolio.

      c. Range of fluctuations in the size of the portfolio.

      d. Average yield of the portfolio.

      e. Impact of cash planning and control procedures (for example, cash forecasting) on timing of short-term investment moves.

7. *Organization considerations.* In order to optimize cash availability for generation of income, has a specific position been assigned responsibility for monitoring all aspects of the organization's cash gathering and disbursing processes?

Typical Improvement Opportunities

Most of the opportunities to significantly improve cash management are implicit in the factors listed above. These opportunities fall into the following action categories:

1. Get the money that is owed to the company into one of its collection or gathering banks faster.
2. Move money faster from gathering banks to the company's disbursing banks.
3. Manage the outflow of funds more skillfully.
4. Reduce excess cash balances.

Getting money quickly to gathering banks   To achieve this objective, the analyst should explore all possibilities for (1) speeding up customer invoicing, (2) reducing the volume of past-due accounts receivable, and (3) optimizing the network of gathering banks used by the company.

Speeding up customer invoicing. For the shipping-billing procedure, the systems staff should make the same sort of detailed analyses of processing times and waiting times that were described earlier for the study of customer service. In addition to substantially speeding up invoice preparation and mailing through the cumulative effect of many small improvements, the analyst may find major opportunities in this area. For example, one study team discovered that, although prevailing industry practice was to bill customers upon shipment of their orders, its accounting department was accumulating shipping papers and preparing only one invoice per customer per month. But analysis showed that the clerical saving resulting from this practice was a microscopically small part of the lost income from delayed collections.

Reducing the volume of overdue accounts receivable. Many companies focus their collection efforts on accounts that are markedly late—30 days or more past the due date. Yet one company found that it could increase its cash resources by over $1 million through an aggressive program of accelerating payments from accounts that were chronically late by only a few days to a few weeks. Its program consisted of

1. Working with treasurers of customer companies simply to make them aware of the problem
2. Modifying the terms of sale to provide for charging interest on late payments

Optimizing the company's network of gathering banks. The final step in getting money to a company "receiving" point as rapidly as possible is to ensure that each of the banks used by the company meets certain requirements. These requirements apply whether the bank is one in which a company field office deposits the payments it receives, or one to which customers are asked to send their payments.

1. The bank should receive 85 to 90 percent of customer payments the day after they are mailed by the customer.
2. It should offer bank wire service.
3. It should be located in the Federal Reserve city serving the area from which collections are being made.
4. The cost of the various services it provides to the company should, of course, be competitive.

When customers' remittances are mailed directly to a gathering bank, they generally go to a lockbox, which is merely a post office box that is opened by the designated bank. Whether a company uses a lockbox or the somewhat slower process of having customers mail remittances to a field office should be based on the bank's per check cost of lockbox processing and the income to be earned from faster cash flow. Thus, the principal factor in determining whether to use lockboxes is the average value of customers' checks.

Accelerating movement of money to disbursing banks Once payments reach the company's gathering bank, the next requirement in a comprehensive cash management program is to move them to one of the disbursing banks in the most cost-effective way. The two principal methods of speeding up inward cash flow at this point are depository transfer checks and wire transfers. Since wire transfers are faster but cost significantly more per remittance than transfer checks, the choice between the two (assuming that the banks are on the wire system) should be based on the incremental investment income to be earned from faster funds transfer versus the incremental cost of wire service.

Managing the outflow of funds In determing whether the outflow of funds is skillfully managed, the analyst should look at opportunities for delaying disbursements and for capitalizing on the float in disbursing accounts.

Delaying disbursements. Although the practice of paying vendors' invoices as late as possible without forfeiting discounts is an obvious means of cash conservation, it is breached as often as it is followed. Hence, analysts should examine this area thoroughly to determine what, in fact, is really happening. Where necessary, they should recommend the establishment of controls to ensure that

1. Checks in payment of discount invoices are mailed not earlier than a prescribed date immediately in advance of discount expiration

2. Checks in payment of nondiscount invoices are mailed at the latest date consistent with maintenance of satisfactory supplier relations

Capitalizing on the float in disbursing accounts. This requires that the size of the float in any given account be accurately forecast so that most of this money can be invested and the account reimbursed only just before checks are presented for payment. This technique is especially applicable where payments from a disbursement account are large and regular and their collection by payees is quite predictable—as in the case of the company payroll, dividend payments, and checks to principal vendors.

Reducing excess cash balances To identify and help capitalize on this opportunity (which is usually substantial in size), the analyst must develop the facts needed to show that:

1. Considering the historical cash flow pattern, present balances (bank balances plus short-term, liquid investments) are higher than required to meet daily operating needs plus a safety margin.

2. Present bank balances provide more income for the banks than is necessary to compensate them for their services to the company.

Once management has accepted these facts and agreed to the recommended course of action, the next step is to review systematically the value of each existing bank account to both the company and the bank. In doing so, the analyst must understand and evaluate such basic aspects of bank economics as the cost of services to the bank and to the company, the bank's reserve requirements, and the minimum average level of deposits that will make the account profitable to the bank at the account's anticipated rate of activity. This sort of analysis generally leads to some consolidation of bank accounts and, more important, to a significant reduction in the average non-income-producing deposits left with many of the banks that the company continues to use.

AREA 4—PUBLIC AGENCY OPERATIONS

An organization in the public sector was chosen as the final example in this chapter to underscore the fact that well-designed systems can help as much to improve the operating performance of a public agency as of a private business. What is more, all societies today are confronted with a compelling need to capitalize on these improvement opportunities. To some extent, this need grows out of the continuously increasing emphasis on productivity in the public sector. But mostly it has evolved from the mounting complexity of public sector operating activities, which range from managing clinics and hospitals to developing new communities, to operating far-flung transit systems.

The example chosen in this area covers the activities of the Michigan State Housing Development Authority, an agency charged with increasing the supply of privately and publicly owned housing available to the low- and moderate-income segment of the state's population. Comparable agencies now exist in thirty states. Their operation represents a formidable management task, not only because of the vast funds entrusted to them, but also because of the diversity of functions they must perform and the number of different participants they must coordinate.

The program of the Michigan Housing Authority will ultimately involve the financing of more than 50,000 housing units at a cost of well over $1 billion. In executing that program the Authority is responsible for acquiring raw land, encouraging private companies to participate in the total housing program, promoting Michigan as a modular home manufacturing center, and developing a uniform state construction code. In addition, in the processing of individual proposals for housing developments, the Authority must either carry out itself, or coordinate the efforts of others in carrying out, such diverse activities as

1. Screening development proposals—evaluating marketability, financing requirements, site desirability, and community acceptance
2. Approving proposals for construction
3. Arranging for mortgage financing
4. Preparing, reviewing, and approving architectural designs and specifications
5. Obtaining construction bids and selecting the builder
6. Supervising construction of the development and monitoring the disbursement of loan funds

7. Recruiting and training a property management team

8. Managing the completed development over its life—achieving planned occupancy levels, controlling operating costs, and keeping the property in good physical condition

9. Evaluating the occupied development as the basis for improving (*a*) the processing of future development proposals, (*b*) Authority standards and decision criteria.

In the execution of this overall process, the Authority must initiate, coordinate, and control the participation of a large number of diverse groups and individual specialists, including development project sponsors, market analysts, community services specialists, site evaluators, housing consultants, architects, mortgage lenders, lawyers, consulting engineers, construction cost estimators, design review teams, builders, sales and rental agents, and property management agents.

Given this complex task, the basic management need to be met is the development of a network of systems that will ensure sound decisions, prompt action, and tight control—without the red tape, delay, and rigidity that often characterize public agency activities. Such systems are, of course, as essential to the accomplishment of the Authority's mission as they are to effective business operation. But in addition, among all public agencies, the need for a high level of operating performance has another important dimension—that of building credibility and avoiding political attacks. Streamlined systems are the public sector executive's best protection against charges of waste, corruption, or favoritism. A well-organized operation, with clear policies and procedures, backed up by adequate documentation of decisions, makes it difficult for dissidents to claim that decisions are made and contracts awarded in ways that do not serve the public interest.

Given this basic imperative, the following sections lay out a series of guidelines for improving the systems that support the day-to-day functions of public agencies.

Symptoms of Poor Operating Performance

Operating inefficiencies in public agencies typically take the form of (1) imbalances between the workload and the work force, and (2) delays in processing transactions and making decisions. The agency's personnel may feel undirected and overloaded, while the chief executive may feel that, despite significant staff increases, more personnel are still required. The outside groups with which the agency deals, whether public or private, may complain that calls are not answered, letters are lost, documentation requirements are unrealistic, and the decision-making process is ponderous.

The analyst's first step should be to quantify these symptoms as much as possible:

1. *Workload versus work force*

   a. What is the range in the workload of each staff over a period of time, and among staffs at the same time?

    b. What has been the pattern of staff additions?

    c. How do staff additions relate to increases in the agency's workload?

    d. Given the agency's operating plan, what are the projected staff levels in 18 to 24 months?

2. *Processing time*

    a. Based on an examination of actual case histories, what has been the overall experience with regard to processing time?

    b. Which steps seem to take the longest?

    c. Are there industry standards against which this performance can be compared?

    d. Does a review of processing delays over time suggest that operating performance is deteriorating?

Analysis and quantification of these and similar symptoms can create strong support for undertaking a comprehensive systems improvement project. In addition, the data gathered will be useful in framing some of the analyses to be conducted in the project itself.

Additional Information To Be Gathered

In addition to the detailed picture of existing processing steps covered in Chap. 9, the analyst who is reviewing the operations of an agency must gather three special types of information:

1. The criteria used for decision making
2. Existing and anticipated bottlenecks in the system
3. Unmet needs—that is, failure of the agency to achieve all its objectives

Decision-making criteria   The public judges most public sector organizations largely on the soundness of the decisions they reach in individual transactions. For this reason, the analyst should probe the following areas to evaluate the criteria that govern the agency's decision making:

1. What are the basic objectives of the agency—that is, what conditions will exist or what results will have been achieved if its performance is outstanding?

2. What are all of the different types of decisions made throughout the agency that favorably or unfavorably affect the achievement of these objectives? What is the relative importance of each type—in other words what is its impact on achievement of the objectives?

3. What criteria (that is, standards, guidelines, decision rules) are used in reaching each type of decision? Are the criteria clearly stated in writing? Have they been designed specifically to facilitate achievement of the agency's objectives?

4. Are adequate decision-making criteria built into *every aspect* of the agency's operations? All too often such criteria are applied at the beginning and end of such a process (say, to project selection and final evaluation), but are not adequately provided for at other points. For example, an analyst studying the Michigan Housing Development process should determine whether the cost criteria applied during the architectural design phase of a project are complete enough to ensure the best balance between aesthetic considerations and construction costs.

Existing or anticipated bottlenecks This aspect of fact-finding is of special importance in a public sector study because of the impact that promptness in handling individual cases can have on outsiders' perceptions of an agency's effectiveness. Here are some of the relevant questions to be asked.

1. What are the steps in the process that most frequently seem to disrupt operations or delay cases?

2. What are the causes of these disruptions and delays and what is the relative importance of each cause? For example, what information required for decision making is frequently delayed, and why?

3. What is the anticipated rate of growth in the agency's workload over the next 2 or 3 years? What effect will the increased volume have in aggravating present bottlenecks and creating new ones?

Unmet needs Sometimes an existing system fails entirely to consider certain cases or meet certain requirements called for in the agency's goals. Under these circumstances, a mere tightening up or streamlining of the existing system would clearly be an incomplete solution. To discover unmet needs, the analyst should dig into the areas covered by the following questions:

1. To what extent is the agency failing to fulfill its overall mission or to achieve any of the subordinate goals into which the mission breaks down? One of the most effective ways of answering this question is to classify a large sample of individual cases (for example, housing developments) according to the specific objective that each is intended to satisfy (for example, upgrading existing urban housing or providing housing opportunities for lower-income families in suburban areas). This step will generally require the analyst to work closely with various decision makers to ensure that the classification structure is correct and that cases are properly classified.

2. Are changes taking place in the agency's effectiveness in meeting objectives? What are they, and which objectives are involved? Do these shifts suggest the need for modifying the objectives themselves?

3. Do the agency's recurring management reports focus top management's attention on the agency's competence in achieving its defined objectives? If not, what are the management information deficiencies and how should they be overcome?

Typical Improvement Opportunities

Most of the opportunities to improve an agency's operating effectiveness are implicit in the preceding sections on diagnosing symptoms and analyzing aspects of the existing system. These opportunities fall into the following categories:

1. *Speed up the processing of individual cases.* Here the analyst should apply all the improvement alternatives set forth in the first section of the chapter on upgrading service to customers.

2. *Improve the quality of the decisions made on individual transactions.* Following are examples of the alternatives the analyst should consider in seeking to capitalize on this opportunity:

   a. Redefine decision criteria to reflect the agency's goals and public purposes more precisely.
   b. Make certain that adequate criteria and information are available at every point in the process at which decisions are made or control is exercised.
   c. Develop and implement a criteria evaluation process by which the adequacy and relevance of existing decision rules are periodically tested and improved.
3. *Strengthen management information and control.*
   a. Overhaul the structure of performance measurement reports to focus top-management attention on how well the agency is accomplishing its basic mission and the underlying objectives.
   b. Develop and implement an operating control system that closely monitors a project through the entire evaluation, decision-making, and execution processes.
   c. Document the agency's systems clearly and completely, and use this material as a basis for training the agency's own staff as well as for coordinating more effectively the efforts of outside groups that participate in the process.

ELEVEN

Cutting Clerical
and Other Administrative
Costs

Confronted with a substantial decline in sales volume, the executive vice president of a manufacturing company set a 20 percent reduction in overhead expense as the objective to be met by his operating executives. Later, in explaining how she made the reduction in her area of responsibility, the company's personnel director said: "I started with a list of all the activities that the 160 people in our department perform. The activities were arranged according to my estimate of their relative importance, and opposite each I entered the number of persons performing it. I then worked up from the bottom of the list, eliminating each activity until I had accounted for 20 percent of our current expense."

Military tacticians would hardly consider this a skillful retreat—and justly so, for cutting people from the payroll by eliminating services requires little, if any, skill. Yet all too often, the sporadic cost reduction programs undertaken by many organizations are either superficial and fruitless, or arbitrary and harmful. In some instances, because of management's failure to face up squarely to a difficult job, the whole cost-cutting effort is focused on such peripheral items as telephone bills, memberships and contributions, and company automobiles. In other instances, management simply slashes costs in the same areas that have been worked over intensively in the past. As the systems staff director of an industrial goods manufacturer remarked: "Every time we dust off and start up our periodic cost reduction drive, we grind the pencils shorter, fly tourist, and beat on the factory some more." In still other cases, companies that resort to the "10 percent cut across-the-board" sooner or later learn that this sort of unskilled surgery cuts away two pounds of corporate muscle for every pound of fat removed. Moreover, there is a good bit of evidence that this technique induces even the leanest, best managed departments to acquire a layer of protective fat.

No management can rely on cost-cutting decrees as a permanent solution to the burgeoning white-collar expense problem dramatized in Fig. 2 (page 23). For one reason, down-the-line managers and supervisors look on arbitrary expense reduction as a temporary tightening of the belt and, consequently, treat it as a short-term measure. Once the pressure from top management is relaxed, high costs almost inevitably return. But—much more important—arbitrary methods of achieving savings seldom eliminate the basic weakness that caused high expense levels in the first place.

In light of these all-too-common shortcomings, four prerequisites must be met if the systems staff is to make anywhere near its full potential contribution to the reduction and control of clerical and other administrative costs. To some extent these prerequisites have been covered in earlier chapters; yet they bear repeating here because they are so central to the effective application of all the other suggestions and guidelines that follow in this chapter.

1. A high savings goal must be set on every administrative cost reduction project undertaken—a goal expressed as a percentage of the number of positions or present cost of the activities to be studied.

2. Top-level executives must make a special effort to get across the idea that they are not interested in a defense of present practice. Instead, they must emphasize that, for the period immediately ahead, they are going to focus their attention on how quickly and how well each of their subordinates contributes to the overall savings goal that has been established.

3. The project team must target its analysis on task lists and work-distribution charts—the fact-gathering devices described in Chap. 9. Concentrating on these documents will do more than anything else to focus the improvement effort on the so-called big ticket items—the major savings opportunities.

4. A rigorous, tough-minded, questioning approach must be applied to every *element* of work and every aspect of the *way* it is being performed. To provide a mental pump primer for systems analysts, this chapter contains well over a hundred experience-based suggestions that illustrate the types of cost reduction opportunities to be pursued. But helpful as these improvement ideas may be, the most significant force in shaping the project team's work is the basic point of view with which team members tackle the study. Their whole way of thinking should be oriented toward achievement of the savings goal. They should *not* ask: "In what ways, if any, can we improve this operation?" or "What steps can we simplify?" or "How many reports can we eliminate?" Rather, they should ask: *"What do we have to do to cut the personnel requirements of this unit by 25 percent?* That is, if the manpower of this unit *had to be reduced* by 25 percent, what would be the safest and least painful way of achieving that objective?" By continuously applying this kind of mental-forcing technique, the analyst is much more likely to come up with major improvements as opposed to merely "scrubbing up" existing operations a bit here and a bit there.

As a guide to studies aimed at cutting clerical and other administrative costs, this chapter is built around the top-management approach set forth in Chap. 5. It offers many examples of possible improvements in each area delineated by the following four questions:

1. Can the activities of this unit be simplified or its work volume reduced by modifying such external factors as policies, organization structure, or the practices or performance of other departments?

2. Is each basic function, task, and operation necessary?

3. Does it duplicate or overlap another function, task, or operation?

4. Can the necessary work be performed in a simpler, faster, or more economical way?

Before these steps are discussed individually, two general observations should be made about them: (1) They overlap to a certain extent, in that more than one of them may uncover the same weakness or opportunity. Nevertheless, since each step will make a substantial contribution to the study, each should be taken separately. (2) To avoid repetition in the statement of principles and techniques, the chapter treats functions, tasks, and operations together, thus combining questions 2, 3, and 4 above. In making an actual analysis, however, the project team should follow a somewhat different sequence. That is, it should explore the need for each. *major function* and investigate basically different ways of performing it (as suggested in Chap. 5) before considering detailed operations.

MODIFYING EXTERNAL FACTORS

As already pointed out, the so-called external factors that affect the volume or flow of work, or the complexity of an operational system, are (1) organization structure and relationships, (2) policies, and (3) the practices or performance of other departments.

Organization

Rather than review all the principles of organization with which systems analysts must deal in their work, this section will focus on some of the specific organizational weaknesses that usually create complexity. The most common are overorganization, lack of clarity in the division of responsibility, lack of standardization among comparable units, and failure to delegate or to decentralize.

Overorganization There is a tendency, particularly at the lower levels of clerical activity, to create an excessive number of separate organizational groups, each performing a relatively small part of a single administrative process. Sometimes this trend toward proliferation springs from a desire to achieve greater specialization or tighter control. Sometimes it represents an effort to give a spurious form of recognition to individuals who have long seniority or particular talents. Whatever the cause, the effects are often high clerical costs, cumbersome procedures, and dilution of responsibility to such an extent that accountability for results is lost.

To uncover a situation of this kind, the analyst should ask these questions: How many separate organization units or subunits participate in the procedure under study? Is the size of each group smaller than one supervisor is capable of directing? How similar are the basic functions performed by these groups? Is there any overlapping or duplication of activities? Within each group, what is the ratio of time spent making ready or becoming familiar with each piece of work to time spent actually performing productive operations? How much time does each group take to record documents passed along to the next unit and to follow their subsequent progress through the routine?

Where this sort of probing reveals profitable opportunities to merge related organization units, some or all of the following benefits can usually be realized:

1. More flexible and economical use of the work force
2. More economical supervision
3. Concentration of responsibility for carrying out a system or procedure in a few key positions
4. Simplification of the whole process through eliminating complex controls; weeding out extra handling, checking, and recording operations; and preventing a circuitous flow of work.

Lack of clarity in the division of responsibility   Most of the problems caused by overorganization can also result from vagueness in the *division of responsibility* for carrying out the major elements of a system or procedure. The obvious effects of failure to delineate responsibility are confusion, overlapping, duplication, and conflict. These conditions in turn produce more serious consequences. Important tasks are poorly done or left undone because "everybody's business is nobody's business." Deficiencies develop in the service rendered to customers. Key decisions are made by persons who have neither the information nor the point of view necessary to ensure sound judgment in those areas.

Lack of standardization among like units   Variations among comparable units are not entirely a matter of organization. They may also occur in the procedures followed as well as in the functions performed. Whenever an administrative process involves the flow of work from several similar units to a common point, analysts should be on the watch for any lack of standardization that creates exceptions in the normal routine. It is not always possible to bring about complete uniformity; some differences in practice may be unavoidable. The majority of variations, however, usually exist only because of tradition or personal preference.

Failure to delegate or to decentralize   A common organizational fault resulting in paperwork bottlenecks is the failure of executives to delegate authority for reviewing and signing documents covering routine transactions of the business. Systems analysts must try to correct this condition whenever they find that it seriously impedes the flow of work or decreases executive effectiveness.

A somewhat similar though less readily soluble problem is that of determining the degree of centralization or decentralization best suited to the performance of a given clerical function. Frequently centralization (in a physical sense as well as in direction and control) is the more economical arrangement. The reason is that the larger volume of work makes possible greater flexibility in the use of personnel, a higher degree of work specialization, less but higher-caliber supervision, and increased mechanization of clerical operations. Centralization also produces tighter control and greater uniformity in the administration of company policies.

Centralization is not always the best arrangement, however. Where a central group attempts to control the day-to-day activities of widely dispersed units, the result is not centralization but a costly hybrid of centralization and decentralization. Communication is either slow or expensive, work is delayed, and unless some duplicate records exist in both the central and the field offices, decisions must be based on insufficient information.

The advantages usually attributed to decentralization are that action is greatly facilitated and decisions are more realistic because they are founded on a better understanding of local conditions and requirements.

Obviously no one answer applies equally to all functions in one company or to any one function in all companies. A firm with 150 branch offices saved 20 percent in the clerical cost of handling accounts receivable by centralizing this function in 12 branch offices strategically located throughout the country. Another company found it desirable, because of competitive service practices, to decentralize its accounts receivable operations to individual branches despite a resulting 15 percent cost increase.

The only guide that can be given is that the analyst's objective should be to achieve the best overall balance among clerical economy, adequacy of control, and quality of service.

Policies

Second among the factors the project team must analyze before tackling individual clerical tasks are the policies that affect the work of the unit under study. In this context, the following excerpt from a *Harvard Business Review* article is germane:

> Every company operates within a framework of formal or informal policies governing the way the company is run. These may cover such varied elements as the geographical market areas to be reached by company products, the means of distributing those products, price leadership, product-line emphasis, personnel promotion and selection criteria, the company's image, the degree of formality between superiors and subordinates in the organization. In many instances, these policies have not been developed as a result of rigorous study but rather to meet the specific needs of the company under specific competitive conditions. Not surprisingly, therefore, many companies . . . have uncovered substantial profit opportuni-

ties by taking a hardheaded look at their established policies. To illustrate:

> *A major financial company had built its reputation on a philosophy of having a strong* local
> *representation in each of its marketing areas. As a result, it had a network of over 200 local*
> *sales offices scattered throughout the United States that performed the selling and accounting*
> *functions for each area, all in an effort to identify the company as a "local" business.*
>
> *Faced with continually rising paperwork costs, management challenged the logic of and*
> *need to do paper processing locally. It found that centralization of paper-handling activities*
> *at fifteen regional centers would reduce costs by over 25 percent—and yet have no real*
> *impact on the company's overall image as a local business. Management would not have*
> *accomplished this substantial improvement in profits if it had merely tried to tighten up the*
> *paperwork-handling function at each of the 200 individual offices.*[1]

In contrast to this sort of far-reaching policy change, the analyst should also look for opportunities to reduce the volume of clerical work by making minor adjustments in policies. For example, can pricing or discount policies be made more uniform without any significant loss of sales income but with a real simplification of invoicing procedures? In a property and casualty insurance company, is it possible to increase the dollar size of the policy that independent agents are permitted to underwrite? In the routine for handling customer complaints, can the company safely reduce the amount of tracing to be done before settlement is made with the customers? In the accounts payable procedure, should the policy for handling overshipments, overcharges, or damage claims be readjusted? In many companies, these discrepancies or claims represent dollar amounts that are smaller than the clerical cost of straightening them out. Can the volume of purchase orders be reduced by adopting a policy of issuing blanket purchase orders to the firm's major vendors? Can purchased items of low unit value be expensed at the time of acquisition instead of charging an inventory account and expensing items individually upon withdrawal from stores? One company found that 35 percent of its stores requisitions covered items that represented less than 5 percent of the value of inventories. For many of these items, the clerical cost of maintaining continuous inventory records, pricing withdrawals, and charging them to jobs was greater than the value of the item withdrawn.

Beyond such changes in policies themselves, can the administration of existing policies be made more uniform so that exceptions to the normal routine are held to a minimum? For example, the activities of the payroll department may be geared to carrying out an established policy for the compensation of sales representatives. But if sales executives are constantly making special arrangements for compensating individual representatives, the payroll unit is forced to set up extra records and controls and to perform much of its work by nonstandard methods. Often the analyst can correct this condition simply by making the executives aware of how frequently it happens and what effect it has on operating costs and speed.

[1] J. Roger Morrison and Richard F. Neuschel, "The Second Squeeze on Profits," *Harvard Business Review,* July–August 1962, p. 64.

Practices or Performance of Other Departments

The efficiency of a system is often affected by the practices or performance of departments that play no direct part in implementing it. Hence, project team members must study this external factor in the same way that they evaluated company policies. Where they find that the volume or flow of work through the process is adversely affected by outsiders, they must look for opportunities to change the practices or improve the performance of the outside groups to gain a better overall result.

An illustration can be found in the experience of a company that tried to cut clerical costs in its stock records, shipping, traffic, and invoicing departments.

The team conducting the study recognized at the outset that the volume of clerical work in these units was determined by the number *of shipments made, not by the size of shipments. Although the company's total volume of business had increased only slightly in recent years, the number of shipments per day had more than doubled and their average size had declined sharply.*

In searching for the cause, the analysts found that a wide variance had developed between delivery promise dates to customers and dates on which merchandise was available for shipment. Some items were produced as much as 3 months behind schedule; others were correspondingly ahead of schedule. To avoid building up excessive finished-goods inventories, the company had resorted to hand-to-mouth shipping and back-ordering.

As a result of this analysis, the company undertook an intensive program to tighten its material procurement, production scheduling, and production control so that output would be geared more closely to composite customer demand. Under those circumstances, the number of partial shipments declined sufficiently to permit a sizable reduction in the clerical force of the four departments in the original survey.

REAPPRAISING THE WORK TO BE DONE

After project team members have fully explored these external factors, they are ready to dig into the details of the work itself. Here their first step should be to determine *what work needs to be done.* This is a matter of critically analyzing each function, task, and operation and of rooting out those parts that prove to be unnecessary or unprofitable. Elimination of this kind is the easiest and most obvious way of simplifying the work of any unit, and it sometimes accomplishes the objective of the study without further effort.

The raw materials for this phase of the development work are the statements of purpose or end use that were obtained during the fact-gathering process. Is the purpose of each function, task, and operation *justifiable?* How does each relate to the objectives of the process as a whole? Does it make a definite contribution to the quality or progress of the work? *Is the value of the contribution greater than the cost? What would happen if the step were eliminated or curtailed?*

In applying these tests of worth or need to an existing process, the analyst will profit by observing three rules:

1. Be sure that the attack is systematic and thorough. Challenge whole functions or major activities before questioning individual tasks or operations. First consider whether the entire function or activity can safely be eliminated. If it cannot, determine whether the volume of work—that is, the number of transactions or the frequency of performance—can be reduced.

2. Do not stop at discovering *why* something is done. Evaluate that reason. Find out whether it is sound. For example, in questioning a body of management reports, determine what specific decisions can be made or action taken on the basis of each piece of information reported. Then make a realistic estimate of the payout of such decisions or action.

3. Take calculated risks. If the need for an activity or its profit-making value is uncertain, eliminate it on a trial basis and see what really happens. Or, as another approach, run controlled tests on less costly alternatives.

As a further guide to the analyst, the remainder of this section contains illustrations of common types of unnecessary or unprofitable clerical work.

Unnecessary or Unprofitable Functions or Tasks

Chapter 5 told how a mail-order house eliminated all shipping-paper files when it discovered that the cost of maintaining them was greater than the cost of replacing lost shipments. A more typical case of unprofitable work is a billing and accounts receivable function that is not geared to the payment practices of customers. Even where customers ordinarily pay against individual invoices, some companies continue to send monthly statements of account that customers neither need nor act upon.

Take the case of sales order handling as an example of one way to reduce unprofitable transactions within a function. Here the systems team should first determine the lowest value of sales order that could be profitably handled, and then take steps to minimize smaller orders or to cut the cost of handling them. Some of the specific alternatives that analysts might consider in meeting this objective are

1. Drop unprofitable accounts with low potential volume.

2. Assign small accounts or divert small orders to a distributor.

3. Encourage larger orders through sales training, incentive compensation, or educational efforts with the appropriate accounts.

4. Impose a service charge on small orders, or make the customer pay the freight bill on orders under a minimum weight or value.

5. Devise shortcuts for preparing orders and keeping sales records on small orders and small accounts.

6. Establish a minimum order value and refuse to accept orders below that amount.

Unnecessary Reports, Records, and Forms

Anyone who has ever looked into the use made of business reports knows how often they are continued long after their usefulness has passed. New reports are sometimes introduced without elimination of the old. One department may discontinue use of a report without notifying the department that prepares it. Special reports requested to meet a temporary need may become assimilated into the regular routine and continued indefinitely. (Chapter 13 contains specific guidelines for testing the value of existing reports.)

The same mortality takes place in the use of forms and records. Even when reports are discontinued, the underlying records are often kept just in case the reports are asked for again. Or a manager or supervisor may set up records simply so that he can respond knowledgeably to any and all questions from higher up—questions that analysis shows are seldom if ever asked.

The waste that can result from such unnecessary documentation sometimes even extends to reports obtained from sources outside the organization. Take life insurance companies, for example. They regularly buy selection information in the form of medical examination reports, retail credit reports, and attending physicians' statements to help their underwriters decide whether to accept, decline, or charge a higher premium for a given applicant. Since these reports can be costly in relation to the annual premium, good business practice requires a sound balance between the cost of the information and the incremental mortality loss that would result without the information. Yet some life insurance companies have never made this sort of value-cost analysis, and many more fail to reexamine their decision rules whenever the cost of obtaining such information goes up significantly—as it has in recent years.

In questioning the necessity of reports, records, and forms, the analyst may be in for some surprises.

One company, for example, found that it could safely eliminate its continuous inventory records on maintenance, repair, and operating supplies by setting up a visual control over each stores item. Under this arrangement, the reorder point on each item is identified by a bin divider, or a line painted around the bin at the reorder level, or the wrapping, taping, or sealing of the quantity representing the reorder point. In addition, at the physical location of each item the company maintains a traveling bin card, on which are recorded

1. *The identifying number and description of the item*
2. *The reorder point*
3. *The reorder quantity*
4. *The last invoice price*
5. *Quantities received*
6. *The physical inventory count*

This sort of record makes possible the pricing of withdrawal requisitions and the accumulation of data on usage for setting reorder quantities. The new procedure has enabled the company to reduce its clerical storekeeping costs by 40 percent and still control its inventory investment, losses, and obsolescence as adequately as in the past.

Unnecessary *copies* of forms or reports are even more common than redundant forms themselves. Consider the company that maintains five complete invoice files, one each in the billing, sales, computer, accounts receivable, and credit departments. Three of these departments seldom refer to invoices and could easily obtain by telephone whatever information they need.

This matter of unnecessary copies will generally repay searching study, for even where analysts cannot save the cost of preparing the whole document, they may substantially reduce the cost of distributing, reading, and filing its copies. Usually these expenses are far greater than the initial cost.

Superfluous Data

Do forms, records, or reports contain more detail than is needed to serve their purpose? What use is actually made of each piece of information? Are money and quantity entered on stock records when only quantity is needed? Are figures shown in pounds and tons, in gallons and barrels, or in cases and dozens when either alone would be sufficient? Where sales control records show the volume of each customer's purchases, are postings limited to quantities shipped by product class? Or do the records contain such superfluous entries as the customer's order number, date the order was received, shipping order number, price, routing, car number?

The amount of *unused* business information that is gathered, recorded, and reported staggers the imagination. In one company a study of seventeen forms, records, and reports prepared by a single department led to a 68 percent reduction in the volume of information recorded. In another company the number of entries on eleven monthly reports was cut from 995 to 281—a reduction of 72 percent.

Situations such as these develop for a variety of reasons. The most common cause is carelessness in defining the specifications of a record or report when it is first set up. People just do not take the time to think through the limits of their purpose or to restrict the information provided to their needs. Sometimes a cautious operator plays safe by recording everything, "just so it will be there if we ever need it." Another cause of superfluous data is the desire to maximize or justify the use of an expensive machine by turning out more and more information.

An important point to remember in this part of the study is that the cost of recording superfluous data is not the only loss. If a report is cluttered up with a lot of unnecessary information, the recipients are likely to disregard it entirely, with the result that the significant facts never come to their attention.

Excessive Control

Does the activity under study include elaborate checks to detect errors that never occur? Is the amount of checking related to the risks involved? What control devices serve only to avoid or fix blame rather than to protect against a real loss? Are unnecessary signatures of approval required on forms?

These questions circumscribe the area in which calculated risks often produce big savings. Following are some illustrations.

1. Can the cost of the credit function be reduced and the processing of orders speeded up by giving a list of approved credit risks to the order department? By establishing a minimum order value below which no credit approval is required?

2. In the case of small claims for automobile property damage, can on-site inspections by claims representatives be reduced or eliminated?

3. Are orders registered for internal control when shipment is made within a day or two after receipt of the order?

4. How many signatures of approval are required on purchase requisitions covering items for which reorder points and reorder quantities have been established?

5. On intracompany transactions, how careful a check is made of mathematical computations? For example, are extensions on labor and material charges recomputed when visual checking for proper placement of decimal points would protect against any serious error?

6. Are figures on reports being balanced to the penny when an overall discrepancy of 1 or 2 percent would not diminish the value of the information to management?

Many such checking and control operations can often be safely eliminated. But in their zeal for cost reduction and work simplification, systems analysts are sometimes inclined to leave too few checks or controls where too many existed before. They can lessen this danger, however, by applying such tests as these: From what actual or potential trouble is the organization trying to protect itself through the application of this control? What is the probability that the trouble will occur? Is it likely to occur often? What are the consequences of its occurrence—for example, what is the average amount of each loss likely to be?

A good illustration is the cost of checking customers' credit when orders are being processed. If the average value of orders is $10, checking can be curtailed much more severely than if the average value is $1,000. Bad-debt losses on a few $1,000 orders can quickly wipe out the savings gained from relaxation of control.

In general, the analyst should apply the same commonsense method of evaluating checks against clerical error. If some doubt exists about the probable volume of error, the line personnel who do the checking should be asked to keep a temporary record of errors by type. In this way, the analyst can measure the cost of the control more precisely against the loss of money or time that would result from permitting errors to go undetected.

Where the need for some kind of check is confirmed, there are still two possibilities for simplification.

1. Would a spot check rather than a check of every transaction be sufficient to maintain the standard of accuracy?

2. Can a substantial part of the checking safely be eliminated by verifying only those items that exceed a certain dollar amount? As an example, refer to the experience of one company that gave up checking all vendors' invoices

amounting to less than $100. (See Chap. 5, the section entitled "What Work Is Worth Doing?")

In evaluating the need for signatures of approval as a prerequisite to action, the analyst should explore two points:

1. What is the nature of the control required? Is it a *qualitative* or a *quantitative* matter? For example, a personnel requisition requires quantitative control since it is concerned with labor requirements. An employment application, on the other hand, requires qualitative control, since it is concerned with the selection of specific individuals. Are the approvals required on these forms related to the type of control needed? Must an employment application, for instance, be approved by high-level executives who neither interview applicants nor are acquainted with the specifications of jobs to be filled?

2. What is the relation between signatures of approval and other means of control over the same action? For example, are as many reviews and approvals of personnel requisitions needed where predetermined quantitative standards have been established and where some postaudit method of control exists (say, budget reports)?

Unnecessary Refinements or Quality Requirements

Closely related to the amount of control that is exercised are standards of work quality and refinements of information. Are production orders completely retyped to make small amendments, when a brief order-modification form would suffice? On multiple-copy forms, are typists permitted to X out mistakes rather than having to erase them? Is it necessary to show cents in financial and operating reports? Can percentage figures be rounded to whole digits, so that calculations need not be made more than one digit beyond the decimal? Would an estimate serve where involved computations are required to develop more precise figures? For example, on purchased items, can the last invoice price or a standard price be used for pricing disbursements, rather than complicated first-in–first-out or average-price calculations? Can small charges for inbound transportation be thrown into an overhead account so that they will not complicate the calculation of inventory unit prices? To reduce the amount of time required to reconcile clock cards with job cards, can arbitrary adjustments be made for discrepancies of a small fraction of an hour? Where indirect factory workers are performing a routine service function, can the accounting distribution of their wages be made weekly on a predetermined percentage basis, rather than by having them submit daily job cards charging their time to the units they serve?

ELIMINATING DUPLICATION

While they are appraising existing functions and operations to determine what work needs to be done, project team members should also be looking for evidence of work duplication. In all likelihood, they will find their greatest savings opportunities here. Duplication is a hardy weed and no field of clerical activity is without it. Executives in all types of organizations talk about it continuously and make sporadic efforts to uproot it, but still it flourishes. One reason is that it is often difficult to detect. It is like the weed that, to the inexperienced gardener, closely resembles the plant to be cultivated.

In their search for evidence of duplication, systems analysts should bear two points in mind:

    1. Duplication is usually partial, not complete, and half obscured rather than clear-cut. Although analysts will occasionally find the same thing being done in exactly the same way in two places, they usually must go beyond that point to discover overlapping or similarity of activities, records, forms, or reports.

    2. Duplication may take several forms. It is not confined to the physical product of clerical work. If analysts are to dig out the opportunities for work elimination or combination which lie in this area, they must look for duplication of (a) function, (b) purpose, (c) operations, and (d) information.

Duplication of Function

An entire major function may be paralleled by a similar function in another division of the organization, or even within the same division. The sales division, for example, may duplicate the work of the production control department by maintaining elaborate production progress statistics and by actively following up the factory to expedite customers' orders.

In digging out all forms of duplication, analysts may often find themselves ranging far afield—sometimes even outside the company. A major title guaranty or title insurance company learned that all the other title companies with which it competed would make available, on a reciprocal basis, their files of completed title searches. Once an initial title search had been completed, subsequent exchanges of that title—if based on the original documentation—required only about one-third the time to review and update. As soon as the company became aware that it could make such an arrangement, it agreed to pool its files with those of other companies, thus saving the work involved in building up most of its own title searches from scratch.

As still another example, a property and casualty insurance company eliminated much duplication of work by authorizing its independent agents to prepare and issue a large percentage of the policy endorsements (that is, policy changes) required by its policyholders. Although this innovation enabled agents to give faster service on endorsements, it did not increase their workload. Under the superseded process they were already filling out a form or writing a letter requesting the company to prepare each required policy change. This particular case, incidentally, typifies a widespread problem in the paperwork of many large organizations—the absurd and wasteful practice of having someone fill out a form that, in effect, simply asks someone else to fill out another form.

Duplication of Purpose

This type of duplication is usually the most difficult kind to recognize, since activities that have a common purpose may be quite different in form. For example, in the study of a shipping and invoicing procedure, the analyst may find that the order department sends to the invoicing unit a copy of each production or shipping order issued. The shipping department, in turn, may number all shipped orders consecutively and send one copy to invoicing, where it is checked off a prenumbered register. In addition, the shipping department may send invoicing a daily manifest of shipments. Each of these different devices is nothing more than a way of protecting against failure to bill a shipment. Any one of the three, by itself, could provide adequate control.

Another common type of duplication of purpose can be found in reports that contain different information but are aimed at controlling the same activity. For instance, to facilitate control of overhead expense, the personnel department may be preparing reports on the size of the work force while the accounting or budget department is preparing operating expense reports. To control selling effort, the sales department may be building up summaries of orders as well as summaries of shipments.

Duplication of Operations

Two routines containing similar operations can often be merged by retaining the necessary parts of each, eliminating the duplication, and reconstructing a single, simplified routine. Here are some examples of the steps the project team can take.

1. In the order processing and invoicing routine, reduce to as few as possible the number of references that must be made to *product* records throughout the whole cycle. Following are the principal reasons for making such references:

   a. To verify or correct the product description, size, grade, or other specifications
   b. To assign correct product code numbers to the items ordered
   c. To price the order
   d. To deduct the items ordered from inventory records
   e. To pull prepunched product cards from a tub file

Ideally, steps *a, b,* and *c*—and occasionally step *d*—can be combined into a single operation. Even step *e* can be included if the tub file also serves as a continuous inventory bank. Or, as an alternative, the order editor can pull product cards from a tub file, and the inventory record can be posted separately.

2. As a closely related example, reduce to one the number of references that must be made to *customer* records during the order processing routine. Customer records may be referred to for product allocation or contract control, order editing, routing of shipments, credit checking, customer coding, or punched-card pulling.

3. Make an analysis of the number of copies of the following documents that are filed throughout the company:

   a. Receiving reports
   b. Incoming inspection reports

    c. Vendors' invoices
    d. Inbound bills of lading
    e. Freight bills
    f. Packing slips
    g. Returned-goods forms
    h. Accounts payable vouchers
    i. Debit notices

Where is each copy filed? In what sequence is it filed? How often is the file referred to, and for what purpose? How much time is spent in maintaining the file? Could not some of these documents be destroyed without filing? Would not one or at most two copies suffice? For example, if purchasing and accounts payable (and possibly traffic) were located close together, could there be one common file of paid invoices and another of closed purchase orders?

In appraising the need for various control operations, project team members will probably find more duplication of *checking* than of any other clerical operation. They may also discover that these extra checks pay for themselves— that the dollar value of errors caught is greater than the cost of catching them. But their search should not end there, for the pyramiding of secondary and tertiary controls is a costly and inefficient substitute for accurate primary control. Hence, the analysts must find out *why* so many errors are being made, or *why* so many are getting past the initial checker. By learning the real causes, they will be able to suggest realistic corrective measures. Better training of the operator or simplification or mechanization of the procedure will often reduce the volume of error to a point where double and triple checks can safely be eliminated.

Duplication of Information

The most common type of duplication is duplication of information. Most business records, reports, or forms contain some information that has already been recorded elsewhere. Separate reports prepared for different purposes may be little more than different arrangements of the same data. They should be consolidated into a single new report that will satisfactorily serve all purposes. The same is true of forms and records, many of which are transcribed wholly or partially from other documents.

Particularly in studying forms and records, analysts should learn to question all *copying* or *posting* operations, since this kind of clerical activity represents duplication in its most elementary form. The team members' objective should be to reduce as far as practical the number of times any given piece of information is written. If a form is copied from another form, they should try (1) to eliminate one of the forms by having additional copies of the other prepared, or (2) to combine both in a redesigned form that will meet all requirements. (Examples of such combinations are given on page 221.) If a record is posted from a form, they should seek to eliminate the record by filing a copy of the original document in its place. In doing so, however, they must be careful not to create more work than they save. A posted record can usually be referred to more quickly than a loose file or a binder of papers. On the other hand, 5 percent usage would not repay posting 100 percent of the transactions. The decision, therefore, should be based on the frequency of reference to the record, the cost of posting to it, and the time that would be lost in referring to source documents.

Where a record cannot be eliminated for practical reasons, the analyst should try to merge it with related records in the same or other departments. An example is miscellaneous cost or expense records that are not properly tied in with the general accounting books. Consolidation of these data can produce not only clerical economies but more meaningful information as well.

Figure 24 shows a useful device for uncovering duplication of information. The obvious advantage of this chart is that it enables the analyst to record a great many related details on a single sheet. Particularly on major studies, therefore, it helps to bring the full extent of existing duplication into sharper focus.

Combination of forms   In tackling duplication of information, bear this rule in mind: Ideally, the form that *initiates* a procedure should be so designed that it will meet the requirements of all departments and eliminate the need for preparing other forms to carry out any part of the procedure. Where the ideal can be achieved, the cost of repetitious typing and checking of the same information is eliminated, the procedure cycle is shortened, and transcription errors disappear.

The following six groups of forms illustrate the possibilities for accomplishing many purposes in a single writing.

GROUP 1

Sales orders	Packing slips
Acknowledgments to customers	Bills of lading
Register of orders	Back orders
Shipping orders	Invoices
Shipping notices	Register of invoices
Shipping labels	Accounts receivable ledger

GROUP 2

Production orders	Material identification tags
Timecards	Cost accounting sheets
Stores requisitions	Inspection reports
Purchase requisitions	

GROUP 3

Purchase requisitions	Receiving reports
Requests for quotation	Incoming inspection reports
Purchase orders	Material identification tags

GROUP 4

Paychecks	Employee earnings records
Payroll journal	

REPORTS

INFORMATION CONTAINED IN REPORTS	DAILY TONNAGE REPORT	DAILY TRAFFIC REPORT	BILLS OF LADING	WEEKLY INVENTORY REPORT	MONTHLY RAW MATERIAL INVENTORY AND DISTRIBUTION	MONTHLY PRODUCTION REPORT	MONTHLY REPORT OF RAW MATERIALS RECEIVED	MONTHLY PLANT REPORT OF SHIPMENTS	MONTHLY COMPUTER REPORT OF SHIPMENTS	MONTHLY RETURNABLE DRUM REPORT
CARS SHIPPED		X	X							
CARS ON SIDING		X								
FINISHED INVENTORY, PRODUCT A				X		X				X
FINISHED INVENTORY, PRODUCT B				X		X				
FINISHED INVENTORY, PRODUCT C				X		X	X			
BEGINNING RAW MATERIALS INVENTORY						X	X			
RAW MATERIALS RECEIVED				X	X	X				
TOTAL RAW MATERIALS USED					X	X				
RAW MATERIALS USED IN EACH PROCESS AND EACH PRODUCT					X					
ENDING RAW MATERIALS INVENTORY				X	X					
INVENTORY DISCREPANCY					X	X				
SHIPMENTS OF PRODUCT A	X					X		X	X	
SHIPMENTS OF PRODUCT B	X					X		X	X	
SHIPMENTS OF PRODUCT C	X					X		X	X	
NUMBER OF DRUMS RETURNED										X
NUMBER OF DRUMS DESTROYED										X
NUMBER OF EMPTY DRUMS ON HAND										X
NUMBER OF DRUMS OUTSTANDING										X

FIG. 24 *A simple chart helps the analyst detect duplication of information.*

GROUP 5

Interplant material transfers   Receiving papers
Shipping papers

GROUP 6

Cash receipts journal   Deposit slips

In studying the system or procedure represented by any one of these groups, the analyst should ask: How many of the documents are now prepared separately? How much identical information do they contain? Can all or most of them be prepared in one writing?

The two most common ways of producing multiple copies of a form are by using duplicating masters or carbon paper—either continuous forms with one-time or reusable carbon, or individual form sets with one-time carbons. In either method, spot carbon, strip carbon, blockout printing, overlay strips, long and short sheets, and other selective devices may be employed to prepare forms of different designs at the same time.

Which multiple-copy method to use depends on the number of copies required and the amount of information that needs to be added or the changes that need to be made after the form has been initiated. A combination order-invoice form serves as a good illustration. If the order-shipping cycle is short, if few partial shipments are made, and if not more than about ten copies are required, a multiple-copy, continuous form containing shipping papers and a fully priced invoice is probably best. On the other hand, if several departments must enter information on the form, if back orders or order changes are frequent, or if the exact quantity to be shipped is not known until shipment has been made, a duplicating master is generally better because of its greater flexibility. For example, in reproducing copies from the master, the operator can clamp strips of master paper over the original master to add new or variable information. In this way, invoices covering back-order shipments can be prepared from the original master by the addition of a special strip containing the invoice number and date, quantities, extensions, and total price of the most recent shipment.

One caution, however: Analysts must be careful not to seek combination for its own sake, but only when it produces demonstrable savings or improvements. This is another way of stating the important principle that duplication is not always avoidable or wasteful. In some situations it may be the simplest way of meeting the requirements.

In combining the order and invoice forms, for example, the analyst must be sure that the advantages of combination are not outweighed by the disadvantages. If the invoice copies cannot be prepared simultaneously with order copies, the advantages of a combination order-invoice duplicating master decline with a decrease in the average number of items per shipment. Thus, to eliminate a small amount of typing per invoice (masthead information and description of one or two items), the savings from the use of continuous-form invoices would be sacrificed and the time of reproducing invoice copies from the duplicating master would be added. Moreover, it is often true that all routine work in connection with a combination order-invoice form must be performed more slowly because of the special care required in invoice preparation.

Checklist of other work combination opportunities   Following are a number of additional clerical activities to which the principles of work combination can profitably be applied. Once again, the list is by no means complete, but as a thought starter it should help systems analysts to identify still other areas that will repay investigation within their own company.

1. Can registers be eliminated by filing a copy of the original document in the same sequence that is used in the register?

2. Are separate accounts receivable and credit records maintained?

3. Can postings to an accounts receivable ledger be eliminated by filing a copy of each unpaid invoice by customer name in an accounts receivable file? (This arrangement is practicable only where customers normally make payments against individual invoices rather than against monthly statements or on account.)

4. Can the sales journal be eliminated by filing invoices chronologically?

5. Can the posting, summarizing, and balancing of an accounts payable register be eliminated by sorting and totaling unpaid invoices by account number at the end of each month?

6. If cash tickets are being prepared for posting credits to accounts receivable, can this operation be eliminated by using customers' remittance checks as the posting media?

7. Are records designed so that they can be used for more than one purpose? For example, if the company has a returnable-container record, is it set up solely to establish customer accountability, or can it also be used as a control tool to increase container turnover? (Increasing turnover is of particular importance when the annual depreciation per container is relatively high in comparison to the value of the product it carries.)

8. In building up statistical data, do departments at each point make full use of subtotals developed elsewhere? For example, at sales headquarters, are order volume statistics built up from branch office summaries instead of from individual orders?

9. Has the maximum consolidation been achieved in the development of sales statistics by product, territory, industry, and customer? Have the sales information requirements of line sales executives been carefully integrated with those of the budgeting, market research, advertising, and production departments?

10. To facilitate the processing of sales orders, are all the fixed information items and special instructions covering each customer contained in one record? Or must order clerks refer to files of previous orders or invoices?

11. Are these customer records maintained by sales headquarters and by the invoicing department as well as by branch offices?

12. Where addressing-machine plates or stencils are used to produce shipping labels, can routing instructions and special shipping or invoicing instructions be included on the plate or stencil?

13. Does the company maintain a continuous order-backlog record, when it could easily obtain this information when needed by running a tape on unfilled orders?

14. Are orders being acknowledged in much the same fashion as they were 50 years ago, even though customers now receive the invoice within a few days of the acknowledgment?

15. Is there any duplication of order scheduling among the order processing, production planning, and factory departments?

16. Are continuous inventory records maintained in the storeroom as well as in the stores-keeping office and the accounting department? At how many points are inventory money balances kept?

17. Are finished-stock records posted twice for each transaction—once when the order is drawn and the second time when it is shipped?

18. Do production records maintained by dispatchers in the plant duplicate records in the production control office?

19. Can the dispatcher's work be combined with timekeeping?

20. Where machine-hours are the controlling element in manufacturing costs, can the computation of labor cost be eliminated by including all direct labor in burden?

21. If monthly maintenance costs are summarized by type of equipment as well as by department, can maintenance job orders be coded in such a way that one adding-machine tape will produce both sets of totals?

22. Can the number of purchase orders be reduced by writing blanket monthly orders for items that are bought frequently?

23. Can a copy of each purchase order be filed by commodity to eliminate the posting of individual and cumulative purchase data to commodity records?

24. Are vendors' invoices checked by both the purchasing and the accounting departments? Are freight bills checked by both traffic and accounting?

25. Do both the personnel and the payroll departments maintain employee history records?

26. Can payroll and personnel records be combined? Or at least, can much of the duplication between them be eliminated? One way of reducing the time required to maintain both records is to set up in the payroll department the original of a payroll change record. When a payroll status change is entered on any employee's card, a copy is then prepared on a photographic copier and sent to the personnel department to replace the superseded copy in its files.

27. Can the posting of records on employee earnings be eliminated by filing a copy of each employee's paycheck or weekly earnings statement? The

individual slips can then be pulled and totaled as needed for government reports, compensation claims, etc.

28. Are various distributions or classifications of data being obtained as by-products of other procedures? For example, is the labor cost distribution obtained as a by-product of payroll preparation? Can distribution of data for sales analysis also be obtained as a by-product of the accounts receivable procedure?

29. Can ledger entries of every type be made as carbon copies of journal entries?

30. Where intracompany forms are first written out by hand and then typed, can one or the other of these writings be eliminated? Can handwritten working papers be reproduced photographically to eliminate the typing of reports?

31. Can several rubber stamps be combined into one? If the form being stamped does not have sufficient space for the impression of a single large stamp, can a slip containing the same impression be stapled to the form?

32. How many different serial numbers are assigned to a sales transaction before it is completed? Are there separate order, shipping, and invoice numbers? How many cross-reference records must be maintained because of these different numbers? Can a single numbering system be designed to serve all purposes?

SIMPLIFYING NECESSARY WORK

The activity under study will have been streamlined considerably by the steps already outlined in this chapter. Policy or information requirements may have been simplified; the organization for performing the activity may have been straightened out; unnecessary work will have been eliminated; duplicate functions or operations will have been combined. Thus, at this point, the project team should have shaken down the work to be done to its irreducible minimum and should now be ready to turn its attention to simplifying the performance of each essential operation.

The following questions typify the kinds of improvements that this part of the analysis can produce:

1. Is the work of a department divided into many minor classifications, each of which has its own special processing routine? If so, operations can usually be speeded up, errors and training time reduced, and personnel used more flexibly by the adoption of one or a very few standard routines.

2. Can the originator of a form supply additional information that would speed up the work in other departments? For example, does he or she prepare sales orders from or check them against customer master cards that contain all the customer information required by the departments that perform subsequent steps in the work cycle?

3. Can terminology on intracompany forms be simplified and standardized so that each form is easier to use?

4. Can the posting of records be simplified by adoption of standard abbreviations?

5. Can the cost of wire communication among company offices be reduced by standard abbreviations or code words for long phrases that occur frequently?

6. Are form letters being used whenever possible in making routine requests, answering inquiries, etc.? For example, the use of a form letter requesting bids or quotations saves not only typing and proofreading time but also the time that buyers spend in dictating, reading, and signing much of their correspondence. Another potential application consists of using a preprinted checkoff form to explain to vendors the reason for deducting any charges against them from accounts payable remittances.

7. Has a simple, uniform classification of commodities been adopted to serve as the foundation for all sales reporting and record keeping?

8. Are codes or other means of classification too large or complex? One company found that it had 56 code numbers to designate the type of order and 189 to designate the type of customer. After the systems staff had simplified these numbers, the coding operation became a matter of memory rather than of constant reference to code books. Also, sales analysis information became more meaningful because it was not broken down into so many minor classifications having limited significance.

9. Can large suppliers be educated to prepare their invoices, packing slips, and shipping papers in such a way that the company can process these documents more easily? Many companies will find that twelve or fifteen suppliers initiate the bulk of their receiving, incoming inspection, and accounts payable paperwork. Where this is true, they can often save clerical time by giving each supplier a written explanation of the documents and information needed to handle its shipments promptly.

10. Have vendors been requested to indicate purchase order numbers on shipping papers? Also, to expedite the receiving process, have they been requested to number all cartons in a shipment in numerical order and to indicate the carton in which the shipping papers have been placed?

11. Checking out discrepancies among vendors' invoices, receiving reports, and purchase orders is one of the most time-consuming operations in invoice verification. Has the department developed statistics on the type and volume of discrepancies and used that information to substantially reduce the number of errors?

12. Does the method of editing customers' orders slow down their processing? If the order editor must enter a good deal of special information, can this operation and the subsequent typing be speeded up by using a supplementary order information form on which the most common instructions are printed and need only be checked by the editor?

13. Where sales orders are written in the field, can order editing at headquarters be reduced by training the originating personnel to use a standard form, uniform nomenclature, and so on?

14. Is every order sent to the traffic department for insertion of routing instructions, or has this process been simplified by the establishment of standard routings for regular customers?

15. If the volume of customer inquiries about order status is large, can it be reduced by having the order department send field sales offices a regular report showing the extent to which shipments are behind schedule on each product?

16. Must every manufacturing order be self-contained or can files of design specifications, bills of materials, and blueprints be maintained in the various shop departments? (This arrangement may not work satisfactorily where the volume of design or methods changes is large.)

17. Where manufacturing orders must contain a number of long, standard clauses covering such matters as testing and crating, use of substitute materials, and inclusion of repair parts, can constant typing of this information be eliminated by the use of rubber stamps or by the attachment of printed sheets on which the applicable clauses have been checked? (The same simplification technique can often be applied to sales orders, shipping papers, or invoices that contain recurring clauses about packing, shipping, or marking; guarantees; substitution of grades; certifications; and so on.)

18. To simplify the development of gross-margin figures by responsibility, can sales be costed out on an average-gross-margin basis rather than on an individual-invoice basis?

19. Have the requisitioning, purchasing, and receiving of items regularly carried in inventory been simplified by the adoption of a traveling requisition card that contains all fixed information on each item, such as name, specifications, code number? Thus, in reordering the items, storeroom clerks need to enter on the card only the quantity and desired delivery date before sending the card to the purchasing department. The advantages of this system are that

    a. It frees the requisitioner from the repetitive typing or writing of recurring information.

    b. It eliminates the need for maintaining records in the purchasing department on commodity purchase histories or sources of supply.

    c. It safeguards against the purchase of incorrect items which might otherwise have resulted from incompleteness in specifications or errors in transcription.

20. Does the accounting system for maintenance, repair, and operating supplies make special provision for low-cost, widely used items? Has the system been simplified by charging to an expense account, at the time of acquisition, any item that meets the following conditions?

    a. Items that have a low unit-of-issue value—say of 35 to 50 cents

    b. Items that are used in so many departments throughout the plant that any method of prorating the accumulated charge-off results in a reasonably equitable distribution

    c. Items that are purchased at fairly regular intervals, which tends to ensure a fairly uniform purchase volume so that no one month is charged with an unusually heavy expense

d. Items that are not typically withdrawn in large quantities (for example, a withdrawal ordinarily does not exceed $10 in value)

On product inventories, can the same practice be adopted on small-cost parts such as bolts, nuts, pins, cap screws, rivets, plugs, washers, and small electrical parts? Or, as an alternative to expensing these items at the time of acquisition, can the company carry them in the inventory account and have the foreman of each department requisition a month's supply of each item at one time? In this way the department that uses them is charged for the items, but the charge is made to a departmental overhead account instead of to many individual orders or manufacturing lots.

21. To reduce payroll verification work, can payroll hours be accumulated by rate and then each rate group extended for an overall check?

22. Can the distribution of maintenance and other service labor costs be simplified by using an average or standard hourly wage rate? Under this arrangement, any unabsorbed balance might remain as a residual charge in the appropriate maintenance department, or it might be charged to general plant expense.

23. Can time reports covering salaried personnel be limited to employees who have worked more or less than the regular schedule of hours?

24. Can the errors in accounting work sheets be more easily detected by means of a standard list of checking operations arranged in a sequence that reflects the probable frequency of error?

25. Does the invoicing procedure take advantage of the following opportunities to simplify invoice pricing?

a. To avoid chain discount computations, print the price list or price book at the amount actually charged to the customer (that is, at the net after discount). Where this cannot be done, at least simplify the extension work by providing a table of chain discount equivalents.

b. Print prices on package merchandise according to the unit of sale (package size) rather than weight (ounces, pounds).

c. Precalculate prices for quantities frequently ordered, and enter them in the price book.

d. If a great many items in the product line have the same unit price (for example, paints of different colors), combine them into a single class on the invoice and make only one price extension for the entire class.

26. Can the maintenance of order-invoice files be simplified by having only one alphabetically arranged customer file and inserting material in it only once for each shipment? The customer's order can ride with the pricing copy and finished copy of the invoice, and then all documents can be filed simultaneously in the customer folder.

27. Is report information standardized for economy and ease of consoli-

dation at higher levels? Are the chart of accounts and methods of recording basic data designed to facilitate the preparation of reports? Can the cost of preparing a report be reduced by changing the underlying records or source documents, or the manual or mechanical methods used?

28. Has an imaginative job been done in developing special tools and work aids that speed up clerical work or reduce errors? A number of these devices were mentioned in Chap. 9. They include codes, tables, charts, visual reference panels, sorting or collating racks, counting or measuring devices, templates for performing or checking clerical operations, and fixtures for positioning work, supplies, or other tools. Many of these items are not available commercially but must be specially designed. For this reason, if analysts see the possibility of applying any such tool, they must generally depend on their own ingenuity to develop it. They will find that the best opportunity for using such devices arises when highly repetitive manual operations—sorting, computing, referring, checking, or stamping—are performed in large volume. This kind of situation should be a signal to look for the possible application of special tools.

CHANGING THE TIMING OF OPERATIONS

While project team members are searching for ways to simplify the performance of individual operations, they should also analyze the *timing* of those operations. Timing in this sense involves three elements: the sequence of operations, their frequency of performance, and delays in the flow of work.

Sequence of Operations

The speed or cost of a procedure can be greatly affected by the order in which its individual steps are performed. A poorly planned sequence may result in extra work or more waiting time between operations. It may require the redoing of work that was initially done too early in the routine. It may lead to the shuttling of papers back and forth in costly rehandling operations. Or, it may increase the total time spent on makeready and put-away work.

The following list illustrates these dangers and suggests a number of ways to surmount them by changing the sequence of operations.

1. Is the timing of checking operations arranged so that the number of checks is held to a minimum? For example, in invoice preparation it is customary to recalculate extensions and totals as a check on the original computations. If this check is made against the invoice work sheet instead of the typed invoice, then the invoice must be checked separately for typographical errors. But if the checker recalculates extensions and totals on the *finished invoice,* then he or she can pick up both computing and typing errors in one operation.

2. Are expensive means of communication being used when, by changing the sequence of operations, less costly means would serve as well? One company spent many thousands of dollars a year to telegraph notices of shipment and

freight car numbers to its customers. Each of its Midwestern plants airmailed a daily list of its shipments to sales headquarters in the East. The headquarters staff in turn sent telegrams to the customers. After following this practice for several years, the company realized that it could achieve the same results and save the expense of wiring by having the plants airmail shipping notices direct to customers on the day shipment was made.

3. Is there any avoidable backtracking of work? At how many different times does each person handle a given document? If more than once, can all the operations of each person be performed in an unbroken sequence? If not, can some operations be reassigned to prevent backtracking?

4. Is sorting being done at the right time or must the sort be destroyed to perform an intermediate operation and then re-created? Analysts should pay close attention to the scheduling of sorting work. They will usually find that sorting more than once into the same category or breakdown can be avoided by changing the sequence of operations.

5. Are reports prepared in a sequence which pyramids upward—that is, can the summaries developed in one report be used as the starting point for another? Here the analyst's purpose should be to reduce to a minimum the number of reports that require the preparer to build up summary information from the original source documents or detail cards. For example, if product sales must be summarized by individual customer, can the results be posted to a peg strip and then crossfooted to obtain product totals by sales representative?

6. Are forms being inserted in typewriters at more than one point in the procedure cycle, even though all the required information could be entered during the first insertion?

7. Where shipments are made from stock, are stock records referred to at the right point in the procedure cycle so that subsequent clerical work is minimized? For example, if limited quantities are carried in stock, the clerk should check the stock record for availability of the items ordered (and post the withdrawal to the record) *before* preparing the order-invoice form. In this way, the need for retyping incorrect invoice forms will be eliminated. But if out-of-stock items are few in number, the clerk can scan incoming orders for these items, back-order them, and proceed with order filling or invoicing without referring to the stock records at all. At the end of the day, the clerk can summarize all invoices by item and make a single withdrawal entry, instead of many, to the card covering each item shipped. Where invoices are summarized daily to get product sales figures, these totals can also serve as summary postings to the stock record.

8. Can physical inventory verification be simplified by making the count of each item at or near its minimum level?

9. By changing the sequence of steps, can related operations be combined, thereby reducing makeready and put-away time? Coding, for example, though usually performed separately, might easily be done during some other operation where the worker has to refer to a file or record suitable for carrying the

code numbers. This technique of combining essential operations by changing their sequence can be widely applied and analysts should keep it in mind as they study individual operations.

Frequency of Performance

In addition to determining whether an operation is performed at the right time in the procedure cycle, the analyst should ask, "Is it done more often than necessary?" This type of inquiry may lead to curtailment of work, even though the operation cannot be eliminated. Following are some examples and suggestions.

1. In one company, each time a shop department required a manufacturing blueprint, reproduction machine operators removed the appropriate tracing from their files, put it through the machine to make a single print, and then refiled the tracing. Analysis showed that 45 percent of the prints requested were reproduced from only 6 percent of the tracings, and that on any given part or subassembly as many as twelve prints might be needed within a 3-month period. On the basis of these findings, quantities of prints to be carried in stock were established for the most active parts. This step materially reduced the frequency of reproduction on many tracings and saved a good bit of makeready and put-away time.

2. The frequency with which the physical inventory is verified can often be reduced with no harmful effects. It may, for example, be safe to count each inventory item only once a year, or only as often as is necessary to keep reasonable control of stock through reconciliation of physical quantities with detailed records and dollar control accounts.

3. In the maintenance of continuous inventory records, the common practice is to compute and enter a new stores balance each time a withdrawal entry is posted to the record. Yet some companies find that they can save clerical labor by limiting withdrawal postings to date, requisition, and quantity, and by computing a new stores balance only after ten or twelve such entries. (Balances are figured more frequently, of course, when the reorder point is being approached.)

4. The frequency of purchase requisitioning can usually be cut by setting larger reorder quantities on fast-moving, relatively low-value items. The objective should be to establish reorder quantities on all stock items at a level that achieves the best overall balance between cost of acquisition (which encourages larger orders) and cost of possession (which encourages smaller orders). The analyst should remember that acquisition costs include not only those incurred in ordering and purchasing, but also those associated with inspection, laboratory certification analyses, receiving, stock handling, recording, and invoice payment.

5. As a closely related type of change, the analyst should determine whether purchase requisitions for similar items can be accumulated for a week so that more items can be entered on an order.

6. A heavy-equipment manufacturing company that received most of its inbound shipments freight-collect decided to avoid daily payments of freight bills by providing its principal carriers with an imprest fund, which it replenished biweekly.

7. A final illustration of this type of change is a reduction in the frequency of report preparation. Can some information that is asked for weekly be prepared monthly, quarterly, or at irregular intervals, and still serve the purpose? (This question may well have been raised during an earlier part of the diagnosis, for when analysts ask, "Is this report necessary?" they presumably also ask, "If it cannot be eliminated, does it need to be prepared so often?")

Delays in the Flow of Work

The various kinds of improvements already discussed in this chapter will do much to accelerate a whole system or procedural cycle. However, after project team members have taken steps to improve the frontline organization structure, eliminate or combine work, and rearrange the sequence of operations, they may find that the process under study still has not been speeded up sufficiently to satisfy the requirements. This shortfall might occur, for example, in (1) the receiving, inspection, and accounts payable routines, which affect the earning of time discounts on suppliers' invoices, and (2) a production planning system that depends on fast summarization of customers' orders as the means of minimizing finished-goods inventories without impairing customer service.

When the need or opportunity for faster action still exists at this stage of the analysis, the best approach is to chart the whole procedure backward from the deadline date. Then determine where the real delays are occurring and when each part of the process must be completed if the final deadline is to be met. The section on customer service in Chap. 10 suggested some shortcuts the analyst should look for in making this diagnosis. Here are some additional suggestions:

1. *Eliminate bottlenecks.* If any obstruction exists in the flow of work after the procedure has been simplified and the methods have been improved, the cause of the bottleneck is likely to be poor distribution of workers among the various jobs, insufficient equipment, or irregular work requiring special treatment. Where routine transactions are being slowed down by nonstandard ones, a possible solution is to set up a separate group for handling only the nonstandard work.

2. *Eliminate peak loads.* Business organizations have devised a variety of techniques to level peak workloads. Two outstanding examples are the cycle billing plan, which has been adopted by most department and specialty stores, and the preparation of factory and office payrolls at different times of the week, which is a common industrial practice. Another example is spreading voluntary deductions from payroll checks over weekly pay periods of each month rather than handling all such deductions in a single week. Thus union dues may be deducted the first week of each month, savings plan contributions the second

week, and so on. Still another technique is that of scheduling purchase requisitions by inventory classes throughout the weeks of each month. These methods of staggering work have wide application and should be explored whenever a group's volume of work fluctuates heavily during the week or month.

3. *Advance the cutoff date.* One way of speeding up month-end financial and management control reports is to close the accounting books earlier. At first glance this may not seem particularly helpful since it does not shorten the period between the cutoff and the availability of reports. It simply moves both steps forward by the same number of days, and those days' transactions are then held over for the next month's reports. But many companies will find that although the great bulk of their transactions are recorded within a few days after the close of the month, their books are being held open for several extra days only to pick up stragglers. These latecomers are usually not large enough in volume to affect the reports. By postponing them until the next month, control information can in fact be given to management more promptly.

4. *Perform part of the work ahead of schedule.* Month-end reports can also be speeded up by doing as much preparatory work as possible before all the figures are available. Some accounting entries, such as depreciation, insurance, weekly payrolls, and other accrued expenses, can be disposed of before the other closing work begins. Parts of report forms can usually be filled out in advance—for example, columnar headings, product or department names, profit and loss or balance sheet captions, and figures used for comparison with actual performance.

In introducing these and other measures to speed up the flow of work, the analyst must be careful that the increase in speed does not cost more than it is worth. In general, the faster that work flows through a procedure, the smaller are the backlogs at any point. And the smaller the backlogs, the greater the risk of incurring idle time because of lack of work. The analyst's job is to achieve the proper balance between these two extremes.

MECHANIZING THE ROUTINE

The mechanization of office routines can make a major contribution to paperwork speed and accuracy, and therefore to the effectiveness of line operating departments (covered in Chap. 10). In addition, it can have a major impact on clerical and other administrative costs, an important potential benefit that will be explored in this section. The discussion here does not attempt to describe the many types of equipment applicable to common clerical operations. Rather, it seeks to establish some guidelines for determining whether a routine should be performed manually or by machine, and, if by machine, what type, make, and quantity of equipment are most suitable. The basic principles in evaluating automation alternatives in general are set forth in this chapter. Application of these principles to the evaluation of computer-based systems is covered in Chap. 14.

In any study where analysts are considering the introduction of mechanical methods, they should bear in mind the following three principles:

1. Mechanical methods are not necessarily cheaper or faster than manual methods. Just because a given operation *can* be performed mechanically is no reason for doing it that way.

2. The choice of methods must be based on *what work needs to be done,* not on what a machine is capable of doing. This point may seem elementary, yet failure to heed it is probably the greatest single cause of unsound or unnecessarily expensive office machine applications.

In witnessing machine demonstrations, office executives are too often influenced by feats of mechanical performance that have little relation to their own particular needs. They are unduly impressed by the speed, capacity, or versatility of a piece of equipment. They develop illusions about the value of new or additional information that the machine is capable of producing. These lures, when coupled with a pervasive desire to be modern, induce in them what another writer refers to as "mechanical hypnotism."

The result is that a good many economic errors are made in the selection of basic methods. Machines are being used where simpler and cheaper manual methods could produce a less elaborate but equally satisfactory result. In other cases the wrong kind of equipment has been chosen—for example machines with greater capacity than is required or machines with multiple functions, special attachments, or automatic features that cannot be fully utilized or that encourage the performance of unnecessary work.

An important point to be stressed in this connection is that office equipment salespeople cannot be expected to figure out the user's requirements. The sales representative's primary function is to make clear what a machine does and how it operates. The user must determine what information the organization needs, how accurate it must be, and how rapidly it must be produced. And decisions must be reached on these points *in advance* if the selection of methods is to be approached in any kind of a rational way.

3. The fact that certain operations in an area are already mechanized should not lead the analyst to assume that all possibilities for mechanization are exhausted. One company, which had been preparing its shipping labels for several years on addressing machines, discovered that it could reduce the work force performing this operation by 50 percent simply by installing automatic paperfeed attachments on the machines. Labels were formerly fed through the machines by hand at an output of 800 per hour; the automatic feed increased production to more than 3,000 per hour.

Once clear decisions have been reached on the information, speed, and accuracy that are needed, the choice between manual and machine methods or between two types of machines should be based on the following factors: comparative costs, comparative speeds, quality of the work produced, flexibility of equipment, ease of operation, effect on the labor force, and reliability of equipment.

Comparative Costs

This factor is usually considered first, since it is of primary importance and since many of the advantages of one method over another are reflected in their costs. The costs of a proposed machine method are composed of the following elements.

1. Recurring costs
   a. Labor
   b. Machine rental
   c. Machine maintenance
   d. Operating supplies, such as printed forms, carbon ribbon, chemicals, film, electric power
   e. Building occupancy or floor space
2. Nonrecurring costs
   a. Machine purchase price depreciated over a predetermined period of years.
   b. Initial setup and conversion costs amortized over the same period of time. These are the costs of making the study and converting from the former method to the new method.

In contrast to these elements, the costs of a manual method are composed of labor, operating supplies, and space.

Of all cost elements, labor is usually the largest, and it is always the most difficult to assess accurately. The only precise way of measuring it is to list every detailed operation, determine the hours required to do each, and multiply them by the hourly rate for the class of worker involved. This approach takes time and effort, but the results are worth the expenditure. Too often the question of labor cost is settled on no more substantial grounds than the assurances of an equipment sales representative that a given machine will "save the work of two operators."

In computing comparative labor costs, analysts must be careful not to attribute economies to the machine that in reality result from other improvements developed simultaneously. In other words, the cost of a proposed new method should never be compared with the cost of the existing method unless the latter is unquestionably as simple and efficient as it can be made. Therefore, to make a sound cost comparison, analysts should first simplify the present routine as much as possible. By doing so, they may find that the proposed new method, which appeared so attractive at first, is clearly uneconomical.

As a further guideline in comparing alternative methods, analysts should remember that proposed methods which exist only on paper are always highly idealized. They seldom reflect the exceptions, the tributaries to the mainstream of paperwork, or the clerical mistakes and inefficiencies that are built into the cost of the present routine. To avoid this error trap, the administrative vice president of one company insists that the estimated costs of proposed methods and procedures be arbitrarily increased by 15 to 20 percent before they are compared with present costs.

In computing the labor cost of an office machine application, analysts should remember two additional points: (1) to include an adequate allowance for setup or makeready time; and (2) to use practical operating speeds, not maximum machine speeds, in setting unit times for various operations. If they do not know what speeds to use in their calculations, they should visit the display room of the equipment manufacturer or distributor and arrange tests of operations similar to those making up the system under study.

Most of the cost elements other than labor can be determined quite readily. The following suggestions will be useful, however, in making sure that all factors are considered:

1. Occupancy costs need not be included in either set of figures, unless the space requirements of one method vary substantially from those of the other.

2. In determining the initial investment on which depreciation is to be computed, the analysts should not forget to include the purchase price of any standby equipment that may be required or the cost of any permanent alterations to the building, such as the erection of soundproof partitions. Questions that arise about the rate of depreciation to use can be answered by members of the accounting or tax department.

3. Whether the new method being considered is a manual or a machine operation, the cost comparison should reflect the nonrecurring expense of making the changeover (for example, the expense of converting records or of retraining personnel).

4. When mechanization saves the full time of one or more persons, another factor to be taken into account is the possible reduction in hidden costs. Under this circumstance the saving may include not only clerical salaries but also payroll taxes, insurance and pension benefits, etc. Where a sizable personnel reduction is possible, the savings estimate should also reflect some decrease in supervision.

As a final guideline in evaluating a proposed new method, analysts should not be content merely with determining that it will, in fact, save some money. They should also express the estimated saving as a rate of return on the investment made. That is, they should calculate the annual saving in operating costs as a percentage of the one-time cost of purchasing the equipment and changing over to the new methods. Then they should ask: "How does this return compare with the company's demanded rate of return on proposed investments in new production facilities? When will we realize this rate of return?"

Comparative Speeds

Theoretically, variations in speed between two methods are reflected in their comparative costs. Once the time requirement for performing a system or procedure has been established, the assumption is that either method will be staffed and equipped to meet that need. Even though the output per person is greater under machine than under manual methods, the unit using manual methods will presumably meet the required deadline by employing more workers. The higher labor costs will, of course, show up in the operating cost figures.

For two reasons, however, this line of reasoning does not always hold true. (1) There is a practical limit beyond which any manual routine cannot be staffed without risking considerable idle time. (2) Not all activities can be speeded up simply by adding more pairs of hands. An illustration is mailing customers' orders from the sales department to the plant compared with teletyping them.

For these reasons, analysts must usually consider the speed of performance under each method. But the important point they should remember is not to judge speed as a thing in itself but only in relation to the requirements to be met. If one method is faster than another but both are fast enough, speed should be given little weight in the decision.

Quality of Work Produced

Under a machine bookkeeping method of posting, the automatic computing and printing of new balance figures are more likely to produce an accurate record than manual posting is. By reproducing successive documents from a single duplicating master, errors in transcription are eliminated. Thus, the quality of work produced is an important intangible in the choice of methods. It is reflected not only in the accuracy of results but also in appearance, completeness of information, and so on.

Flexibility

To determine how quickly and thoroughly a mechanical method can be adapted to changing requirements, the analyst should ask questions like the following.

How flexible will the machine installation be in handling peak loads? Is it likely to become a bottleneck because of a limited number of machines and trained operators, or because the work cannot be broken down and distributed among many people as manual work can? How frequently are these peaks likely to occur, and what will their probable size be? Are the machines sufficiently flexible to handle unavoidable exceptions to the normal routine?

Under the various methods being considered, how much flexibility will the company have in cutting the costs of the system or procedure if volume should drop sharply? How easily can typical requirements be met under each method? For example, for each of the various types of duplicating masters available on the market, how easy is it to add information manually, to correct the master, or to block out information in reproducing copies? If a major change should be required, how fast could it be made? As an illustration, for each alternative method of payroll preparation, how quickly could a general rate increase be incorporated in the master record from which payrolls are prepared?

Ease of Operation

How difficult the machine is to operate may have been reflected in the operating cost figures. Ordinarily the more specialized the skills required of the operator, the higher will be the wage rate. But in addition to the cost aspect of this factor, the analyst should also consider the possibility of labor turnover and the time required to train new operators.

Effect on the Labor Force

How is the clerical force likely to react to mechanization of a manual operation? In weighing this factor, the analyst should heed its long-range implications more than its immediate ones. Most workers feel some momentary uneasiness about learning new techniques, or some fear that mechanization means large-scale labor displacement. These reactions will soon pass, however, once the newness of the system wears off. Thus the matter of real concern is what attitudes the operators will have toward the new method when they become accustomed to it. They may be neutral, favorable, or unfavorable. For example, they are likely to feel some dissatisfaction if the new method is noisier or more tiring than the old, or if it uses inks or dyes that soil the hands or clothing. On the other hand, workers usually favor a new method that gives them a salable skill or eliminates some of the drudgery of repetitive manual operations.

Reliability of Equipment

Finally, in considering mechanical methods, the analysts should try to find out as much as possible about the mechanical reliability of the equipment. If they have had no previous experience with the manufacturer's equipment, they should be sure to talk with several other users about their machine breakdown experience and the promptness of the supplier's repair service.

CAPITALIZING ON OTHER COST-CUTTING OPPORTUNITIES

To conclude their work on a project aimed at reducing clerical and other administrative costs, project team members should turn their attention to the four specialized areas of office layout, forms design, filing systems, and supervision of clerical labor. These factors may have been considered somewhat tangentially in earlier parts of the study but now they need to be attacked directly because of the impact they can have on office costs.

Revising the Office Layout

By means of the organizational changes and the work elimination or combination already discussed, the systems staff has presumably cut down the *number* of times work moves from desk to desk and from department to department. Having reduced the number of moves to a minimum, project team members must now see what they can do to shorten these moves. In other words, they must revise the office layout to cut down the *distance* traveled by the work and by the workers. This is, of course, not the only objective of layout studies. Others include conservation of space, improved supervision of the work force, isolation of noisy equipment, and provision of better natural lighting. But ordinarily the analyst's principal purpose in rearranging the layout is to speed up the flow of work through the routine and to reduce the time lost by workers in moving about the office. This objective can be achieved through the following steps.

   1. Eliminate crisscrossing and backtracking of work
   2. Locate each worker as close as practicable to
      a. The person from whom he or she receives work
      b. The office machines, reference facilities, or files that must be used or the persons who must be consulted in performing the work
      c. The person to whom the worker gives completed work

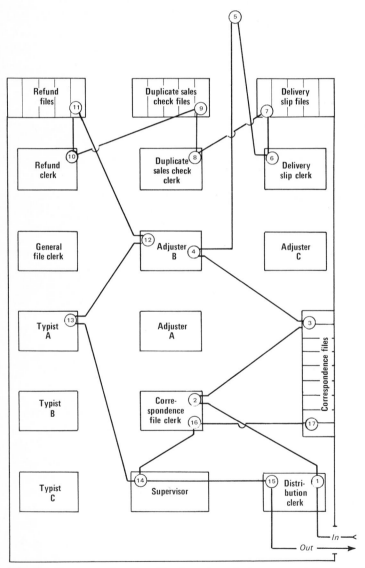

FIG. 25 *By rearranging the office layout, analysts can speed the flow of work.*

Figure 25 illustrates the application of these steps to an actual layout. Compare this "after" chart with the "before" chart of the same area on page 175 to see the ways in which movement can be eliminated by rearranging facilities. The flow lines on the two charts show that in the processing of a typical complaint, the distance traveled was cut from 277 to 182 feet, a reduction of approximately 35 percent.

The same techniques of charting and analysis can be profitably used in making departmental location studies. Figure 26 shows in graphic form the volume of interfloor personnel traffic within the office of a company occupying a multistory building. The study revealed that during an average week, the employees of different departments had about 1,300 personal contacts. Of these contacts, about 900 or almost 70 percent were between floors—from one to eight floors apart.

On the basis of this analysis, the company gave up its space on one floor of the building, acquired additional space on another floor which it already partially occupied, and made a number of changes in the location of various departments. The combined effect of the changes was to reduce the number of interfloor contacts by 65 percent. Moreover, the new arrangement helped to improve interdepartmental coordination by facilitating important contacts that had been neglected in the past.

Redesigning Forms

The next step in the wrap-up phase of the project refocuses attention on the printed forms used in the procedure under study. When this point is reached, the analyst has already determined what existing forms or copies of forms or information on forms can be eliminated. The remaining job is to improve the design of the essential forms in a way that facilitates their preparation and subsequent use. The design of every form has an important effect on the time required to fill it out initially, distribute it, add information to it, transcribe information from it, interpret it, file it, and find it. The number of minutes saved in processing a single copy of any given form may not seem consequential, but we have only to think of the tremendous volume of paperwork in any organization to realize what a cumulative impact better forms design can have.

Following is a checklist of the questions analysts should ask in testing the design of existing forms as well as proposed new forms.

1. *Is the form clear?* Is its title sufficiently definitive to indicate its exact purpose or use. Are all its captions self-explanatory, or is there likely to be some question about the specific information required? Is it clear who should fill in each space, box, column, or line? Is the routing or distribution of each copy plainly evident? Are the various copies identified by numbers, titles, or differently colored papers? If the form is used by many people throughout the organization, does it contain printed instructions wherever necessary to promote understanding and uniformity in its preparation? As an illustration, Fig. 27 shows the printed instructions appearing on the back of one company's employee expense report form. A note on the face of the form calls attention to these instructions.

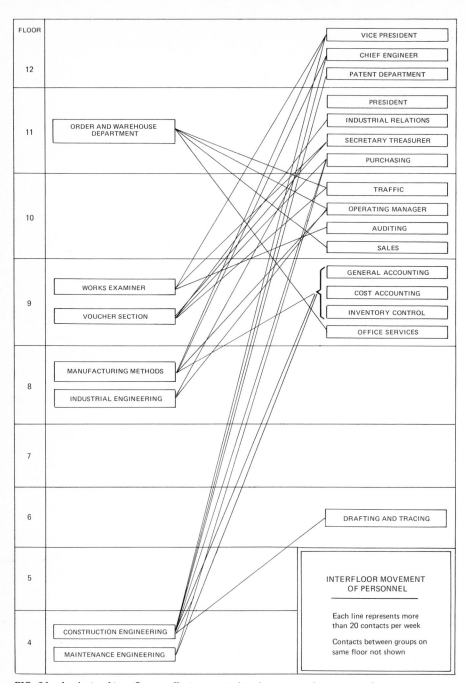

FIG. 26 *Analysis of interfloor traffic is essential in department location studies.*

INSTRUCTIONS FOR FILLING OUT THE EXPENSE REPORT

General

1. If you incur expenses daily, send in your expense report at the close of each calendar week. If you incur expenses only now and then, hand in your report not later than one week following the date of the expenditure.
2. All major items of expense must be supported by receipts: Attach to your report such receipts as those received for hotel service, for railroad and airline transportation, etc.
3. Make a copy of the report for yourself—you'll find it a convenience for answering questions or for checking against the accounting department's records.

Dates

1. The date in the upper right-hand corner of this report is to be the one on which you hand in your report.
2. After "from" and "to" on the second line, put in the first and last dates of the period for which you are reporting expenditures. All expenses for that period of time must be included.
3. Date column: Show the date on which each item of expense was incurred. If more than one item was incurred on the same date, don't repeat the date.

Location

Use this column to show the name of each city visited. Be sure the correct location is indicated for each item of expense. However, when several items were incurred in one city, don't repeat the location.

Nature of work

Use this column to state, in general terms, the nature of each work assignment. Show the calls made; when you make calls on several firms, list the name of each firm.

Description of expenses

1. In this column, itemize and briefly describe each item of expense. Don't combine different kinds of expenses in one item.
2. When you report transportation expenses, specify the kind of transportation—bus, train, airplane, automobile, etc.
3. When you report transportation by automobile, support the amount claimed by showing the number of miles traveled multiplied by the mileage rate.
4. Even though the personnel traffic representative paid for your transportation tickets, enter that information in this column. For example, "R.R.—New York City and Washington through P.T.R., $19." Do not enter the cost of these tickets in the "Amount" column.
5. Airline transportation purchased on scrip cards should be listed in this column. For example, "Washington and New York City—scrip." Do not enter the cost of this transportation in the "Amount" column.

Amount

Enter in this column those amounts of money you actually paid out.

Distribution of expense

Enter in this block the correct account number. (The account number consists of a two-digit control account number plus a two-digit subaccount number. Control account numbers applicable to each department are listed in the "Departmental Responsibilities" section of the Accounting SPI Manual; subaccount numbers are listed in the "Expense Subaccount" section of the Accounting SPI Manual.) If the items on this report are to be charged against an authorization order, be sure the correct A.O. number and product code number are entered. (For list of product code numbers, see "Profit and Loss" section of Accounting SPI Manual.

FIG. 27 *Printed instructions help employees fill out forms properly.*

2. *Is the amount of writing held to a minimum?* Have words or clauses that must be used repeatedly been printed on the form so that only the variable information needs to be filled in? This technique can be adopted even though the printed information does not apply to every transaction covered by the form. For example, a company having few items in its product line might print the names of all products on its order, invoice, and bill-of-lading forms. Thus, the items covered by any one copy of these forms would be specified by the quantity figures entered ahead of certain product names. An equally common way of reducing the amount of writing is to provide boxes for checking any of several printed captions rather than to print questions that require word entries.

3. *Do the location and sequence of information facilitate preparation and use of the form?* If the information on a form is copied from or posted to another form, are the items on both forms arranged in the same sequence? For example, if customer master information cards are used in typing sales orders, the card record should be virtually a facsimile of the order form. Or if data on the form are to be keypunched into cards or tapes, the form should lend itself to good punching sequence. Where more than one person enters information on or uses the form, are the items that concern each person grouped together to speed up writing, analysis, or reference? (This requirement cannot always be fully met, since the information referred to by one user may be entered by more than one clerk. In such a case, preference should be given to the person who would lose the most time by poor arrangement.) If each space on a form does not always need to be filled in, are the most frequently used spaces at the top of the form? Is the file reference information positioned according to the method of filing or binding?

4. *Is the form well spaced?* Is enough room allowed for each entry so that the form can be completed without crowding, unintelligible abbreviations, or incomplete information? Are the vertical dimensions of spaces related to the method of filling in the form—say, by hand or machine? If the form is to be typewritten, has the printing been arranged to minimize the number of starting points and thus permit maximum use of the tabular key? If the form is to be bound at the sides or top, are adequate binding margins provided?

5. *Is the size of the form right?* Is it so small that it is difficult to handle or likely to be lost? Is it too large for convenient filing? Is it a standard size that can be printed economically? If it is to be inserted in an envelope, are its size and shape such that no fold or only one fold is necessary?

6. *What other features have been, or could be, added to facilitate preparation or use of the form?* Where multiple-copy forms are used in large volume, are they printed as continuous forms to eliminate repeated insertion and alignment in the typewriter? Where desirable, are spaces on the form numbered to facilitate reference or to relate the parts of the form to written instructions covering its use? Are colored papers used to indicate where or when the form was initiated? Has the distribution of copies been planned so that the persons who use the form most receive the original or clear carbon copies? If the form is to be mailed, is it designed for insertion in a window envelope?

Improving Filing Systems

Filing operations generally receive little attention in systems and procedures studies. Yet 10 to 15 percent of the entire clerical cost of a process may be spent on filing or on searching files for the documents the system has produced. By simplifying filing and finding operations, the analyst can not only save clerical and supervisory time but also encourage the use of valuable information that lies buried in the company's files.

The following paragraphs suggest some of the key points the analyst should consider in reviewing the method of filing individual documents—whether they are cards, printed forms, or correspondence. The reader who wishes to examine the principles and techniques of good filing in greater detail will find a number of excellent books devoted to that subject.

Since the purpose of any file is to facilitate the finding of a document when it is needed, the type of filing system used should be governed by two factors: (1) the information by which the filed material is most likely to be identified, and (2) the probable frequency of reference to the file. The second factor has, of course, an economic bearing on the problem, for the more frequently a file is referred to, the more money can profitably be spent on clerical labor and equipment to maintain it.

1. *Simplicity.* The first requirement of any filing system is that it be simple to understand and operate. This means that any document should be easy to classify for filing and easy to find after filing. Numerical filing, for example, is usually faster and more accurate than alphabetic filing and should, therefore, be used whenever either alternative is feasible. The analyst should also be on the watch for involved methods of classifying or indexing, complex cross-reference techniques, or other special features that are likely to increase the volume of error or require the use of highly trained operators.

2. *Indexing.* The speed of finding a given file position (either to insert or remove a document, or to post to a card record) is determined almost entirely by how well the file is indexed. The most complete form of indexing is a card file or loose-leaf binder in which the reference margin of every card or sheet is visible. The cost of this kind of equipment is usually justified whenever a large volume of references or postings is made to each form before it is permanently removed from the file.

In so-called "blind files," such as tub files, wheel-attached cards, and drawers of addressing-machine plates or stencils, visible indexes should be provided for every twenty to thirty units, depending on how fast the file is expanding. Fewer index dividers may be needed in a conventional file of manila folders, provided that the reference tab on each folder is clearly identified and visible.

3. *Posting to record files.* In studying files of records to which postings are made, the analyst should consider whether the type of file is suited to the method of posting. If the record is posted by hand, can the entry be made without removing the card or sheet from the file? If posting is done by machine, can the form be removed easily?

4. *Folder filing.* Even when the finding of individual folders has been speeded up by better indexing of file drawers, much time can be wasted in filing documents in the folder or removing them from the folder. Ideally, the file clerk should be able to perform these operations without lifting the folder from the file drawer. For this reason, the practice of fastening documents to folders should be avoided wherever possible. As long as a folder remains in the file, its contents are not likely to be lost or disarranged. When it is removed to be sent to another department, the material can be quickly clamped to the back of the folder by a binder clip.

Other aids to fast filing are individual hanging folders and hanging cloth jackets, each holding four or five manila folders. With either device, folders remain erect with their reference tabs plainly visible, and at the same time the file can be kept loose enough to permit fast drop filing.

In examining the individual folders that make up the file, the analyst should also consider in what sequence papers are placed in the folders. The possible alternatives are (1) to drop each document at the front or back of the folder, and (2) to interfile the document in some predetermined order among the papers already in the folder. Obviously the first alternative results in faster filing but slower finding than the second. The choice, therefore, should be determined by the probable frequency of reference to the file.

As an illustration, assume the following conditions:

    a. That a correspondence file contains folders not for ididual corre-spondents but for small alphabetic breaks, such as Aa–Ag, Ah–Al

    b. That *interfiling* every letter within the proper folder takes 5 seconds more than drop filing

    c. That finding a letter in an unalphabetized folder takes 20 seconds longer than finding a letter in an alphabetized folder

    d. That an average of 10 percent of the letters in all folders are referred to after being filed

Under this set of conditions, drop filing would obviously be the more economical practice, since the increased finding time for every 100 letters in the file would be only 200 seconds, while alphabetic interfiling of these 100 letters would have taken 500 more seconds than drop filing. The net saving of time decreases, however, as the percentage of references increases, and the advan-tage swings the other way when more than 25 percent of the letters are referred to.

    5. *Filing or posting sequence.* Finally, the analyst should find out whether the documents to be filed or posted to a card record are first sorted in the same sequence as that in which the file or record is arranged. Unless the file or record is small, this practice will usually save more time in filing or posting than that lost in presorting.

Strengthening Management of the Clerical Labor Force

Once the work to be done has been fully streamlined and the most efficient methods have been developed for performing the essential work, the final prerequisite to minimizing clerical and other administrative costs lies in ensur-ing that the ideal system now developed is carried out effectively by a skillfully managed clerical force. Meeting this prerequisite requires (1) optimizing the division of clerical labor, and (2) supervising the hour-by-hour work of the clerical force.

Optimizing the division of clerical labor   At this point, project team members should ordinarily prepare work-distribution charts (see Fig. 15, page 165) covering the activities and tasks to be performed after all the recommended changes have been adopted. These charts will help them develop further recommendations on how many people will be needed and how the various tasks should be divided among them. Although this step will not be necessary if few major changes are being recommended, the analyst should at least reexamine the original work-distribution charts to see what improvements, if any, can be made in the existing division of labor.

Determining the best division of tasks among a clerical force is no easy matter, for in most situations the analyst will find strong arguments favoring both specialization and flexibility in the use of personnel. Since these arguments have different weights in different situations, all that can be done here is to suggest an approach that will help the analyst reach well-balanced decisions.

As a starting point, the project team should apply two tests to the existing or proposed division of labor:

1. Is the most economical grade of labor employed on each task? In any office operation, some tasks require the exercise of judgment or the use of specialized skills. Others require extreme care, because the results of error can be costly. Still others are easy to learn and demand little concentration or skill. To obtain maximum economy in the use of clerical labor, tasks must be grouped according to skill and experience requirements so that salary levels reflect the real value of work performed.

2. Are several different tasks requiring the same grade of labor being performed by each of several clerks? Suppose, for example, that three tasks within one organization unit require different types of eye-hand coordination or manual dexterity, but that all three require an equal *grade* of skill and thus command the same salary. If three operators are needed for the total work, should each task be fully assigned to one operator, or be divided among the three? At least in principle the first alternative should be chosen, since greater specialization ordinarily means greater proficiency. In addition, less time is wasted when each operator can stick with one task instead of jumping from one to another.

But these two guides are deceptively precise; in actual practice few problems in the division of work are as clear-cut. The detailed clerical operations of any one unit may require many fine gradations of skill or experience, so that it is difficult to say where the content of one job should end and another begin. Moreover, in applying these principles of specialization, the analyst must also weigh the arguments favoring diversification of work:

1. Workers need a change of pace to minimize job fatigue or monotony.

2. Time can be saved when a transaction passes through fewer hands. The more time each operator must spend in becoming familiar with a given document, the more support there is for having the same person perform several related steps on a batch of documents instead of a single step on all documents.

3. Flexibility is gained by diversification. In smaller organizations, diversifying duties makes possible more effective use of the work force when some employees are absent or when the workload shifts.

4. Diversification reduces the need for special controls. The fewer the persons handling any one document, the less will be the temptation to set up special records to control its progress through the routine.

Supervising the hour-by-hour work of the clerical force Although this activity is not strictly a part of systems analysis or design, it is dealt with here because it plays an important role in the control of clerical and other administrative costs. Even with well-designed paperwork systems, an organization can have high office costs under the following conditions.

1. Sometimes the supervisory force is really not *managing* clerical labor but rather, is *doing the work of the group.* The only distinction between the supervisors and the group is that the supervisors handle the difficult, special, or troublesome cases while the workers handle the routine ones.

2. The supervisory force may not be approaching its job in a *planned, positive* way designed to optimize worker productivity. Rather, it takes the passive approach of answering workers' questions and handling trouble as it arises.

Against this background, the key understanding to be built into the consciousness of a clerical supervisor is that she has a costly *set of resources* to manage—resources consisting of workers, equipment, and space. Her job is to use these resources in a way that achieves optimum efficiency or productivity over the long run. Basically, this means that she must do everything necessary to eliminate idle time beyond the workers' control, and to train and encourage her group to work more effectively. How the skillful clerical supervisor discharges these two responsibilities is apparent in the *way she spends her time* and the *conditions or events she keeps an eye on.*

The effective clerical supervisor spends most of her time on the following tasks.

1. Scheduling work and ensuring that the clerical complement in each unit is at the right level in relation to the anticipated workload—meaning that the available minutes match the forecast of needed minutes.

2. Watching and controlling the hour-by-hour flow of work and the availability of work at each work station. To keep on top of the work flow, she must spend a substantial part of her time each day out on the floor—not in her office.

3. Detecting and eliminating bottlenecks and minimizing the impact of work peaks.

4. Using work measurement standards to coach, retrain, upgrade, motivate, and otherwise increase the workers' productivity.

5. Checking on the quality of work (accuracy, completeness, length of time taken to handle a transaction, etc.).

6. Analyzing monthly reports on work efficiency and relating them to the personnel available to perform each function.

7. Monitoring the training of new workers.

To do these tasks well, the supervisor of a clerical unit needs specific information on four points.

1. The persons absent (scheduled or unscheduled) and tardy (including the amount of each tardiness).

2. The current workload. The supervisor should keep informed of any unusual work peaks during the course of the day. In addition, she should be

notified at the day's end of the number of unprocessed pieces of work of each type (that is, the work backlog for each type of transaction).

3. The skills available to perform each clerical activity making up the work of her unit. The supervisor should keep an up-to-date inventory that shows each job in the unit and the names of workers who have been trained to perform it. This inventory of skills is her key reference document in planning temporary reassignments of workers.

4. The kinds of transactions that may, if necessary, be postponed or backlogged, and the maximum length of time that each type may be postponed.

Armed with this sort of information, the clerical supervisor is in a position to keep a variable work force and a variable workload in balance. To maintain this balance at the lowest cost, she will ordinarily take the following steps in the sequence listed. (Some of these steps were covered in the section of Chap. 10 on speeding up service to customers. They are repeated here because they are as relevant to the reduction of clerical costs as they are to speeding up the flow of work.)

1. Shift workers temporarily to other jobs. This step requires that the supervisor plan and carry out a successful program for cross-training employees. The practice of shifting workers within the department can be extended to borrowing workers from other departments. For example, if cash receipts are heavy at particular periods of the month, the accounts payable unit may borrow personnel from the credit department during the peak period. Since the credit department ordinarily delays collection follow-up work until heavy cash receipts are posted, this solution often works to the advantage of both units.

2. Backlog postponable work.

3. Recruit a small corps of part-time workers to help out on the one or two heaviest workdays of each week.

4. Work overtime.

5. Hire temporary help from one of the clerical labor service organizations.

TWELVE

Improving Top-Management Effectiveness through Better Decision-making and Control Systems

With one notable difference, the analysis and the design or overhaul of a *general-management* system require the same steps as the *operational* systems described in Chap. 10. For both types of systems the analyst must

1. Define the end-result objectives of the system.
2. Identify the activities that must be performed in a superior way to achieve the defined objectives.
3. Determine the detailed steps required to carry out those activities.
4. Evaluate how well the steps are performed and, by this process, highlight shortcomings and improvement opportunities.
5. Identify alternative ways of correcting each shortcoming or capitalizing on each opportunity.
6. Evaluate each alternative and develop the recommended improvement program.

The one notable difference is this: To improve a general-management system, the analyst must bring to the task an understanding of the top-management decision-making process, which is by no means so precise or so readily definable as the functional activity underlying an operational system. It is, for example, a much more complex task to improve the information and processes by which management makes tradeoff decisions among competing requests for capital expenditures than it is to improve the systems that control inventory management or service to customers.

To illustrate the approach to be taken and the special problems involved in improving general-management systems, we have chosen the process of *stra-*

tegic analysis and decision making. Our reason is that this process concerns the most crucial activity that the members of general management perform. In addition, the design of an effective strategic planning system demands the highest levels of competence and sophistication likely to be required on any general-management systems project.

THE NATURE OF STRATEGIC PLANNING

As a beginning point, the nature of the strategic planning task can best be delineated by showing how it fits into the total responsibilities of a general-management executive—that is, the head of a company or of a self-contained subsidiary or product division. Expressed in its simplest form, the job of any such top-management executive consists of the two elements shown in Fig. 28. The first is the executive's *operating* responsibility—his ongoing job of guiding the enterprise in becoming more effective and efficient at what it is now doing. The second is his entrepreneurial or strategic role—his responsibility for ensuring the company's future by leapfrogging competition, moving in new directions, and adapting skillfully to changing external forces.

In this context, corporate strategy consists of the broad plans by which management expects to preserve or significantly change the character of the company. It is made up of those major moves that will defend the company against hostile forces or enable it to grow significantly in size, improve its profitability, strengthen its industry position, or stabilize its operations in a way that minimizes the impact of wide cyclical swings. Development of a viable strategy typically involves fundamental policy decisions—as contrasted with continuing improvements in operating efficiency. Further, the effectiveness of this developmental process depends on a number of factors: first, an analyti-

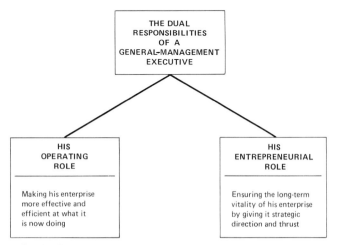

FIG. 28 *A general-management executive plays two roles.*

cally based understanding of the *environment* within which the company will, or may have to, operate; second, a shrewd perception of *competitive pressures and reactions;* and finally, an imaginative conception of the *strategic alternatives* that are within the grasp of a strong, informed, farsighted management.

The most important determinant of effective strategic planning is the thought process behind it. To this end, top management must educate and persuade the key members of the management group to think about their business the way a security analyst or business economist looks at an enterprise. What situation is the company in? What are the major factors that will determine whether it will be successful 10 years hence or, indeed, whether it will even be around? What are the critical things that must be done to keep this large, prosperous, and growing business equally strong and dynamic in the future, instead of allowing it to become inbred and moribund?

In asking these sorts of searching questions about their own company, top-level executives sometimes mistakenly assume that because the company is large and growing and strong today, its continued success is assured. Nothing could be further from the truth, for the concept of a *corporate* life cycle is as valid as that of the human life cycle. And the phenomena of decline, decay, and even mortality are as grimly present among business enterprises as they are among human beings.

Several examples reinforce this blunt fact of business life. Of the 100 top industrial companies in the United States in 1919, only 43 are among the top 100 today, and 30 have disappeared entirely. Of the companies in the original list of 100 that still survive, the ones that ranked eighth and seventeenth in size among the corporate giants in 1919 are now seventy-second and one hundred fifty-seventh, respectively. The same inexorable pattern of decline appears in other countries—although statistics are not readily available as far back as 50 years. In fact, it was only as recently as 1956 that *Fortune* magazine began to publish its annual listing of the 100 largest industrial companies outside the United States. But even in the relatively short time since then, 53 of the 100 companies on the original list have disappeared from this elite group as others have surged forward to take their places.

These dramatic changes in the fortunes of the world's leading industrial enterprises did not just happen. The period since the end of World War I is unquestionably too long a time for chance to have been the only factor behind their respective performances. Sustained effectiveness (or the lack of it) in strategic planning and execution has clearly played a leading role.

Still deeper insight into the character and scope of strategic planning can be gained from some real-life examples of corporate strategic moves. The following list is taken from the experiences of leading businesses in five different countries over the past decade.

1. Growth through acquisition of new products or product lines for the company's present markets or for different ones.

2. Growth in earnings per share through acquisition of, or merger with, other businesses. These moves have taken several forms:

    a. Horizontal expansion—addition of interrelated or complementary products or services

    b. Vertical integration—either backward to assure the company's sources of raw materials or forward to strengthen its market position and profitability by moving closer to the ultimate end user

    c. Diversification into new businesses

3. Acquisition of a specialty chemical manufacturing company to reduce the acquiring company's dependence on highly competitive, price-sensitive commodity chemicals.

4. Divestment of an existing business that had ceased to be compatible (in terms of technology and markets served) with the other components of the company's evolving product line. In another case, divestment of an existing business in which the company had neither the technological nor marketing strength to achieve profitability within the foreseeable future.

5. Substantial long-term shift in the allocation of resources (capital investment as well as research and development effort) from one existing business (industrial chemicals) to another (plastics).

6. Basic shift from heavy investment for the future growth of an existing business to "harvesting" the business and maximizing the cash flow available for the company's other businesses.

7. Major changes in marketing policy or structure (for example, entering wholly new markets, shifting to new channels of distribution).

8. Adoption of a basically different approach to product pricing aimed at capturing a carefully selected and more profitable segment of the market.

9. Shifts in the geographical area within which the company operates— domestically or overseas.

10. Basic change in research and development emphasis from improving existing products to developing new ones for different markets. In another instance, concentration of the company's research and development resources on a single technology in its major product line and abandonment of two other alternative technologies.

11. Development of overseas sources for the manufacture of components or finished products.

12. Basic alteration of the company's financial structure—say, in ownership or capitalization.

THE ROLE OF SYSTEMS IN STRATEGIC PLANNING

Given the entrepreneurial flair, the innovative drive, and the critical judgment that must underlie the development of meaningful strategies, we may logically ask: What role can formal systemization play in this process? Is not strategic planning such a highly individualized activity, so dependent on bold, imaginative, visionary thinking, that it will only be stultified if it is reduced to a prescribed series of procedural steps?

In partial answer to these questions, it has been our experience that companies which do not subject themselves to the discipline of a recurring strategic analysis and decision-making process are typically the ones that do the poorest job of strategic planning. On the other hand, it is equally true that the mere existence of a formal system will not ensure high-quality strategic output. Despite this fact, more than a few organizations have made the mistake of assuming that because they have a well-established, highly codified, consistently followed strategic process, the generation of competitively superior strategies will inevitably result.

The real need, therefore, is to recognize that there are both limitations and advantages to the use of some sort of formal system in carrying out the strategic planning activity. Managements must understand what they should *not* expect systems to accomplish in the strategy area as well as what these tools *can* contribute if used intelligently.

What Systems Cannot Do

Two of the most critical steps in the development of "win" strategies are the imaginative identification of a wide range of strategic options and the ultimate choice that is made among them. A well-designed system can facilitate these creative activities, but it cannot, of course, supplant them. No amount of fact-finding or analysis—however thorough and discerning—will ever replace the need for wisdom, courage, and future-oriented leadership in making tough decisions on where the enterprise should be headed or how scarce resources should be allocated among strategic alternatives.

What Systems Can Do

Having acknowledged this limitation, we emphasize with equal certainty that a well-designed, skillfully used system *can make the difference*—the difference between an expedient, haphazard approach to strategic planning and a well-considered, consistently productive one. An effective system makes this contribution by ensuring that three prerequisites to superior strategic performance are met: thoroughness of effort, continuity of effort, and unity of effort.

Thoroughness of effort A rigorous, well-documented system helps to make strategic planning much more comprehensive and searching, because it not only details the critical steps to be performed but also establishes high standards of analysis to be met. This sort of prescription is needed to ensure that all strategic opportunities and threats (in current operations, related businesses, and totally new fields) are identified and soundly evaluated.

Continuity of effort System is the forcing technique needed to make sure that strategic planning is done regularly and not left to happenstance. Some organizations tend to regard strategic moves as the product of an occasional bold thrust or creative stroke that, like invention or discovery, cannot be programmed. But all too often this notion simply serves as an excuse for failing to anticipate and prepare for change and, instead, merely reacting to it when it begins to pose a threat that compels attention. In this context, one of the principal values of a formal strategic planning system is that it forces key executives to regularly update their assessment of the strategic outlook for their business and, in that light, to reaffirm or redirect their own course.

Unity of effort   The final value of a formal system for strategic planning is that it helps to marshal the talents of many members of the management team and increase their contributions to the process. In a small enterprise, top-management leadership and example can serve as the marshaling force. But in large organizations, where top-management visibility is low, leadership and example are not enough. Here the involvement of many members of management can best be brought about by the discipline of a system that prescribes for everyone the steps to be followed, the factors to be considered, the kinds of analyses to be made, and the decisions to be reached.

This systematic involvement of large numbers of the management team produces two important benefits. First, it engenders better strategic plans, because it almost always brings to the surface a larger number of strategic alternatives to be evaluated. But more significant over the long run, it facilitates the career development of managers throughout the organization. It helps functionally oriented executives to more quickly comprehend the general-management process, and in that way builds strategic planning into the very core of an organization's operations.

Summary

The real issue, therefore, is not whether to have some sort of formal strategic planning system or no system at all. The real issue is whether the system that an organization does have will be just another paperwork exercise or set of perfunctory procedural routines, or whether it will be an effective means of orchestrating the many decision-making steps, the different analytical cuts, the numerous management disciplines, and the entrepreneurial talent of the enterprise.

DESIGN OF A STRATEGIC PLANNING SYSTEM

In undertaking the design of a general-management system that meets the foregoing specifications, the analyst needs to think in terms of two parameters: (1) the broad factors that influence the scope and complexity of any organization's approach to strategic planning, and (2) the individual steps around which the system should be built.

Broad Factors to Consider

The three factors that influence the scope and complexity of any organization's approach to strategic planning are:
1. The basic purposes that must be satisfied for effective planning of any kind

   2. The predominant characteristics that distinguish the company or industry

   3. The organizational makeup of the company

Purposes of effective planning   Any type of business planning, whether strategic or operational, must meet the following four tests if it is to have a constructive influence on the way an organization is managed.

   1. Plans should be aimed at achieving *increments*—at producing significantly better results than the organization would have achieved if it did no planning at all. In other words, plans must be made to challenge or stretch the organization.

   2. Plans that are made should be carried out. They should *not* be forecasts of what the organization thinks will happen, but statements of what the organization is going to spend money and effort to *make happen.*

   3. Plans should consist not just of end-result objectives (say, a certain level of earnings or growth in market share), but also of concrete action programs and timetables spelling out how the organization will achieve those objectives.

   4. Planning should not be regarded as a staff function. To be sure, staff personnel can play a facilitating or supporting role. But if planning is to be a dynamic force in the management of an enterprise, it must be viewed by everyone in the organization as an inextricable and critically important part of the line manager's job.

Distinguishing characteristics of company or industry   The complexity of the strategic planning process and the amount of attention given to the various fact-finding, analytical, and decision-making steps within that process will be influenced by company and industry characteristics. Here are some examples.

   1. Type of market (consumer versus industrial)

   2. Stage of the product life cycle (early growth versus maturity or decline)

   3. Operating scope (single function versus vertical integration)

   4. Geographic spread (local versus multinational)

   5. Critical resource (technology versus capital)

   6. Profit-making cycle (for example, one-season, high-fashion merchandise versus life insurance)

   7. Degree of regulation (public utility versus manufacturing company)

Organizational makeup   The final broad factor that will determine the character and complexity of the strategic planning process is the diversity of the enterprise and its form of organization. If the company is organized around a single business, top management takes the initiative and provides the leadership needed to carry out all aspects of the process: making strategic analyses, identifying and evaluating alternatives, reaching final decisions, and allocating resources.

But if the company is a multibusiness enterprise, top management's strategic role is different. Such an enterprise may consist of a parent company and a group of subsidiaries, or a company organized into a series of relatively self-contained product divisions. In either case, much of the job of initiating strategic recommendations shifts to the managements of the individual businesses. Under this arrangement, therefore, the strategy role of top management of the total enterprise consists of the following kinds of activities.

1. Developing and promulgating the company's overall objectives covering such elements as the kind of company it should be, the markets it should serve, and the financial (e.g., liquidity and earnings per share) and the nonfinancial goals it should reach

2. Providing strategic planning guidelines to the business divisions, including:

    a. Economic forecasts (general business conditions and outlook)

    b. Forecasts and analyses of significant social, political, and technological threats or opportunities

    c. Capital or other resource constraints that need to be considered in strategic planning

3. Identifying and evaluating strategic needs or opportunities in fields so different from the company's present businesses that none of the product divisions could reasonably be expected to perceive these alternatives

4. Reviewing, challenging, and testing the validity of each division's proposed strategies

5. Determining how well the divisions' proposed strategies, taken together, will meet the company's financial and nonfinancial objectives, and what should be done about the gap, if any

6. Choosing a portfolio of strategies—divisional and corporate—that best meet the company's overall objectives and make the optimum use of available resources: capital, managerial, and technological

7. Monitoring company performance in executing agreed-upon strategic plans

Individual Steps in the Strategic Planning Process

Notwithstanding the variations in scope and complexity that result from the foregoing conditions, all strategic planning systems need to make provision for the following major steps:

1. Defining the organization's end-result objectives

2. Developing the guidelines and assumptions that will govern the current iteration of the strategic planning process

3. Evaluating the long-term attractiveness of each business that the company is now in

4. Identifying the strengths and weaknesses of the company and its position in each of its businesses

5. Determining the size of the gap between the company's end-result objectives and the results likely to be achieved from pursuit of the existing product and market strategies

6. Identifying a full range of alternative strategies for closing the gap

7. Developing contingency plans for each strategic alternative being considered

8. Making a searching evaluation of each of these strategic options and the assumptions on which it is based

9. Reaching final decisions on the strategy to be followed in each of the company's businesses and allocating resources accordingly

Defining the organization's overall objectives   As a starting point, top management must define and communicate to the other members of management the financial and nonfinancial goals that the enterprise is dedicated to achieving. Financial goals are ordinarily expressed in terms of growth rate in earnings per share, targets for return on equity and return on assets, cash-flow objectives, dividend policy, and debt-equity policy. Nonfinancial objectives, in turn, may include such elements as

1. Parameters on industry concentration—for example, limits or targets on the amount of earnings that the company wants to obtain from certain industries, and extent of participation in those industries. For example, will the company be vertically integrated within its product lines, or will it simply be a one-stage manufacturing and marketing company?

2. Parameters on geographic concentration. These might be expressed as limits or targets on the amount of earnings the company wants to obtain from various world economic blocs.

3. Parameters on risk exposure—say, acceptable levels of capital investment in, and projected earnings from, certain high-risk industries or new product ventures.

4. Guidelines on corporate values or corporate responsibilities (for example, affirmative action programs).

However these corporate goals may be expressed, they are usually of such a fundamental character that they remain fixed over long periods of time.

Developing strategic planning guidelines and assumptions   Compared with the definition of the organization's end-result objectives, the second step in the strategic planning process is one that must be taken each time the process is repeated. It is the step in which top management develops and communicates the guidelines and assumptions on which the current strategic analyses and recommendations should be based. It typically includes the following elements.

1. A forecast of the economic conditions that will affect the company as a whole (for example, levels of business activity, inflation and its impact on the company)

2. Identification and evaluation of significant sociopolitical trends and developments (for example, the changing makeup of the country's labor force and resulting future labor relations requirements, increased governmental regulation, consumerism, threats against corporate bigness)

3. Assumptions about the availability of capital (for example, a 3- to 5-year

forecast of capital resources, including operating cash flow and additional capital to be obtained from external sources)

4. Identification of any special corporationwide objectives that are to be given emphasis in the development of current strategic plans (for example, a drive to significantly expand the company's international position)

Evaluating the long-term attractiveness of each of the company's businesses   Against the background of the corporationwide considerations set forth in the two preceding steps, we now get down to the task of developing strategic recommendations for each of the company's current businesses. To provide a sound basis for these recommendations, we must first make a searching reevaluation of each existing business, its major improvement opportunities, and the wisdom of staying in it. Given what we now know, would we enter each of our current product and market areas today? If we would not, we must then decide whether we should stay in them, for how long, and under what circumstances. And given a decision to remain, we must select a strategy to make the most of the opportunities.

This assessment of the long-term attractiveness of each business should be based on an evaluation of the significant environmental forces affecting the industry. These include sociopolitical forces and economic trends.

Sociopolitical forces.   Having earlier identified the environmental forces that may act upon the corporation as a whole, we must now single out those sociopolitical threats and opportunities that are of special significance to each individual business. These forces may take the form of greater government intervention—for example, pollution and waste control, consumer protection, or increased antitrust activity. Or they may grow out of such social changes as increased affluence and leisure time, the suburbanization of the population, the changing role of women, the dominance of the young and the elderly in the marketplace, and the demand for new financial services.

One of the most striking examples of such forces is the explosive effect that automotive safety requirements and air pollution constraints have had on the world automobile industry. Another is the dramatic long-term shift from businesses to consumers as the primary source of demand deposits—the free money supply of commercial banks.

Economic trends.   This part of the environmental analysis consists of identifying and determining the long-term significance of changes taking place in the basic economics of the industry. The task of spotting these changes calls for gathering and analyzing the following kinds of facts.

1. The trend in the profitability of companies making up the industry
2. Changes in the rate of growth in demand—domestic and foreign
3. The trend in price levels in relation to major elements of cost; the present and probable future pressures on product prices and costs
4. Shifts in the sensitivity of sales volume to price
5. Changes in the size and utilization of industry capacity
6. Threats of process obsolescence in the industry

7. Barriers to entry into the industry
8. New-product developments, changes in the degree of product differentiation, and threats of product obsolescence
9. Any threat of a substantial rise in imports from other countries
10. Prospects for vertical integration by major customers

Identifying the company's strengths and weaknesses   Having completed an evaluation of the long-term attractiveness of the industry as a whole, the planner must now assess the company's strengths and weaknesses and their effect on its present and probable future position within the industry. Since strengths and weaknesses are relative matters, the performance of the company must be compared with that of each competitor on such factors as product superiority, technological capability, and manufacturing and distribution costs.

Product superiority.   Which company in the industry has the image of holding product leadership? On what product features is this image based? If our company is not the leader, how far behind the leader are we? What would we have to do to catch up? What would this cost and what would it be worth? How important to our customers is product differentiation in contrast to price? How important is service to customers, and how do we compare with our competitors on this element of performance?

Technological position.   How important is technological leadership in the industry? Can our company be a technological follower and still achieve its earnings goal? How does the technology of each competitor compare with ours? In the area of product development, does our company originate a larger number of profitable new products than any of our competitors? If we are not competitive technologically, where throughout the world are the technologies we need and how can we acquire them?

Cost performance. How do our unit costs (of production, selling, and distribution) compare with those of our competitors? Of the total of each of these costs, what portion is fixed and what variable? What important facilities do our competitors appear to be acquiring that could significantly improve their cost performance (for example, purchase of new equipment from a common supplier, construction of a new plant)?

Determining the size of the strategic gap   Armed with an evaluation of the prospects for the industry and the company's strengths and weaknesses within that industry, the planner is now in a position to make an informed estimate of the size of the so-called strategic gap. This is the difference between the contribution that the business division is expected to make to the corporate financial goal and the profit results likely to be achieved if the division simply continues to pursue its present product and market strategies. The size of the gap, if there is any, will determine the character and magnitude of the remaining steps in the process. Estimating the gap gives the planner a realistic measure of the earnings that must be achieved through changes in strategic direction—whether through a basically different approach to pricing, a substantially increased investment to stimulate growth of the present business, the development of new products, entrance into new markets with existing or new products, or entrance into a wholly new business.

Identifying alternative strategies This step, more than any other, determines the quality of an organization's strategic planning. Even among companies that have adopted a formal approach to strategic analysis and decision making, relatively few are sufficiently imaginative and resourceful to develop a wide range of strategic options. They tend to settle on a single course of action, which often amounts to little more than striving harder at what they have been doing in the past.

This common shortcoming is not easy to avoid, for the development of a reasonably comprehensive group of viable strategic alternatives is the most difficult and creative aspect of the whole planning process. Carrying it out successfully requires that the planners look at each of the company's *present businesses* in fresh new ways, and then explore a full range of *new business* opportunities—new products, new markets, or both.

Identifying strategic opportunities in present products and markets. At this stage the planners must remove the blinders that tend to keep their sights fixed on what is now being done. The critical question they must ask is *not* how the company can fine-tune what it is doing, but "Within the framework of our present product-market structure, what can we do differently?" Following are two approaches illustrating the ways of thinking and methods of analysis that planners must apply to their present businesses.

1. In light of our analysis of the industry's long-term prospects and our company's strengths and weaknesses, which of three basic strategies should we pursue:

    a. Should the company drive to achieve or maintain industry leadership as a means of building future years' earnings? This strategy might require heavy financial commitments in order to

        (1) Expand capacity ahead of sales

        (2) Maintain technological leadership

        (3) Price to expand market size and share

        (4) Acquire other companies in the industry, where feasible

    b. Should the company drive to maximize earnings over the short to medium term (say, 3 to 5 years)? This strategy in turn may require such supporting tactics as

        (1) Pricing to hold the company's market share and to keep capacity fully utilized

        (2) Maintaining a competitive level of innovation, but focusing product development increasingly on design changes that cut manufacturing costs

        (3) Remaining a low-cost producer through offshore sourcing

    c. Should the company retrench or harvest that division of the business to achieve maximum cash flow over the short run? This strategy might require such moves as

(1) Eliminating product development expense
(2) Minimizing new capital investment and concentrating it on fast-payback projects
(3) Withdrawing from unprofitable market segments and disposing of facilities opportunistically
(4) Forgoing part of the company's market share in profitable market segments in order to maintain prices and profits

2. What strategies does each of our competitors appear to be pursuing and what implications do these moves have for our company? If we are not the industry leader now, what specific things must we do to overtake competition in pricing? In technology? In service to customers? In marketing effectiveness? If we are the leader today, what must we do to maintain that position? What is the pursuit of each of these objectives likely to cost us, what are the chances of success, and what long-term profit improvement, if any, will this action produce which we would not otherwise realize?

As an illustration of the kinds of strategic issues that can be surfaced by the foregoing lines of inquiry, the strategic planning staff of one manufacturing company recently posed the following questions to its division management:

MAJOR CORPORATE RESOURCES	MAJOR ENVIRONMENTAL OPPORTUNITIES AND THREATS		
	A. Rapid growth of new market in Europe	B. Sharp increase in labor costs	C. (Etc.)
Capital	• Acquisition or joint venture in Europe		
Management			
Process Technology	• License European manufacturer • Build own plant in Europe	• Increase offshore sourcing capacity	
Product Cost Leadership		• Increase automation	
Product Technology		• Increase prices while retaining market share	
Marketing Skill			

FIG. 29 *The strategic planning process should match the key resources of an organization to the opportunities and threats it faces.*

Given the continuing attractiveness of our industry and our own growing technological competence, should we make the research and development investment needed to achieve significant product differentiation and higher margins? Or should we play safe by sticking to our traditional limited product line with high sales volumes and low unit costs?

Identifying opportunities to move into new fields. In searching for attractive opportunities to move into wholly new areas (new products or businesses, different markets, or both), here are some analytical steps that planners should take to stimulate and stretch their thinking.

1. *Determine how the key resources of the company could be used to deal with each opportunity or threat identified in the environmental analysis.* Building on two of the earlier steps in the strategic analysis, this step leads the planner to consider how the real strengths of the organization can be used to capitalize on opportunities to move into new fields or to withstand the threats facing the business. The matrix in Fig. 29 illustrates the application of this approach to a specific situation.

The foregoing approach can also be used to advantage even in situations where the analysis of environmental forces does not, in itself, lead to identification of specific opportunities. As an example, one property and casualty insurance company defined its major resources or fields of distinctive competence as including extensive business and personal contacts; safety engineering capability; actuarial, underwriting, and reinsurance competence; and surplus cash. In light of this analysis, the top-management team then singled out the following opportunities for diversification through the application of these resources.

BUSINESS AND PERSONAL CONTACTS

Life insurance sales	Stock brokerage
Mutual fund sales	Advertising agency
Title insurance company	Travel agency
Data processing services	

SAFETY ENGINEERING CAPABILITY

Safety consulting	Nuclear safety services
Appraisal firm	Safety equipment manufacturing

ACTUARIAL, UNDERWRITING, AND REINSURANCE COMPETENCE

Actuarial consulting	Reinsurance company

SURPLUS CASH

Premium financing	Factoring
Personal finance company	Car rental
Real estate	Airplane financing
Computer leasing	

2. *Review upstream and downstream possibilities.* This review should lead to the identification of all significant opportunities for integrating backward to the company's basic raw materials and forward to the end user. (As an example, American Motors is essentially an assembler of purchased components and a marketer, while Ford Motor Company is integrated backward to basic steel production.) But even in companies that are already well integrated, the

planner should still pursue this line of inquiry with the objective of determining whether *all* strategic options at each production stage have been fully evaluated. For instance, in a company manufacturing basic, intermediate, and specialty chemicals, has the most profitable balance been struck between selling intermediates to other chemical processors and upgrading them into the company's own specialty products?

3. *Review research projects now underway or contemplated.* The key question to be answered here is: Does the company have the technological capability to launch, or has it already launched, promising research projects that will enable it to diversify away from its existing business base through a major *new-product-development program?*

4. *Look for opportunities to transplant strategic concepts from one business to another.* This step requires that planners develop a deep understanding of the basic strategies underlying the superior performance of the world's most successful enterprises. For example, a manufacturer of expensive high-technology equipment might consider applying IBM's leasing strategy, which has been one of the cornerstones of that company's success during the great growth period of the computer industry.

Developing contingency plans Rarely do all the forecasts, assumptions, and action programs on which strategic plans are based work out exactly as originally developed. For this reason the next critical step in the process is to ask a series of "What if?" questions to identify the possible risks inherent in each alternative strategy. These are events that can significantly upset achievement of the objective that the strategy is aimed at accomplishing. They might, for example, include a possible strike, a 25 percent smaller or larger sales demand than that forecast in the strategic plan, a successful new-product development in a competitive technology, a sharp price drop by a major competitor, or a change in tariff levels leading to a rapid increase in imports.

Since the range of potential contingencies is always great, planners should not try to "cover the waterfront." Rather, they should focus selectively on the two, three, or four developments that are the most likely to occur and that can have the worst impact on the company if they do. To prepare for these contingencies in a rigorous, substantive way, the planners should take two steps:

1. Make as realistic an estimate as possible of
   a. The odds that the contingency will occur
   b. Its possible severity (what it could do to the division's profits)
   c. The amount of lead time the company will have between the first warning signals and the need to respond to the contingency

2. Develop the sort of action plans that will enable the company to react quickly and effectively if the event should occur

If analysis indicates that the likelihood of a given contingency is high and its predicted impact severe, the company will presumably not be content with developing plans for responding when the event occurs, but will take preventive steps to avoid or minimize the risk. But for most possible contingencies, the full need will be met by creating either a detailed fallback plan, or a checklist of steps to be taken in quickly analyzing and developing an appropriate response to any event in a given class of contingencies.

The wide-ranging value of this sort of contingency planning is well illustrated in an article by Louis Gerstner:

> An electronic components company depended on a single large customer for 30 percent of its sales. Management simply asked: "What if this key customer should integrate backward?" There was no visible reason to believe that such a move was in the offing, and the question would probably not have surfaced as a serious issue without the forcing device of required contingency planning. But, in this instance, developing the contingency plan led to two real benefits. First, it brought out the need for some preventive medicine, and this became a continuing part of the company's relationship with its big customer. Second, it led to a detailed economic analysis of the risks and disadvantages to the customer of backward integration. One year later that analysis was instrumental in convincing the customer that a tentative step he had been about to take toward integration would be unwise.[1]

Testing the soundness of strategic proposals The next step in a formal strategic planning system should be that of checking each proposed strategy for acceptability, completeness, and validity. Since strategic moves often involve large and irrevocable commitments of resources, this is the point in the planning process at which entrepreneurial thinking and innovation must give way to hardheaded pragmatism. Before top management makes final decisions that will shape the future of the enterprise, it must ensure that each of the alternative strategies from which it will choose is a well-thought-out, well-supported proposal. And although this sort of challenging review should be applied in some measure to every option being considered, it should be applied most searchingly to proposals that represent *major* departures from former strategies and that require the largest commitments of resources.

Acceptability. Assuming that each alternative strategy achieves the results predicted for it, how close will that level of performance come to meeting the earnings contribution that top management expects of the division? How compatible is the strategy with the corporation's other goals (for example, with its nonfinancial objectives)?

Completeness. The test of completeness lies in posing the following two sets of questions.

1. In light of our analysis of the attractiveness of the industry and of our company's strengths and weaknesses, have we identified *all* of the reasonable alternatives that we ought to consider? How effectively do these alternatives respond to the major competitive opportunities and threats confronting the company?

2. For each strategic alternative, have we developed a clear written statement of the following components of the strategy and the way in which they were generated?

[1]Louis V. Gerstner, Jr., "Can Strategic Planning Pay Off," *Business Horizons,* December 1972, p. 11.

> *a.* Estimated market share, sales volume, and earnings
> *b.* Assumptions on which these estimates are based (for example, market growth rate, price levels, imports from Japan)
> *c.* Detailed action plan and timetable for carrying out the strategy
> *d.* Character and magnitude of risks (including possible actions of competitors)
> *e.* Contingency plans
> *f.* Resource requirements—for example, capital and scarce human resources (managerial and technical)

Validity. The final step in the testing process consists of challenging the credibility of the assumptions, estimates, and other building blocks that make up each strategy being considered. Following are some questions to be asked and analyses to be made in this validity-testing process:

1. How realistic is each of the *quantitative projections* in light of the company's own past performance? This question should be asked persistently about each forecast forming a part of the strategy—for example, the forecasts of overall market size, sales prices, the company's market share, its sales volume, and its unit costs.

2. Whenever the foregoing comparisons reveal significant discontinuities between the company's past performance and the strategic forecast, those making the validity test must then ask: How concrete and adequate are the *action programs* through which we expect to achieve the marked improvements specified in the strategic plan? For example, if our strategy is based on in-house development of a major new product, has management of our research and development department done a rigorous job of quantifying the probable results of the project (new-product performance features, production start date, capital requirements, and costs to manufacture)?

3. In light of our past performance in implementing major new undertakings, how realistic is the *timetable* for carrying out each strategic alternative?

4. What substantiation have the planners developed for their assumptions on possible *moves of competitors?*

5. If a proposed strategy calls for entering a new field consisting of new products and new markets, how thoroughly have the planners identified the *new resources of knowledge and skill* (managerial and technical) that will be needed to succeed in the field? What evidence is there that we have been able to develop such incremental human resources from within the company in the past? How detailed are the program and timetable for developing or acquiring this talent?

6. How thoroughly have planners identified and dealt with *intangible considerations* that can be crucial in the entry into new markets? For example, in moving to establish manufacturing operations in Europe, these intangibles may include tax concessions, the trade union atmosphere, governmental attitudes, and cartel arrangements.

Making final strategic decisions Top management comes now to the critical step in which it must consider all the recommended strategies that have been developed in various parts of the organization and select from them the courses of action to be followed. Of all the steps in the planning process, this one lends itself least to systemization. It is basically an intellectual activity, requiring of key executives a quality of judgment that is at once visionary yet rational, bold yet exacting.

Notwithstanding this fact, however, the likelihood that management will reach the best strategic decisions can almost always be increased if the criteria to be used for evaluation and decision making are well thought out and put into writing. These criteria should be designed (1) to facilitate the comparison of recommended strategies with each other, and (2) to achieve a sound balance in the portfolio of strategies that are finally adopted.

Comparing alternative strategies. As a prelude to making final decisions, top management must reach a judgment on the *relative* attractiveness of each proposed strategy. This requires the comparison of all alternatives in terms of such criteria as the following.

1. Compatibility with the company's overall objectives.
2. Funds required (capital and operating expenditures).
3. Projected cash flow.
4. Projected profitability.
5. Technical resources required and potential technical risks.
6. The competitive position that will result, and the potential competitive risks (including international competitive threats).
7. Management capability and other human resources needed, and the likelihood that the organization can acquire, recruit, or develop them.
8. Interactions with other businesses of the company, and the fit with other strategies.
9. Probability of success—that is, an estimate of the company's chances of achieving the predicted sales and earnings. This probability might be expressed as

    a. High—a more than 80 percent chance of achieving the target
    b. Medium—50 to 80 percent chance
    c. Low—less than 50 percent chance

Achieving balance among strategies adopted. Although the mix of strategies that represents the best overall balance will vary considerably from company to company, two guidelines for achieving a balanced portfolio are applicable in most situations.

1. The combination of strategies finally selected must represent a sound balance between opportunity and risk. In other words, management must fix on the best distribution between higher-risk, high-return alternatives and play-it-safe, lower-return ones.
2. In addition, the portfolio of strategies finally adopted should achieve a sound balance between short-term and long-term earnings. In a multibusiness enterprise, this balance can best be struck by a mixture of

    a. Rapid-growth businesses that will provide high future earnings while requiring heavy capital investment today and contributing only modestly to current earnings
    b. Businesses generating high current earnings

  c. Businesses generating large amounts of excess cash for reinvestment elsewhere

Summary

Against the background of these individual steps in strategic analysis and decision making, systems analysts should never lose sight of one key point: The strategic planning systems they design must encourage *substance,* not *form.* The great danger is that this whole process will become cursed with "format fixation"—that instead of stimulating and marshaling the organization's entrepreneurial thinking, it will degenerate into a bureaucracy of forms, documentation requirements, and procedural routines. Where this happens, the systems that are developed tend to straitjacket those managers who can make imaginative strategic analyses and to cover up for those who cannot.

The principal way in which systems analysts can help their organization avoid such debilitation is by designing strategic planning systems that focus not on detailed procedural steps but on such broader, fundamental elements as (1) the basic objectives of the process, (2) techniques and checklists that help to ensure searching analysis, (3) criteria for evaluating alternatives and making decisions, and (4) information systems for measuring the strategic performance of the company against its competitors.

OTHER TYPES OF GENERAL-MANAGEMENT SYSTEMS

Most of the other general-management systems found in large-scale organizations have been identified in earlier chapters. To recapitulate, they include the following.

  1. *Operational* planning and control systems. These typically consist of one year's profit planning and budgeting and the network of management reports that measure actual versus planned operating results.

  2. Systems for planning and developing the *human resources* of the organization, including systems for (*a*) forecasting personnel requirements, (*b*) maintaining an inventory of managerial and professional personnel, (*c*) evaluating performance, (*d*) making compensation decisions, (*e*) identifying candidates for promotion and moving them along defined career paths.

  3. Systems for managing *capital expenditures,* including capital budgeting, review and approval of appropriations, and project control.

The important remaining point to be made is that although these three categories of systems cover very different aspects of the general-management function, they need to be designed in a way that reflects a common management philosophy. Not only must they be consistent with each other, but they must be strongly and visibly supportive of each other.

This need is illustrated by the experience of a large manufacturing company that carried out a major program to strengthen its systems for strategic planning.

Top executives assigned the highest priority to the program and took every step that they believed was necessary to ensure success of the effort. The chief executive personally launched the program and actively participated in it until its completion. In addition, he assigned some of the company's most talented line and staff members to work full time on the project. And when it was over, he sponsored a formal management training program in the strategic planning techniques that had been developed.

But the company overlooked one critical ingredient: In the design and administration of its management development and motivation systems, it had unwittingly built a short-term orientation into its managerial value system. As part of its formal executive development effort, it had implemented a comprehensive career-path program for its management group. Every manager with any significant potential was moved to a new position every 3 years. In addition, the company's management compensation plan provided that a sizable portion of each manager's annual compensation was to be based on his or her performance and paid as a lump-sum incentive bonus. But since no manager remained in a job for more than 3 years, performance had to be measured in terms of short-term, quantifiable results. As a result, many managers were—consciously or unconsciously—motivated to pursue courses of action that had the effect of "milking" their businesses instead of capitalizing on their long-term strategic potential.

Another example of inconsistency in general-management systems can be found in the traditional approach to capital development decisions that many companies follow. Under this approach, all capital appropriation requests are ordinarily filled regardless of the business division or product-line component that originates them, provided only that they meet a prescribed financial criterion such as demanded rate of return. And whenever the total of the capital funds requested exceeds the funds available, the required rate of return is simply raised and once again applied as a single, insensitive decision criterion to all requests.

This approach to decision making on capital expenditures debases the strategic planning process. By taking into account only the one conventional financial requirement, it permits the flow of funds to a mature business at the same or an even faster rate than to a high-potential business that is just moving into the rapid-growth stage of the product life cycle. Thus, by the application of this single-number criterion, all the other considerations that were painstakingly identified and evaluated in the strategic planning process go for naught.

In summary, the point to be stressed is this: Whenever systems analysts work on the evaluation or improvement of any general-management system, they must also make a high-spot diagnosis of all other general-management systems to ensure that their current efforts help to achieve or maintain the level of unity needed among such systems.

THIRTEEN

Strengthening Management Information Systems

One of the most pervasive buzz words in the lexicon of the present management generation is the acronym MIS. It is a rare conference of management, accounting, or computer personnel that does not generate a good bit of ritualistic oratory on such variants of the theme as "The Total Management Information System," "The MIS Revolution," "Information Sciences Make the Future Manageable," "MIS—Why Not Now?" and "MIS—The Answer to Your Company's Data Explosion."

Further proof of the interest in MIS comes from the *Harvard Business Review*. One of the most popular of its bound series of related-article reprints is the *Management Information Series,* which contains fourteen of the MIS articles that have appeared in that journal over an 8-year period.

WHY THE INTEREST IN MIS?

The manifest interest that executives in all types of endeavor show in the field of MIS, and the lofty claims made for it, are not hard to explain. Among the many causes, three stand out.

1. *The long-term growth in size and complexity of many business organizations.* This trend has increasingly forced management to depend on formal written summaries of operating results rather than on informal, on-the-spot observations or judgments that were possible when businesses were smaller.

2. *The expanding need for better ways of measuring executive performance.* This need, in turn, has been sparked by the intensive use of three management

techniques: decentralization of operations, management development programs, and incentive compensation for executives. Although the three are not closely related, they have a common thread: To be successfully applied, each requires the use of objective methods for measuring executive performance.

For example, if decentralization is to mean delegation rather than abdication, then the executive who used to run the whole show himself must learn instead how to evaluate the way someone else is running it. Again, one of the key steps in many formal management development programs is determination of the profit-making components of individual jobs. This step helps to show executives exactly what is expected of them and thus builds commonly understood bases for evaluating results achieved. Finally, one of the most significant characteristics of successful executive compensation plans is that they achieve the maximum motivational value by ensuring that each person's incentive pay is a real reflection of performance.

Thus, all three of these techniques call for reevaluation of formal management information systems, for they are one of the principal means through which performance results are summarized, measured, and made known to those who must interpret them.

3. *The widespread application of the computer.* The exponential growth in the use of EDP equipment, which was described in Chap. 2, is unquestionably the greatest single cause of today's interest in upgrading an organization's formal structure of management information. As an illustration, in discussing his company's acquisition of a third-generation computer, an executive of a chemical company said, "We'll be happy if we break even on this new EDP investment insofar as clerical cost is concerned. Our real purpose is to sharpen our management controls by speeding up the availability of existing information and by developing important additional information not now available." Another company defined the purpose of its computer installation as "the production of management information that will increase profits." Still another described its EDP objective as that of "increasing the ability of management to make decisions sooner from more readily available and greater amounts of information."

These illustrative statements underscore the fact that executives view the computer as both a competitive opportunity and a competitive threat—an opportunity to generate the sort of decision-making and control information that will significantly improve their own organization's performance, and a threat to the extent that competitive organizations will gain the MIS advantage much sooner than they do.

THE PROBLEM AND THE OPPORTUNITY

In spite of the sustained interest in MIS and the sense of need that underlies it, most managements (general-management executives as well as heads of major functions) still do not have the information they need to effectively plan and control the results for which they are responsible. This is a harsh conclusion and it needs, therefore, to be reinforced by some specifics in the two areas where formal information serves as a management tool: controlling current operations and planning the future.

Evidence of an MIS Shortfall

The newspapers of May 1974 carried the story of a major New York bank that had "sustained losses of $14 million and a possible additional loss of $25 million in its foreign-currency trading."[1] According to the press accounts, these losses had occurred *without the knowledge of the top officers of the bank.*

Other managements might well feel smug in the conviction that no such looseness or lack of control exists in their organizations. And yet while they may be sufficiently on top of things to prevent any such shock losses as the one just mentioned, they may still be unaware that they are simply not getting the critical information they need to optimize the performance of their enterprise.

As an example, the top managements of companies in the life insurance industry are, in a very real sense, in the business of *managing assumptions*—that is, the assumptions on which their policy premium rates are set. These include forecasts of (1) mortality (or other types of claims), (2) persistency of policies, (3) expenses (both first-year and renewal expenses), and (4) investment yield. The relationship of actual results to these sorts of assumptions on a given type of contract obviously has a significant effect on policy dividend rates, on the need for expense control and tighter underwriting of new business, and on the pricing of future contracts. Yet relatively few companies in the life insurance industry have built into their management information structures a regular tracking system that shows—at least annually—how actual results compare with each of the foregoing assumptions on every product, and that suggests what ought to be done about the differences.

As another illustration, among commercial banks the range of profitability (expressed as a rate of return on assets) is extremely wide. One of the causes of this variation is the fact that there are large differences in profitability among the various customer services provided by the banks, and also among the several market segments served (for example, national or multinational accounts, local businesses, individuals with high net worth, and consumers). Notwithstanding these differences, relatively few commercial banks have yet developed a management information system that regularly shows the comparative profitability of their products, market segments, or major individual accounts. And without such information, bank managements must labor under the severest handicap in trying to improve their profit performance through greater market and product selectivity.

These illustrations could be repeated almost endlessly in other industries and types of endeavor. For despite the widespread use of the computer and the pressure for better MIS, it is a rare executive indeed who is fully satisfied even

---

[1] *The New York Times,* May 13, 1974, p. 1.

with the information he or she is getting for control of current operations, to say nothing of that needed for planning strategic moves.

Causes of the MIS Shortfall

If an organization is to succeed in overcoming these weaknesses, its management team must first understand why they exist at all. Since the development and use of better information tools are of such significance to modern management, why do most organizations still have so far to go in fashioning these tools and helping their executives learn to use them skillfully?

There is no simple answer to this question. In fact, there are at least five conditions that contribute to the MIS shortfall, and taken together, they constitute a formidable barrier. These conditions are described in the sections that follow.

Information systems are skewed toward external reporting requirements   The financial information systems of most companies have been shaped primarily to meet external reporting requirements, which happen to bear little relationship to the information needed to manage the business. This fact of corporate life is especially relevant among regulated industries—banks, insurance companies, common carriers, telecommunications companies, public utilities, and the like. Here are two examples.

1. In an article on the telecommunications industry, *Business Week* stated: "In a competitive environment, AT&T's accounting could be its Achilles' heel. For years, Bell has run its operations under accounting procedures originally developed before the turn of the century and later cast in bronze as the standard system of accounts . . . imposed by regulatory agencies."[2]

2. The principal financial reporting vehicle of the life insurance industry in the United States is a document prescribed by state insurance departments and referred to as "the convention statement." Because this report calls for the submission of extremely detailed schedules of information, it tends to give life insurance executives the illusion that they possess an extensive system of operational and financial controls. But the convention statement is based essentially on a solvency or liquidation accounting approach rather than a going-business approach. And for that reason, it places minimum emphasis on information needed to run the business.

This skew toward external reporting is by no means peculiar to the information systems of regulated industries. Accountants in all businesses, regulated and nonregulated alike, have always tended to organize their record keeping and reporting around external demands, whether these are imposed by tax authorities, the SEC, shareholders, lending banks, trade associations, or any other.

MIS lacks adequate nonfinancial information   In addition to their strong external focus, traditional accounting and financial reporting systems are expressed

[2]"Who Will Supply the Office of the Future?" *Business Week,* July 27, 1974, p. 45.

primarily in dollars. All too often, however, dollars are simply a means of reporting the *final results* achieved. These results are important, of course, and must form part of any organization's MIS. But if management is to influence the end results reported in the form of dollars, it must, in addition, have recurring information on the *causes* of dollars. This part of any management information system is typically expressed in nonfinancial terms. But because the accounting system has served for so long as management's primary source of formal information, the structure of underlying nonfinancial information is still woefully deficient in most organizations.

To further emphasize the point, financial and nonfinancial information can be likened to the two blades of a scissors. Either, by itself, is far from being half as good as the two together. Both are essential and must be shaped to support each other if a useful management tool is to result.

The computer has not filled the void   Because of the common but mistaken notion that MIS willy-nilly means computer-generated outputs, the development of improved management information has, in many organizations, been assigned or left by default to computer personnel. It is in no way a criticism of these specialists to point out that although they understand well what the computer can do, it is an exceptional computer expert who has an in-depth understanding of the decision-making and control aspects of managing the enterprise.

For this reason, organizations that have left the development of their management information systems to computer personnel have often become preoccupied with development of all-purpose data banks, EDP transaction-processing systems, automated display devices, and other alleged electronic "advances," but have accomplished little or nothing in response to the gut question: How can we use this high-cost, high-potential equipment to significantly upgrade our management effectiveness?

MIS improvement drives are often superficial   In organizations that have made a direct frontal attack on the management information problem, the improvements achieved are often of little consequence because the scope of the attack is much too narrow. One of the most common approaches might be characterized as a "housecleaning" drive aimed at scrubbing up the structure of existing formal reports. Such an approach consists essentially of two steps. First, some of the most questionable reports are discontinued without notice and then reinstated only if the former recipients complain loudly and convincingly enough. Second, copies of all other reports are gathered and analyzed for duplication, following which the recipient of each copy is asked whether he or she could get along without it. On the reports that have to be kept—and they typically turn out to be over 95 percent—further efforts are made to simplify the contents, rearrange columns, present some of the information in chart form, and the like.

This approach seldom produces any major, enduring benefits because it suffers from three defects.

1. It is self-limiting in the sense that it ties the mind to the makeup of existing reports. Its whole focus is on patching up what the organization now has.

2. It provides no reasoned basis for determining what information is needed. The principal criteria for discontinuing or keeping reports are personal preference and prejudice.

3. It is not dynamic. It does not go deep enough in trying to rebuild the structure of management information into a real profit-making tool. It is a negative approach aimed primarily at cutting office expense through elimination of unnecessary or duplicate information.

In contrast to this typical approach, the job of report improvement should be regarded as much more than just removing clutter and confusion from existing reports. Such housecleaning, to be sure, is part of the job. But the main objective should be to make certain that the report structure as a whole gives management the critical information needed for running the business and for determining how each segment is contributing to the company's overall goals.

Management does not know what information it needs   To compound these failures of staff or specialist personnel to meet the real information needs of the organization, the final basic reason for the MIS shortfall is that top management and the heads of major divisions have not been very successful in rescuing themselves from the fruitless efforts of others. This results, in many instances, because there are still large gaps in management's understanding of the decision-making process itself. Given these gaps, many management teams simply do not have a rational, disciplined, and universally applied approach to defining the total body of information they need to manage the enterprise as a whole or any major piece of it.

In some of the same or still other organizations, managements have failed to take the initiative in this whole field because they believe that once the organization has a large-scale EDP system installed and functioning, the emergence of a substantially improved set of information systems is just a matter of time. Their line of reasoning is that management in the past has had to act with inadequate facts and with facts that arrived too late. With the widespread application of the computer, this problem, they contend should largely disappear since management should now be able to get all the facts it needs and get them quickly. Hence, say these proponents, the possibilities for error in management judgment should be sharply reduced. As a consequence, the willingness of management to utilize these new information-generating devices should, as one enthusiast put it, "surely separate the leaders from the laggards in modern-day management."

This way of thinking is based on the assumption that management has always known what specific facts it needs but has simply not had the tools required to process, summarize, and report them at a reasonable cost and in time to be of value. One advocate of this point of view draws a parallel with another field of decision making by referring to similar problems that meteorologists have faced in developing weather predictions. He points out that for some time enough has been known about the factors affecting the weather to make accurate predictions possible. The main drawback has been the tremendous amount of calculation involved, which formerly would have taken so long that the prediction, once obtained, would have been only of historical interest. But now with the advent of high-speed computation, practical manipulation of the long-known facts has become a possibility.

The fallacy of applying this parallel to business lies in the assumption that we already know an equal amount about the factors affecting business growth, profitability, and competitive strength, and that all we need to do now is determine how large-scale computers can be used effectively in performing the "tremendous amount of calculation involved" in organizing and presenting this existing knowledge.

Unfortunately the problem is not so simple. The real need in applying computers to the management of large-scale enterprises is a matter not so much of manipulating existing knowledge as of identifying much more clearly the factors that affect business health and the specific ways in which each factor exerts its influence in any given situation.

THE TOP-MANAGEMENT APPROACH
TO MIS IMPROVEMENT

In light of this delineation of problem and opportunity, the rest of this chapter is aimed at:

1. Providing a systematic basis for determining what kinds of recurring information each level of management needs for planning and control.

2. Establishing principles for building this information into an integrated structure of management and operating reports that are easy to understand, interpret, and use.

Determining What Needs to Be Planned
and Controlled

The starting point in determining what recurring information each level of management needs is to understand clearly what the basic purposes of formal reports are and how they relate to an organization's underlying data base and information-generating mechanisms.

These concepts and relationships are portrayed in the schematic diagram appearing in Fig. 30.

The key conclusions to be drawn from this chart are as follows:

1. Management's information needs cannot be met simply by building up a comprehensive data base—which, in its extreme form, means building into the computer storage system every bit of information that might conceivably be needed or asked for by anyone in the future. No matter how skillfully developed, a data bank of external and internal information is still nothing more than a warehouse of raw material—and the most unprocessed of raw material, at that.

PLANNING

	STRATEGIC	OPERATIONAL	PERFORMANCE MEASUREMENT AND CONTROL

MANAGEMENT PLANNING AND CONTROL SYSTEMS	**For example**
	• Basically new products
	• Market coverage strategy
	• Distribution channels (agent versus direct)
	• Investment strategy
	• Diversification
	  — Related fields
	  — Unrelated fields
		Typically a one-year profit plan covering objectives and action programs on
		• Premium income
		• Underwriting gain
		• Operating expense
		• Investment yield and gain
		• Personnel development
			• Strategic performance
			• Operational performance

MANAGEMENT INFORMATION SYSTEMS	• To evaluate the performance and outlook of the company and each of its major competitors • To identify improvement opportunities and needs, as well as external threats • To identify and evaluate the payout and risks of alternative courses of action		• To compare actual with planned results and programs • To fix responsibility for variances from plan • To "localize" or facilitate identification of the reasons for variances • To take corrective action
DATA CONVERSION TOOLS	Accounting Systems Computer Specifications for Management and Operating Reports Advanced Analytical Techniques (Operations Research)		
DATA BASE	EXTERNAL OR ENVIRONMENTAL DATA Competitive Economic Governmental (Regulatory) Social	INTERNAL DATA Marketing Expense Underwriting Investment Claims Personnel	

FIG. 30 *How raw data and management information relate to the planning and control processes of a property and casualty insurance company.*

2. Converting the raw material in a data bank into useful management information generally requires the power of the computer and sometimes the use of advanced analytical or operations research techniques.

3. But neither the building of a data bank nor its conversion into useful management information systems can be done intelligently unless the analyst first defines the specific purposes of the information. This means identifying the key strategic and operating decisions to be made and the critical elements of performance to be controlled.

Any such statement of factors to be planned and controlled must be expressed in terms of (1) common financial indices, and (2) the key factors governing success in the industry or other type of endeavor.

Common financial indices   These elements, most of which are components of the return-on-investment equation, include the following.

    1. Income
        a. Sales revenue
            (1) Sales volume
            (2) Product mix
            (3) Selling price
        b. Investment and other income
    2. Costs and expenses
        a. Direct
        b. Indirect
    3. Cash flow
    4. Assets
        a. Facilities
        b. Working capital
        c. Other assets

Since these variables are much the same in all make-and-sell businesses, the analyst's task here is to determine the significance of each within his or her company. This assessment should cover the relative importance of each variable, the level at which it is influenced or controlled, and the detail in which it should be measured. For example, control of direct manufacturing costs may be relatively less important in a pharmaceutical company than it is in some other companies, whereas control of research and development expenditures may be relatively more important. The product mix of sales will be extremely important in some businesses, unimportant in others. During periods of increasing shortages and rising costs of new capital, cash flow performance and liquidity can replace growth in earnings per share as the most important measure of overall corporate results.

To illustrate this process, Fig. 31 shows a list of the key financial variables for one company's marketing activities. This diagram was developed in an attempt to answer two questions: What are the basic elements of performance to be controlled in this situation—that is, what is the best indicator of overall marketing performance? And what are the financial components that make up the total result?

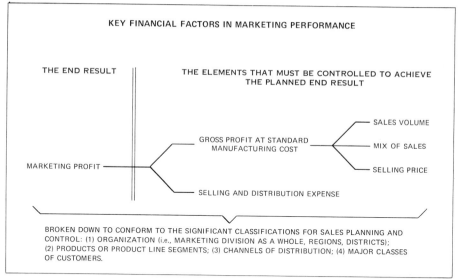

FIG. 31 *Marketing results depend on how well key financial elements are controlled.*

Key factors for success in the industry The next step in determining what needs to be planned and controlled is to isolate the critical factors on which success depends in that particular industry. For any business—large or small—these will ordinarily consist of not more than four to six factors. They are the activities or aspects of the business that have a major impact not only on short-term profits but also on long-term growth and competitive strength. As such, they are the more important because they represent the only means through which the key financial variables just discussed can be influenced or changed.

Obviously these factors (unlike the common financial indices) will differ widely from industry to industry. For example, in the automobile industry they include car styling and maintenance of a hard-hitting dealer organization; in the chemical business, the development of new products or new uses for existing products. For a distributor of major appliances, effective management of field inventories is critical; for a life insurance company, the key is development of an adequate supply of competent branch or agency managers. For manufacturers of such commodities as industrial fasteners or fractional horsepower motors, manufacturing cost performance is a survival criterion; for other businesses, it may be substantially less important. In a business where much of the so-called value added is bought (as in aircraft manufacture) purchasing competence and the ability to manage vendors effectively can have primary profit-making significance. In a business where margins are thin, such as a supermarket chain, skill in product pricing can be the major determinant of competitive success.

The story is much the same for other types of businesses, whether they are construction companies, savings banks, textile converters, public utilities, department stores, newspaper publishers, or common carriers. For each such industry there is a handful of key factors controlling success. And for no two industries is this list of factors exactly the same. Indeed, even within one company, the critical factors will vary from time to time with shifts in industry position, product superiority, distribution methods, economic conditions, availability of raw materials, and the like.

Because of these wide variances, all that can be done here is to suggest a way of thinking that should help the analyst identify the key factors in any given situation. Specifically, the following questions should serve this purpose.

1. What things have to be done exceptionally well to *win* in this industry? Particularly, what must we do well today to ensure leadership in profit results and competitive vitality in the years ahead?

2. What are the factors that have caused or could cause companies in the industry to fail?

3. What are the unique strengths of the company's principal competitors?

4. What are the risks of product or process obsolescence? How likely are they to occur, and how critical could they be if they did occur?

5. What are the things that have to be done to increase sales volume? What are the ways in which a company in this industry increases its share of market? What impact on profits could each of these ways of growing have?

6. What are the major elements of cost, and what are the ways of reducing each?

7. What are the big profit leverage points in the industry—that is, what would be the comparative impact on profits of an equal amount of management effort expanded on each of a whole series of possible improvement opportunities?

8. For each major functional segment of the business, what are the key recurring decisions that have to be made? How much impact on profits could a good or bad decision in each of these categories have? In what way—if any— can the performance of this function give the company a competitive advantage?

Once the key factors for success in the industry have been identified, the next step in determining what needs to be planned and controlled is to explode each factor into its components. This means identifying the activities that the company must carry out in a competitively superior way to ensure that it is consistently among the top companies in the industry in results achieved on each of the key success factors.

Identifying these component activities is not easy. It requires a great deal of penetrating analysis, extended discussions with key executives, study of competitors' operations, and similar steps—all aimed at acquiring a sophisticated understanding of the profit economics of the business. But however demanding and difficult the process may be, it is the only way of singling out the elements of performance on which the organization, and therefore its management information systems, should concentrate.

As an example of this whole process, the key factors for success in the property and casualty insurance industry consist of:

1. Optimizing[3] the company's loss ratio by line of business (for example, personal insurance lines such as homeowners' and personal automobile coverage).

2. Strengthening the company's so-called "agency plant." This means acquiring and maintaining sound relationships with a network of independent agents who consistently generate a significant volume of high-quality business for the company.

3. Reducing and controlling operating expense.

4. Accelerating cash flow; for example, speeding up premium collections so that these funds can be included in the company's investment portfolio more quickly.

5. Maximizing long-term investment yield and gains.

To extend the example further, Fig. 32 shows how these five key success factors might be broken down into the network of underlying activities that must be carried out well to ensure superior performance on each of the key factors themselves.

Thus far our focus has been on the business as a whole. But a very much similar approach can be followed to identify the factors governing the contribution that each department makes to the total enterprise. Specifically, the questions to be asked here are

1. In what specific ways can this department contribute to the company's "bottom-line" results?

2. What conditions will exist or what results will have been achieved if the performance of the department is outstanding? What are the things that have to be done well to achieve this kind of end result?

As an example, such an analysis was made of the purchasing department in a large company.

> The analysis showed that only two of the major elements to be planned and controlled were covered by the company's accounting system: (1) the amounts of money paid for purchased components and materials, and (2) the expense of running the purchasing department.
>
> But because the company was largely engaged in the assembly of purchased parts and components, other elements of purchasing performance not reported in the accounting or budgetary control systems also had a major bearing on short-term operating results. For example, through proper selection and follow-up of vendors, buyers can ensure the kind of delivery performance that avoids costly emergency purchases or production interruptions on the one hand, and excessive inventory buildups on the other. Through careful monitoring of vendors' quality performance, the company can weed out vendors with a high rejection rate at incoming inspection and also avoid acceptance of items that

[3]The term "optimizing" is used here instead of "minimizing" because the *lowest* loss ratio (percentage of total premium income on the line of business that is paid out in claims) is seldom, if ever, the best loss ratio. The reason is that the lowest possible loss ratio would result in the company's issuing so few policies of this type that the line of business would make very little contribution to the flow of funds available for investment and therefore for generation of investment income.

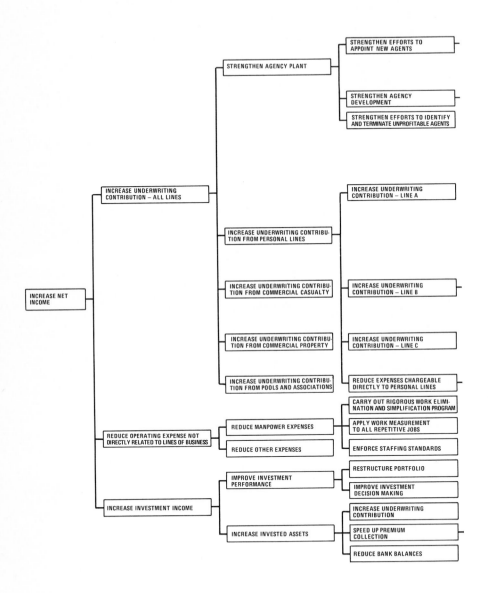

FIG. 32 *Underlying the key factors for success of a property and casualty insurance company is a host of activities that must be performed in a competitively superior way.*

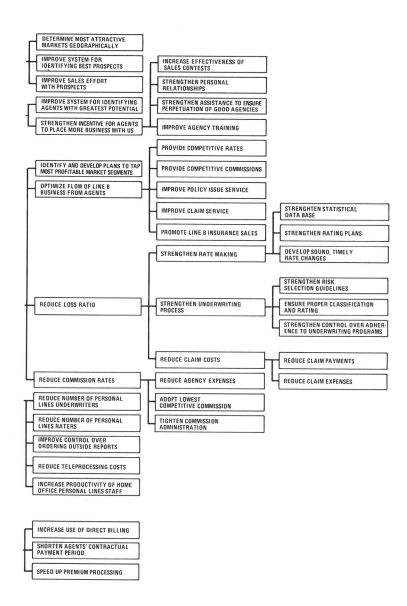

cause difficulty during assembly or premature breakdown of the company's finished products after sale.

This analysis led to the following checklist of performance factors for planning and controlling the company's purchasing activity.

DEPARTMENTAL OPERATING COSTS

1. Is the departmental budget based on up-to-date engineered standards—that is, has a comprehensive program of work elimination, work simplification, and work measurement been carried out during the past 3 years?

2. How does actual departmental expense compare with budget?

CONSEQUENTIAL COSTS AND BENEFITS

1. Price performance: On key items purchased, have prices paid been consistently at, above, or below the market?

2. Results of purchasing research: What better or less costly substitute materials have been developed or savings opportunities uncovered through research into different quantities, types of suppliers, types of packaging, etc.?

3. Vendor delivery performance: Have delivery dates been secured from suppliers? How well have they been met? What have been the consequences of delivery failures?

4. Vendor quality performance: What have been the consequences of rejection or failure of purchased items—for example, production interruptions, rework costs, and complaints or product returns from our customers?

Developing the Information Needed for Planning and Control

Once the major elements to be planned and controlled have been identified, the next step in a fundamental approach to improving an organization's MIS is to determine the specific information needed for reaching decisions on each element and measuring how well it is performed. In carrying out this phase of the diagnosis, the analyst will find it profitable to think in terms of the basic characteristics of (1) a soundly conceived management information system as a whole, and (2) the individual planning and control reports that comprise the system. These characteristics are set forth in the three sections that follow.

The overall management information system   A well-designed system of management information as a whole has the following characteristics.

1. It focuses attention on the key factors for success in the industry and the activities that produce results in each of these key-factor areas.

2. It provides all executives and managers with the information they need for

   a. *Planning* the end results for which they are responsible and the concrete action programs through which they will achieve these results.

   b. *Measuring* their own and their subordinates' performance and taking corrective action.

3. The system reflects the fact that the kind of information needed for planning is different from the kind needed for performance measurement and control.

4. It is complete in the sense that it covers
   a. Nonfinancial as well as financial factors and elements of performance.
   b. Long-term goals and programs as well as short-term ones.

5. It has an *operating focus* rather than an accounting and financial focus. This means that information exists not just because traditional accounting practices require it, but because it is specifically designed to help top-management executives as well as operating managers to do their jobs better.

Planning reports Here are the characteristics of a comprehensive, well-designed body of planning information.

1. It is structured to help achieve the following purposes (which represent basic steps in both the strategic and the operational planning processes):
   a. To assess the performance and outlook of the company and each of its major competitors (for example, as to market share, profitability, manufacturing costs, technological position)
   b. To identify and evaluate external threats to the company or the industry as a whole
   c. To identify opportunities and needs for making basic improvements in overall company performance, the performance of any given unit, the profitability of a product, etc.
   d. To bring all units up to the level of the best
   e. To estimate the payout and risks of alternative courses of action (usually alternative uses for the available resources of manpower and money)

2. Planning information is expressed in terms that make possible *meaningful* comparisons
   a. Between performance of the same organization unit in the current year and in prior years
   b. Among like organization units (say, among branch offices, plants, or operating divisions of the same company)
   c. Among products and market segments or classes of customers

Control reports In contrast to planning reports, well-designed management control reports have the following characteristics.

1. They follow lines of organization and are structured primarily to reflect the responsibilities of individual positions. This means that the results reported should be segregated according to the individual most responsible for control over them.

2. For each separate organizational unit of the business, they cover all the important controllable elements of performance that affect the group's total contribution to companywide goals. As a corollary, allocated or prorated costs over which executives have no control should not be combined in their reports

with the costs over which they do have control. (The surest way of avoiding this danger is to exclude allocated costs from control reports altogether. If that is not possible, such costs should at least be segregated or earmarked to indicate that the executive is not expected to control them.)

3. Control reports facilitate evaluation of what has happened by measuring what was actually achieved against what *should have been achieved,* not against what was achieved during a previous period. In other words, they compare actual results with planned results, not with past results.

4. In addition, control reports show causes of variances from planned results. Where possible, these causes should be isolated in the reported figures themselves. For example, variances from planned marketing profit should be broken down to show the portion of the overall variance caused by sales volume, mix of sales, selling price, and selling expense. Where causes cannot be expressed quantitatively, they should be reported and interpreted in an accompanying narrative explanation.

5. Control reports represent an *integrated* structure in which the information furnished to all levels of management is tied together and simply becomes more condensed as higher levels of management are reached. This means that the control report structure ought to have unity. One should be able to follow logically and step by step from one report to another and from summary to supplementary or subsidiary reports.

6. Control reports represent the hard core of information needed for performance measurement, rather than a flood of information designed merely to keep everyone informed.

7. They are consistent with the company's organization philosophy in terms of the positions to which they are distributed and the amount of detail presented to each level of management. (Receipt of the same report by more than two levels of management—although not conclusive—at least points to the possibility of faulty distribution.)

8. Finally, control reports cover relatively short periods of time (for example, 1 month) and are issued promptly enough to facilitate corrective action.

The distinction between planning and control reports   From the foregoing, we see that control reports have two distinguishing features.

1. They compare actual performance with planned results or predetermined unit standards instead of with past-period results.

2. They cover relatively short periods of time so that corrective action can be taken promptly. Ordinarily the time span ranges from daily for frontline control (of work spoilage, for example) to monthly for overall control by higher levels of management.

On each of these two features, planning reports differ materially from control reports. To be really significant, the information needed in formulating operating plans and policies must cover comparatively long periods of time. Also, the primary emphasis in planning reports is on the long-term trend of actual performance. Thus, planning reports look to the past, comparing the

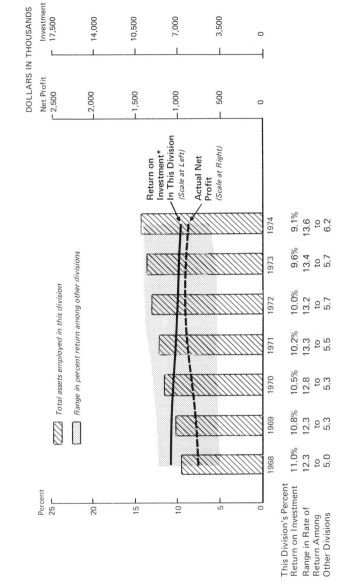

FIG. 33 *Planning reports consider the past and present in projecting the future.*

results of the same organization unit or group of similar units over a period of years to highlight such factors as growth or decline in industry position, profit margins, return on investment, and unit costs or productivity.

In summary, therefore, planning reports focus greater attention on historical performance in order to develop a meaningful picture of trends. And they are usually not prepared or reviewed as frequently as control reports.

These distinctions are illustrated in Fig. 33, an example of a planning report, (page 287) and Fig. 34, an example of a control report.

Evaluating an Individual Report

Up to this point the discussion has focused on building a model *system* of management information without paying much attention to the details of existing reports. Sooner or later, however, the analyst will have to evaluate existing reports and decide what disposition to make of them and how each relates to the model system being constructed.

This phase of the work is often critical because the proposed discontinuance of information now being received is much more likely to incite resistance than the recommendation that new information be provided. The following series of questions suggests a way of probing deeply to get at the real worth of an existing report.

1. What specific decisions can be made or action can be taken on the basis of the information contained in this report? That is, if the results appearing in the report were significantly different from what was expected, what specific decisions would be called for or what action would be taken?

2. What is this report designed to *protect against*? That is, what event or development is it intended to detect or forestall? How likely is this to happen, and what would be its real consequences?

This line of analysis can help greatly to get rid of some "sacred cows." As an illustration, take the experience of a systems analyst in a large industrial organization who was engaged in a study of the company's purchasing department. In the course of the work, the analyst raised a question about the purpose of a monthly report that was sent to the director of purchasing. For each of the six geographically dispersed purchasing departments reporting to that executive, the report showed the value of purchase orders placed during the preceding month, broken down by fifteen major commodity classifications. In reply to the analyst's question, the executive said, "Why, this is one of my most important reports. It shows me the volume of activity in each of the purchasing units for which I am responsible."

An important point to recognize in this response is that the purchasing director did what others commonly do in attempting to explain the purpose of a report: He simply described the *contents* of the report without referring to its purpose or end use at all. Stated another way, he said what the report told him, not what he did or could do with it.

Persistent digging by the analyst showed that the only conceivable purpose of the report was to alert the executive when the volume of activity in any of the six purchasing units had changed sufficiently to suggest the need for changing

SUMMARY OF OPERATING PROFIT PERFORMANCE

MONTH   YEAR

THIS MONTH			YEAR-TO-DATE	
AMOUNT	PERCENT TO NET SALES	PROFIT AND LOSS ITEM	PERCENT TO NET SALES	AMOUNT
$		NET SALES		$
		Less: Standard Cost To Manufacture Goods Sold		
		GROSS MARGIN REALIZED AT STANDARD COST		
		Less: Actual Selling and Distribution Expense		
		MERCHANDISING PROFIT		
		Less: Unfavorable or *Plus:* (Favorable) Variances from Standard Manufacturing Cost		
		Less: General, Administrative, and Research Expenses		
		ACTUAL OPERATING PROFIT		
		Operating Profit Objective		
		Amount Over or (Under) Objective		
		NONOPERATING INCOME AND EXPENSE (NET)		
		PROFIT BEFORE INCOME TAXES		
		FEDERAL AND STATE INCOME TAXES		
		NET PROFIT		

REASONS FOR VARIANCE FROM OPERATING PROFIT OBJECTIVE

		FAVORABLE OR (UNFAVORABLE) VARIANCES FROM PLANNED MERCHANDISING PROFIT CAUSED BY		
		Volume or Mix of Sales		
		Sales Price		
		Sales Discounts and Allowances		
		Selling and Distribution Expense		
		Total Merchandising Variance		
		FAVORABLE OR (UNFAVORABLE) VARIANCES FROM PLANNED MANUFACTURING COST CAUSED BY		
		Labor Usage		
		Labor Price		
		Material Usage		
		Material Price		
		Variable Overhead — Amount Spent (Over) or Under Flexible Budget		
		Fixed Overhead		
		Amount Spent (Over) or Under Fixed Budget		
		Amount Over — or (Underabsorbed) Because Production Volume Was Up or Down		
		Total Manufacturing Variance		
		VARIANCES IN GENERAL AND ADMINISTRATIVE EXPENSE		
		VARIANCES IN RESEARCH AND DEVELOPMENT EXPENSE		
		Total Variance from Operating Profit Objective		

Distribution: *President, Divisional Vice Presidents, Controller, and Treasurer.*

FIG. 34 *Control reports compare actual results against predetermined standards or objectives.*

the size of its staff. But for the following reasons, the existing report could not even serve this purpose well.

1. The personnel complement of a service unit like purchasing can ordinarily be adjusted only as longer-term changes take place in its workload. But the report under consideration contained no meaningful picture of trends; it

showed only the dollar value of purchases for the preceding month and the year to date.

2. The dollar information in the report was built up from vendors' invoices which the six purchasing units had approved for payment. But on many purchase orders, vendors' deliveries were made and invoices processed several weeks and often months after the purchase order was placed. Thus, the dollar figures in the report were not a good indicator of current work volume.

3. In addition, the dollar value of purchases is not the best measure of purchasing activity, since no more time is ordinarily required to purchase $500 worth of a given operating supply than $100 worth. For this reason, the *number of different items* purchased—by significant commodity class—is a much better indicator of activity.

The various purchasing units were already maintaining statistics on number of purchase requisitions processed, broken down by commodity class. Because of this and the other study findings, the analyst was finally able to secure discontinuance of the report in question. In its place, he developed a simple long-term trend chart that showed to the director of purchasing the fluctuations in volume of purchase requisitions placed by each unit under him.

Effective Techniques for Presenting Reports

Once an MIS model has been constructed by the process of identifying the information needed to plan and control the critical elements of performance, the remaining task is to use these basic raw materials in the design of finished reports. This is essentially a matter of arranging the desired information in a form that can be easily understood by an executive who is not figure-oriented. Admittedly, the concern here is with form, not substance, but a great many business executives will testify that it has a major bearing on the usefulness of reports—and therefore on the extent to which they are used.

Following are some guidelines that will help the analyst carry out this part of the MIS improvement program.

1. Be sure that every report is a simple and concise statement of only the pertinent facts so that the recipient can see immediately what he needs to know without being confused or without having significant information obscured by excessive detail. To this end, here are steps the analyst can take to avoid the "bed sheet" type of figure reports that discourage use:

    a. Be sure that the amount of detail is limited to that which can be controlled. For example, reports for frontline sales supervisors do not need to show dollars as well as product units when prices are set by headquarters executives.

    b. Eliminate all computed figures that may have been necessary to build up a schedule but are not needed to understand the end result.

    c. Report only actual performance and variance from plans. To include the budget amount in the report serves only to complicate it and to divert attention from the important facts.

    d. Eliminate cents and round large figures to the nearest thousand or million dollars.

    e. Review the account classifications or product breakdowns to see whether they are producing amounts smaller than management needs in order to control expenses or sales results effectively.

2. Express every report in the language of the user. Be particularly careful to avoid technical accounting jargon, such as "burden variance debit." Also be sure that report titles, columnar headings, and item names or descriptions are so clear that no one *could* misunderstand them.

3. Use presentation techniques that highlight significant developments or out-of-line performance. The basic objective should be to focus the reader's attention on matters requiring decision, action, or further investigation. The recipient should be able to grasp the significant facts and relationships quickly without first having to engage in an extensive "mining" operation. The most common ways of highlighting matters of importance are as follows.

    a. Compute and show deviations from plan (that is, from standards or budget) so that the reader does not have to make the calculations mentally.

    b. If the significance of the data is not readily apparent to the executive who will use the report, then call attention to the important facts. That is, if the report is not clearly self-explanatory, distill its significance for the user. One way of doing this is to red-circle the important variances. This is satisfactory where no further explanation is needed to make the report meaningful. Another method is to provide a narrative interpretation, either in footnotes or in a supplementary report.

    c. Use graphic methods of presentation wherever feasible—not only to reveal trends, but also to make the entire report easier to understand. Remember, however, that mere charting does not by itself ensure clarity. Like a statistical report, a chart that attempts to give too much information can be more confusing than helpful.

CONCLUSION

This chapter has not intended to suggest that any structure of management information or system of control can replace the need for judgment, vision, resourcefulness, skill in motivating men, entrepreneurial drive, or any other executive quality. There is much evidence, however, that successful application of the approach outlined here can help to bring these executive qualities into fuller play by keeping executives better informed and by focusing their attention on the factors that have a major bearing on profit results.

As an illustration of what can be accomplished by applying this approach, one company that used it was able to reduce its recurring reports from 560 to 140 and the number of copies of reports being distributed from over 2,800 to 460. An interesting sidelight is that the number of reports received by the president was reduced from 37 to 8.

But reduction in the quantity of information was not the most important result achieved. The big gains—in terms of their bearing on management effectiveness—were scored in the quality of the information produced. The president of the company put it this way: "I would guess that as a result of what we have done, almost 75 percent of the figures have been taken off my desk. And yet, for the first time, we have confidence that all the important bases are covered and that we are getting the vital information we need to run and build our business."

FOURTEEN

Optimizing Development of Computer-based Systems

Up to this point, we have dealt only tangentially with the role of the computer in the execution of operational and general-management systems. Chapter 2 referred to the contribution that a formal systems improvement program can make in helping an organization capitalize on the full potential of the computer. Chapter 3 cited the hazards of electronic myopia and the danger of assuming that the systemization of modern office operations is synonymous with computerization. Chapter 13 noted that in many companies the computer is still not being used in a way that meets management's need for better decision-making and control information.

But aside from such passing references as these, relatively little coverage has been given so far to the impact of the computer on the systems function. The reason for delaying treatment of so important a piece of the whole until now is twofold.

First, however powerful and versatile a tool the computer may be, it is still only a *method* of carrying out systems—a means and not an end. In this light, it is important to establish a solid foundation of substance before dealing with form: to develop a concern for the relevance and utility of a system before becoming preoccupied with methods of carrying it out, whether they are manual, mechanical, or electronic. This is why the preceding chapters have focused so much attention on such elements as the end-result objectives of a system, the activities that must be performed well to ensure achievement of those objectives, and the approach to determining the real profit-making value of everything that is now being done.

293

Second, most aspects of computer systems management are not only highly specialized but also closely related to each other. For this reason, the subjects of EDP systems design and computer use can best be dealt with altogether in a single chapter rather than spread throughout the book as pieces of several other chapters.

Having underscored the fact that the computer is only a *method* of carrying out systems, we hasten to add that achieving the optimum use of this particular methodology is, for most organizations, a goal worth the highest level of effort. At least four reasons support this conviction.

1. *The penalty for failure is very great.* Chapter 2 noted that because the benefits to be gained from skillful use of the computer are so far-reaching, organizations that fail to master this tool inevitably fall behind in the competitive race.

2. *In most organizations, the total cost of computer operations is huge and growing.* Today it is not uncommon for this function to cost between $10 million and $25 million a year in medium- to large-sized companies. Clearly no organization can justify an annual expense of such magnitude without having both a full awareness of the values flowing from that expenditure and a program for increasing those values.

3. *Designing and installing most individual computer-based systems is an expensive undertaking, requiring a long lead time.* Because the penalty cost of faulty selections is so high, those responsible for computer systems management must select potential computer applications with extreme care. The critical significance of this point is intensified when one considers the lead time involved. As Richard Young explains:

> The typical large systems project takes two to four years from the first investigation and proposal until it is programmed and ready for implementation. If many people or many different locations are involved, a year or two more may be required for full implementation. It may then take three to four years thereafter for the system to show an adequate return on the effort invested; that is, the pay-back period may be three to four years after full installation. Thus the typical EDP system should be designed to operate effectively under the conditions that will exist five to ten years after the system is first proposed.[1]

In the face of lead times as long as these, no organization can afford to approach the selection of new computer applications with anything less than the best talent it can muster.

4. *Finally, in most companies, the needed human resources are scarce.* Personnel required for the development of new computer applications are in especially short supply. The number of attractive new EDP applications invariably outstrips the resources any organization has or can acquire in a reasonable period of time. This fact highlights, once again, the need for committing the highest possible level of management talent to planning and controlling the whole computer program.

[1]Richard C. Young, "Systems and Data Processing Departments Need Long-range Planning," *Computers and Automation*, May 1967, p. 30.

Compelling as these four reasons may be, some companies have done a vastly better job than others in harnessing the power of the computer. To measure the differences between companies and to identify the reasons for them, a McKinsey research study (referred to in Chap. 2) sought answers to the following kinds of questions: What companies are leading the field in EDP performance results? How did they get where they are today? What are the distinguishing features of their computer efforts?

The study revealed that, without exception, the front-running companies owed their superior performance to the approach they were taking to the four critical aspects of computer management:

1. Setting objectives for the EDP effort
2. Fixing responsibility for EDP results
3. Planning the slate of computer applications
4. Planning and controlling the design of individual EDP systems

SETTING OBJECTIVES FOR THE EDP EFFORT

In all the companies that are realizing outstanding *economic* results from computer applications, top management is simply unwilling to settle for anything less. In the not-so-successful companies, many managers exhibit a tendency to keep the computer at arm's length for fear of exposing their technical inadequacies. The president of one company frankly admits: "If a stockholder were to walk into my office today and ask me what was going on in each of our departments, I could answer his question confidently for every group except our computer department."

In contrast to this sort of abdication, top managements of the most successful computer users have established the *rule of high expectations*. They have made unmistakably clear their determination to focus the organization's whole EDP effort on the *profitability of the computer as a business investment*. They stoutly insist that every computer development project be subjected to the same rigorous scrutiny and subsequent measurement that they have always applied to major capital expenditures.

To further reinforce this overall economic goal, they have imposed two other requirements: first, that their organization periodically evaluate where it stands and where it is headed in its use of the computer; second, that it establish high standards for the conduct of every phase of computer systems development.

Taking Stock of Computer Accomplishments and Outlook

As part of its annual planning process, the top management of one large multinational company requires that the computer department incorporate in its written plans the sort of stocktaking called for by the following questions.

1. How effective are our present EDP systems? Are they providing management with the information it needs for effective decision making and control? Have we realized all the savings and other benefits we originally

anticipated from the applications now installed on our computer? What, in summary, are our winning and losing computer applications? Is our present EDP effort soundly organized, managed, and staffed? Where does the company stand in relation to the "state of the art" and computer results achieved by other companies? What steps can we take in the immediate future to improve the timeliness, accuracy, and profit contribution of existing systems? What mistakes have we made and what lessons have we learned that will help us develop better EDP plans this year and in the future?

2. Are our plans for future data processing projects well conceived and realistic? Do they have a high profit improvement potential? Are these planned projects directed at the key problems of our business? Are the proposed data processing equipment and personnel adequate to support the planned applications? Is there sufficient managerial control of existing operations and of the development and implementation of new systems?

Setting Standards for Key Aspects of Computer Systems Development

As a second means of making the rule of high expectations more explicit, top managements of the most successful computer users insist that rigorous criteria be developed and codified to govern the performance of every aspect of the computer systems program. These aspects are covered extensively in the remaining sections of this chapter. It will suffice here simply to say that what top management should press for is a set of definitive qualitative standards that will help ensure consistently superior performance in each of the following phases of computer management.

1. Identification of the highest potential applications for computer systems development

2. Cost-effective design, testing, and installation of new computer systems

3. Postinstallation evaluation to ensure that forecast benefits are achieved or exceeded.

FIXING RESPONSIBILITY FOR EDP RESULTS

A second distinguishing characteristic of the most successful computer users is that top management assigns responsibility for EDP results to the *line and functional executives* served by the organization's computer systems and then sees to it that they exercise this responsibility.

In contrast, executives of organizations with less successful computer operations wittingly or unwittingly assign responsibility for computer results to EDP technicians. Although they would not dream of turning any other type of major project over to a contractor without their own deep personal involvement, they will do so with computer systems that affect the whole fabric of their business operation and management process. In saying this, we do not mean to imply that EDP professionals should play no part in computer systems design. Quite the contrary. They can make an important contribution and clearly should be partners in the whole development process. At the same time, however, no organization should run the risk of relying exclusively or even primarily on computer specialists for EDP systems selection, design, and evaluation. For one reason, the design of the most productive computer systems today requires not only knowledge of what the computer can do but also knowledge of the profit economics of the business and its key decision-making and control requirements. No one would seriously contend that an organization should count on its EDP technicians to bring this understanding of the business to bear. Yet such a dependence arises out of default whenever top management fails to hold line and functional executives accountable for EDP results.

The second reason for not relying primarily on computer-oriented specialists is that many of them, having gained their experience during an exciting period of growth in computer technology, are often more interested in developing technically challenging and sophisticated applications than they are in improving the management of the business. The point is well illustrated by the experience of one large manufacturing company whose computer staff developed a cathode-ray tube system for instantly presenting data generated by an outmoded cost accounting system. As the company's executive vice president said, "This new system is a technical triumph but a commercial failure. And I have to admit that responsibility for this failure rests with our operating managers, not our computer people. We just never did match their technical virtuosity with our own knowledge of where faster, tighter controls are needed to improve our profit results."

In summary, most of the top-performing computer users take either one of two organizational approaches: In addition to the usual operations research specialists and other professionals, some companies assign to the corporate computer staff an experienced representative of each major functional division of the business. Other companies, following the project approach to computer systems development, use project teams that include employees temporarily transferred from operating departments. But whichever approach these companies adopt, departmental and divisional managers know that top management will *insist on economic results*—and that in the final analysis, *they* are the ones who will be held responsible for achieving those results.

PLANNING THE SLATE OF COMPUTER APPLICATIONS

The task of optimizing use of the computer in any organization can be approached in two ways. The first is basically a technical approach aimed at ensuring that the computer performs the existing operations in the most efficient way. The second is an analytical approach aimed at making sure that the work done by the computer is relevant and meets the real needs of the enterprise.

The first approach—although obviously essential—does not have the leverage of the second. It is concerned with selecting and fine-tuning hardware and software, using this disk drive rather than that, eliminating unnecessary program steps, improving techniques for detecting errors, balancing work loads, and deploying programmers and machine operators effectively. Since these activities require a high level of technical competence, they should be left primarily to data processing managers—provided, however, that general management has some adequate basis for measuring performance.

The second approach moves beyond technical preoccupation. It is concerned with planning the slate of computer systems applications that are to be developed. Here lies the real key to successful use of the computer. If projects are poorly chosen, the organization can manage their design and installation outstandingly well and still lose money on them. But if they are well chosen, the organization can often tolerate a fair amount of slippage between forecast and realized benefits and still achieve a high rate of return on each application.

The process of planning the slate of computer applications breaks down into the following closely related phases.

1. Identifying potential systems development applications
2. Analyzing and evaluating each proposed application
3. Selecting the projects that will make up the total systems development program

Identifying Potential Applications

Planning an organization's slate of computer systems development projects begins with the imaginative identification of the potential applications from which the ultimate choices must be made. In most organizations, this phase of the overall process is haphazard and unorganized. In addition, all too often, it is left largely to EDP professionals, who lack the background needed to determine where the most profitable applications can be found.

In contrast, successful identification of projects is an ordered, mind-stretching process that meets the following two tests:

1. It is concerned with fundamentals.
2. Responsibility for it is clearly fixed.

Attention to fundamentals Those responsible for project identification must school themselves to think independently and searchingly. They must not limit their sights to typical applications that come quickly to mind, to the existing network of manual systems, or to the computer systems that other companies have developed. Rather they must bring to this process the same sort of thinking and questioning that should be applied to structuring a management information system (Chap. 13). Specifically, in looking for high-potential computer applications, they need to ask: (1) What are the key factors controlling success in our industry? What must we do well to be a competitive leader? (2) What costs and revenues would be affected most by improved decision making and tighter control?

The fruits of this kind of disciplined thinking are many and varied. Following are some examples of profit-making applications taken from successful computer users.

1. A fertilizer manufacturer has developed a computer-based simulation model that helps production management answer such questions as: How much should we plan to manufacture, ship and store at each plant location to minimize our total costs of production, distribution, and storage over a 1-year period?

2. A major oil company has developed computer-based techniques of seismic analysis to assist in the planning of exploratory drilling.

3. Several railroads have greatly increased their utilization of freight cars and reduced the need for additional investment in rolling stock by creating control centers that keep up-to-the-minute records of the location and condition of each car.

4. A leading consumer-goods company has developed an EDP system for interpreting test marketing results faster and more reliably through rapid access to comprehensive historical information on sales by product, package, promotional technique, and the like.

5. A manufacturer of high-fashion clothing, with national outlets and multiple plants, has developed a computer forecasting model for establishing preliminary cutting schedules at the beginning of each season. This application has proved so successful in matching production to demand that a new project is now under way to apply computer forecasting methods to the development of purchasing plans.

6. An international oil company has developed computer models of its crude oil and finished-product distribution systems which have enabled it to increase the utilization of its marine tanker fleet and to cut ship-chartering costs.

7. A nationwide retail chain is using a central computer-based dispatch system to reduce branch-store inventories by cutting the stock replenishment cycle.

Even though the foregoing types of applications can have a great influence on costs, revenues, and service—and consequently on profits—computer systems planners must recognize that such applications are all examples of *operational* systems—for example, marketing, production, and logistical systems. Although most high-potential new applications are likely to be found in operational areas, systems designers must not confine themselves to those areas. They should also look for possible applications in the *general-management* area. Difficult though it may be to adapt a general-management decision system to the computer, the potential benefits of doing so are almost certain to be correspondingly great. Moreover, there can be little doubt that as the EDP leaders achieve more maturity in their computer-based operational systems applications, they will turn increasing attention to the development of more advanced general-management systems.

As an illustration of the great promise for computer use in the general-management field, one organization has developed a computer-based planning model that it is using successfully to analyze and forecast

1. The financial results of acquisitions, mergers, and divestments and of the alternative methods of financing them

2. The outcome of alternative strategies for raising capital funds, including internal financing through cash flow, reduction in dividends, or sale and leaseback of facilities; short-term borrowing; and the sale of equity, convertible, or fixed-income investments

3. The financial and operating performance of a multidivisional, multinational company under a series of different assumptions on currency values, inflation rates, product demand and price levels, investment in added plant capacity, and investment in new market development

4. The economic impact of alternative strategies for new product development

5. The financial results of possible competitive moves such as reducing or increasing prices, reducing or increasing market share.

Clear assignment of responsibility In fixing responsibility for identifying potential systems development projects, two requirements must be met. First, for the reasons already discussed, a representative of each functional division of the organization must participate in the process along with EDP personnel. Second, a single individual must be charged with leading the group, and especially with guiding it in the development of interfunctional and general-management systems proposals. In one company, this responsibility is assigned to the corporate systems and procedures director. In another, it is assigned to the head of the central computer department, but—in the discharge of this special function—he and the project-identification team report to a computer planning committee made up of four functional executives.

Evaluating Potential Computer Applications

The basic purpose of this phase of the overall computer systems development process is to gather and analyze the information needed to evaluate each potential application and either reject it or recommend it for inclusion in the approved project slate. This phase consists of five steps.

1. Defining the project
2. Identifying alternative systems approaches to carrying out the function
3. Estimating benefits
4. Estimating development and operating costs
5. Determining feasibility

Although these steps are generally carried out in the sequence indicated, the analyst should regard them as a checklist rather than as an inflexible schedule. At many points in the analysis, it may be desirable to back up and repeat a task. For example, the analysis of costs and benefits for one approach may indicate that another approach should be developed. In addition, all five steps are not always carried to the end for each project. Rather, the analyst should evaluate the information gathered at each step as a basis for recommending whether to continue or abandon the project.

Defining the project The first step in developing a sound basis for evaluation is to clearly define the objectives and scope of the project.

Project objectives. To be meaningful, such objectives must be expressed in quantitative terms wherever possible, as well as in terms of the kind of improvement sought. Moreover, the statement must be sufficiently precise so that, following installation of the new system, management can readily determine whether the project is, in fact, achieving the objective for which it was originally approved.

This need for precision is often disregarded. Computer systems planners tend to express project objectives in vague, general terms such as: "To reduce the costs of our operations," "To automate the issuance of personal automobile and homeowners' policies," "To develop a model of our corporate financial system," or "To provide better information for decision making." None of these statements has any value as a basis for reaching a "go, no-go" decision on a project. Although systems planners begin to come closer to the mark when they state that the objective of a project is to improve a specific type of decision, even then they need to stipulate *in what way* and *to what degree.* If the project involves the design of a simulation model, they must identify what activity will be performed better, and *how much better,* through the use of the model.

Examples of valid objectives are (1) to reduce warehouse inventories by 20 percent and hold out-of-stock conditions to 2 percent; and (2) to reduce average processing time on commercial property and casualty claims from 60 to 30 days.

Project scope. The scope must be stated clearly if the proposed undertaking is to be understood by members of management who must approve it and by the EDP and user-department personnel who must carry it out. Further, if the statement of scope is to be helpful in ultimately controlling the execution of the project—its design, testing, and implementation—it must include a description of at least five elements:

1. The activities that are included in the proposed system and those that are specifically excluded

2. The relationship of this project to other projects or systems—whether existing, under development, or requested for the future

3. The changes in existing systems that will be required by adoption of the proposed project

4. The effect of this project on departments other than the requesting or user department

5. The departments that either supply data to the system or receive information from it

Identifying alternative systems approaches The objectives set for the project may be reached in a number of ways. This section of the project proposal spells out all such alternative approaches. They may represent variations in the processes of the user department, variations in the EDP techniques to be employed, or variations in the degree of automation. They may even represent differences of opinion between EDP and user personnel on how the problem can best be solved. No matter what the source, the planning team should include on its list all worthwhile systems alternatives.

Estimating benefits

The senior vice president for administration of a large consumer-goods manufacturing company was recently asked to approve a slate of three project proposals for the development of new computer-based systems. Following is a brief description of each of these projects.

1. To design a model of the company's physical distribution system—a model that, the company hoped, would be used in both long-range planning and daily management of operations. The justification offered for this project was simply: "Cost data on the present distribution system are scanty and out of date."

2. To design a revised sales-call reporting system. Under this proposal, the computer was to analyze sales representatives' routes as well as product- and customer-profitability. Based on this analysis, the computer would then prepare weekly instructions to sales representatives that specified (a) customers to call on, (b) the sequence of calls, (c) targeted sales by product, and (d) weekly sales quotas. The justification cited here was merely: "Improved utilization of sales representatives' time and better customer service."

3. To design a computer-based strategic management information system. No details were given on the approach to be taken in determining management information needs, on the cost of collecting required information and making it available, or on the proposed techniques for helping and motivating managers to use the new information. The justification for this project was only that "It will be a basic research project"—as indeed it would have been.

These three proposals were listed, without specific cost or benefit estimates, in the annually updated EDP plan submitted as part of the computer department's budget request for the year. Even more significant, these three projects alone would have accounted for over 80 percent of the resources available for new systems development.

This example typifies the problem that many managements face today in approving computer systems development programs. The project proposals that are advanced lack an explicit statement of the economic results expected from them. Admittedly, the task of estimating benefits is always complex. And now that the emphasis in computer use has shifted from cost reduction applications to management decision-making and control applications, estimates are even more difficult to make. Certainly the potential value of an EDP system aimed at cutting direct costs or investment requirements is much easier to measure than that of a system designed to improve information for management decisions. Because of this increased difficulty, proposers of new computer systems tend more and more to resort to generality: "The *intangible* benefits of this proposed application, although impossible to measure, are unquestionably much larger than the costs."

On this matter of intangible benefits, Harvey Golub, one of the long-time leaders in the theory and practice of management uses of the computer, has stated: "Only measurable benefits are real; intangible benefits are not measurable; intangible benefits, therefore, are not real."[2]

---

[2]Harvey Golub refers to this statement (in an unpublished manuscript) as one of "Golub's Laws of Computerdom." Others of his so-called laws are quoted in later sections of the chapter.

Because intangible benefits are difficult to measure does not mean that computer systems planners should ignore them. To do so would clearly be foolhardy, particularly since many of the major benefits of the computer still waiting to be exploited are among those most difficult to quantify. What we do maintain, however, is that difficult though the task may be, it can be done with much more rigor and tough-mindedness than most analysts have shown to date.

As an example, if one of the principal outputs claimed for a new system is "improved information," the systems analyst should ask the user department (1) to specify the decisions that will be affected and the precise ways in which the new information will improve the making of those decisions, and (2) to estimate the profit-making value of a 10 or 20 percent improvement in the quality of such decisions. Even though this process cannot be carried out with finite exactness, it should produce insights and judgments that will greatly improve the evaluation of proposed computer systems projects.

Estimating development and operating costs   Having determined the range of potential benefits, the analyst should next estimate the development and operating costs of the system, including those that will be incurred by both the user and the computer departments. This estimating process is often poorly performed simply because EDP professionals as a group do not plan or control project work effectively. This widespread weakness, in turn, has produced a long history of cost and schedule overruns.

For example, a review of one company's history of EDP applications over the past 5 years revealed the following facts. Out of twenty-five major systems developed, all but one—that is, twenty-four out of twenty-five—had exceeded the original project estimates of time and costs by 25 to 300 percent. What is more, the one exception had been subcontracted.

Development costs.   The major elements of development costs are
1. Systems analysis and design
2. Programming, testing, and documentation
3. Computer program test time
4. File creation and conversion
5. Training
6. Parallel operation and start-up

Among the foregoing, systems analysis and programming account for the highest proportion of cost, but the user department will also incur significant cost. For this reason, it is important that the potential users understand and help develop the cost estimates.

Development costs should be estimated in terms of worker-months of effort, by skill category, for both the computer and the user departments. Realistic estimates should also be made of total developmental *time*. The addition of these two estimates of effort and elapsed development time will produce a statement of the total personnel requirements.

Operating costs. Once the new system is developed and installed, the major elements of operating costs are

1. Data capture and preparation
2. Other clerical support activities
3. Operation of the computer and peripheral equipment
4. Maintenance programming
5. Supplies

These costs should be estimated on an annual basis. They should cover all activities from initial data capture through dissemination of the final reports to the user, including all support activities necessary for the system. Again, the user department should be deeply involved in developing such estimates, since it is likely to incur a large share of the operating costs.

Determining feasibility   If project evaluation stopped at this point, the analyst would have a single, quantified measure of potential project value, expressed in terms of payback period, return on investment, net present value, and so on. These potential benefits must now be balanced by an estimate of the probability of achieving them, or conversely, of the risk of *not* achieving them—in short, an analysis of feasibility. This assessment of the chance of success, or project feasibility, has three major components: economic, operational, and technical. Because it is impossible to measure feasibility precisely, the values assigned are usually indicated in such terms as "high," "medium," or "low."

Economic feasibility. This phase of the feasibility evaluation is aimed at measuring the likelihood that the estimated economic results of the project will actually be realized. Will the system, when developed, produce the promised benefits without significantly exceeding the estimated development time and estimated development and operating costs? In determining economic feasibility, the analyst should consider the following factors.

1. *Relationship of costs to benefits.* This relationship can be expressed in terms of payback period, cash flow, or return on investment. For example, a short payback period indicates high feasibility, since large percentage changes in costs or benefits are less likely to negate the economic value of the project.

2. *Total system development time.* For projects with long development times, economic feasibility is lowered, because the problem that the project is designed to solve may change during the development period.

3. *Economic life of the system.* The shorter the economic life of a project, the lower its economic feasibility. A project that is expected to have value for 1 year is less feasible than one having a possible economic life of 10 years, even though the total net benefits are the same. If development and implementation are delayed, the life of the project may be sharply curtailed and its economic benefits substantially reduced.

4. *Confidence in the estimates of benefits and costs* as evidenced by
   a. The range in the estimates of benefits and costs: the narrower the range, the higher the level of confidence
   b. The ease or difficulty of quantifying the estimated benefits

Operational feasibility. In this step, the analyst must evaluate the likelihood that, if the system is developed, it will be successfully implemented and used. Will operating personnel adapt to the new system, or will they resist or ignore it? Operational feasibility is measured on the basis of four factors.

1. *User attitudes and acceptance.* Operational feasibility generally increases when the management of the user department requests and supports a project. Even though a project may have originated outside the department, active support of the entire user department is essential. If user acceptance is low, the project probably will not produce anticipated benefits, regardless of the soundness or sophistication of the system.

2. *Changes in policies, organization, or managerial behavior.* If the proposed system will require changes in (*a*) existing policies, (*b*) organizational structure or reporting relationships, or (*c*) managerial values or behavior patterns, its operational feasibility will be correspondingly reduced.

3. *Changes in present operating routines.* The amount and type of change required in existing procedures affect operational feasibility because they determine training requirements. Whether those requirements are high or low depends on (*a*) the number of persons that need to be retrained, (*b*) the magnitude of the change needed, and (*c*) the degree to which skills must be upgraded.

4. *Implementation time.* The longer the implementation time, the higher the risk and lower the operational feasibility. This is because (*a*) experienced and trained individuals on whom the company is counting may become impatient or disillusioned and leave, and (*b*) user interest may be difficult to maintain.

The consequences of ignoring operational feasibility can be severe. A striking example is found in the experience of a medium-sized specialty chemical company. Its products had varying profitability rates, and many of them could be used interchangeably in certain applications. Staff members of the company thought that if sales representatives knew the gross margins on various products and grades, they would automatically sell the more profitable products. So the staff designed and installed a system that provided this information. But as it turned out, the pattern of sales did not change at all. The reason was that the salespeople were compensated on the basis of gross volume. Hence, whenever they had a choice between two products, they naturally chose to sell the one that generated the most sales dollars, not the one that contributed most to company profits. The sales manager was unwilling to change the compensation scheme, and the new computer system was consequently thrown out.

Technical feasibility. In the past, this aspect of the feasibility evaluation process was concerned with the capability of the hardware and software required, how relatively new and untried they were, and so on. But now, with the continuing advances that have been made in hardware capability and software developments, the focus of technical feasibility evaluation has shifted in most organizations to different questions: What level of skills will this proposed application require in systems analysis and design, operations research, and programming? Do we have these skills in the organization now? Can we readily acquire them?

A good example of the relevance of these questions is found in the experience of a commercial bank that recently wrote off, as a sunk cost, an investment of over $5 million in the development of a trust accounting package. The prime cause of failure appeared to be its decision to install, simultaneously, four massive modules of inter-woven coding which constituted a single software package. Although this maneuver might have been appropriate in a sophisticated environment of highly skilled and experienced programmers, the approach failed at the bank because the EDP department lacked the necessary manpower skills. A more modest, evolutionary approach would have offered far greater likelihood of success. [3]

Choosing the Systems Development Projects To Be Undertaken

Once those responsible for computer systems planning have completed their evaluations of potential development projects, they are ready to compare projects and recommend the ones that should be rejected and the ones that should be undertaken—and in what sequence. Final decisions on the project slate should be made by top management or its delegant—say, by the computer planning committee referred to earlier in the chapter—not by the EDP manager or systems staff. The reason is that the selection process usually requires the assignment of priorities to the projects proposed by functional departments. The result may be that one department is asked to wait for a year or two while a computer system is first developed for another.

Because project selection and the building of a project slate demand considerable judgment and compromise, the process cannot be reduced to a step-by-step routine or to a simple numerical ranking of projects. It can, however, be facilitated by a set of guidelines covering the essential considerations to be weighed in reaching final judgments. Following are four guidelines that will meet this need:

1. *Rank projects according to the ratio of benefits to cost, and then according to feasibility.* Ranking of projects is relatively easy if it involves only their position on a scale showing return on investment or some other measurement of benefit-cost relationship. But it becomes more difficult if feasibility is also considered. Assume, for example, that the benefit-cost ratio of one project is 50 percent above the same ratio for another project, but that the first project has a lower feasibility ranking than the second. Such situations are not at all uncommon in computer systems selection, and they call for the best combination of operating and technical judgment that the company can command.

2. *Be sure to take into account the interdependence of projects.* This factor is, of course, one of the major considerations in establishing the sequence in which projects will be undertaken. In addition, project interdependence must always be evaluated whenever any proposed project is rejected. Even though a project may not be justified on its own merits, its completion may be a prerequisite to the pursuit of other more attractive projects.

[3]F. Warren McFarlan, "Management Audit of the EDP Department," *Harvard Business Review,* May–June 1973, p. 140.

3. *Make project tradeoffs in light of EDP department constraints.* Each project considered for the project slate will require a different amount of EDP department resources—personnel (by skill category) and computer hours. While the resources required on each project can be measured in terms of cost, the total of a given resource constitutes a practical constraint because it cannot be expanded quickly. An example is a project that requires a large amount of computer time for data conversion when such time is scarce. In that instance, another project which is slightly less desirable in terms of benefit-cost ratio and feasibility may have to be substituted, because it makes better use of available EDP resources. In making such tradeoffs between projects, the following needs and benefits should be considered.

    a. Requirements for skilled systems analysts or specialized computer programmers.

    b. Requirements for large numbers of programmers.

    c. Computer machine-hours required for data conversion.

    d. The benefits that the project will generate within the EDP department. These may include a reduction in computer hours required, an increase in program efficiency, and smaller file size.

4. *Maintain a balanced mix of projects.* While giving the greatest weight to economic return and feasibility in their selection decisions, executives should also seek to maintain a balance among the projects that make up the slate. To arrive at a proper balance, they should consider

    a. *Departments served.* If a high level of interest in, and commitment to, effective use of the computer is to be maintained throughout the organization, EDP development efforts at any one time should ideally be spread over at least three functional areas.

    b. *Types of benefits.* Similarly, a balance should be maintained among projects aimed at achieving different objectives—say, clerical cost reduction, manufacturing cost reduction, improved management decision making.

    c. *Development times and risks.* Finally, the total slate should represent a sound mix of short-term, quick-payout projects and high-payback ones, which usually entail higher risks and longer development times.

PLANNING AND CONTROLLING THE DESIGN OF INDIVIDUAL EDP SYSTEMS

All the detailed action steps, guidelines, and ground rules set forth in Chap. 8, "Planning and Controlling the Individual Systems Project," are applicable to projects involving the design and testing of EDP systems. But in addition, another set of special requirements must be met and warnings heeded if computer systems projects are to stay out of trouble. The reasons for these extra precautions are implicit in Harvey Golub's second law of computerdom: "No major computer systems project is ever completed on time, within budgets, or with the same staff that started it, and none does exactly what it is supposed to do. Yours will not be the first to avoid these shortfalls."

Even though this statement may have been exaggerated for the sake of emphasis, our own experience is that it comes very close to the mark. What is more, the reasons for the persistent occurrence of large cost and schedule overruns on computer systems projects are easy to understand. They spring from the following conditions.

1. The sheer size of many EDP development projects and the long time they take.

2. The uniqueness of most such projects. The majority represent a first-time design experience for the organization seeking to carry them out.

3. The close coordination required between computer professionals and personnel of user departments.

4. The fact (already mentioned) that EDP specialists as a group do not adequately plan and control project work.

Because of these conditions and because of the history of overruns on computer systems projects, a great deal more than an ordinary project management effort is needed if any such undertaking is to be completed on time and within the budget. Specifically, three aspects of project management must be carried out more thoroughly and more rigorously than is ordinarily required.

1. Planning the project
2. Controlling project progress
3. Controlling in-process changes in systems design

Planning the Project

The key requirement to be met in computer systems project planning is that the total job be broken down in much greater detail than is needed on other less complex projects of shorter duration. This breakdown should consist of four segments.

1. *A comprehensive list of the identifiable pieces of work making up the project.* The major tasks will ordinarily include the following.

Systems specifications	Conversion system design
Systems design—noncomputer	Conversion system training
Systems design—computer	Conversion implementation
Program specifications	Pilot test evaluation
Forms design	Parallel test evaluation
Program logic	Manual operations procedures
Program coding	and training
Program logic debugging	Machine operations procedures
Creation and evaluation of	and training
program test data	
Creation and evaluation of	
system test data	

The project planner should then (*a*) divide each of these tasks into its component parts or subtasks, (*b*) estimate the total time (worker-months) required of each skill needed to carry out a subtask, (*c*) assign responsibility for each subtask to a specific person, and (*d*) gain agreement on the time estimates and responsibility assignments from the person who is accountable for each major task.

2. *Milestone events.* These are the so-called products or outputs that will be generated throughout the life of the project and that will serve as the principal bases for measuring accomplishments.

3. *A work-sequenced timetable.* The timetable should be presented in a form that recognizes the relationships between tasks and subtasks and therefore schedules them in proper sequence. It might take the form of a Gantt chart, PERT chart, or merely a precedence listing with start and completion dates.

4. *A personnel schedule.* This part of the plan should summarize the amount of worker effort in each skill category needed on the project in each month of its duration. It should also include the requirements for personnel from the user department or departments.

Controlling Project Progress

Since, for all the reasons already given, the planning of computer systems development projects tends to be imperfect, formal controls over project progress take on special significance. If these controls are to be fully effective, they must meet the following requirements.

1. Procedures must be developed for accumulating and regularly reporting actual time spent, by skill category, on each subtask.

2. Regular progress reviews should be held at frequent intervals and should involve the entire project team, including user-department personnel.

3. At each review session, the project leader must help the team probe deeply to

    a. Identify problems and anticipate slippages.

    b. Develop concrete plans and timetables for overcoming slippages.

    c. Reevaluate the likelihood of achieving project objectives and the present attractiveness of those objectives in light of any significant escalation in development costs or in the time to completion.

4. In addition to regular project progress reviews, frequent informal face-to-face meetings among the persons involved are a critical part of good project management. Programmers should meet with systems analysts, analysts with task leaders, leaders with the project team manager, the project manager with user-department executives, and lower-level analysts with lower-level supervisors in the user departments. Face-to-face meetings are the best possible means of identifying problems in the early stages, pinning them down, and resolving them, thus reducing the number of crises that will arise later on.

Controlling In-Process Changes in Systems Design

One of the most common problems in major computer systems development projects is that of controlling the tendency to continuously expand the scope of the project or introduce an endless succession of midstream changes in system design or programming. The price of giving way to this tendency is suggested in Harvey Golub's third law: "If project content is allowed to change freely, the rate of change will exceed the rate of progress."

Two real-life examples will serve to validate the point.

A property and casualty insurance company undertook the development of a computer-based claims-processing system. The estimated cost of the project was $15 million and the estimated time to completion was 2 years. At the end of 3 years of systems design and programming effort, the project team estimated that 3 more years would be required to complete the project. At that point, top management reached the unavoidable conclusion that the project would literally never be completed. The project team had been making so many changes in the scope and content of the work that the added work being created by these changes consistently exceeded the work that was being completed.

The second example comes from the experience of an international airline that was developing an on-line reservation system using a software package that had been produced by one of the computer equipment manufacturers.

In the course of testing the system, the project team accumulated a backlog of 133 technical changes that it wanted to make before installation. To make these changes the team was ready to delay installation for a full year. As it turned out, when the changes were thoroughly examined, only 5 of the 133 were really required to operate the system effectively. So the rest were deferred until the system was installed and were then scheduled as an orderly series of modifications.

PART FOUR

Cashing in on Systems Improvement Recommendations

FIFTEEN

Presenting and Selling Recommendations

The fourth phase of a systems project is that of presenting recommendations and obtaining the acceptance or authorization necessary to put them into effect. This is the critical point in every study. As Chap. 6 emphasized, the contribution of a systems staff will be judged not by the number or quality of the recommendations it develops, but by the number and importance of the constructive changes made as a result of its work. For this reason, the presentation of recommendations should be preceded by the same careful thought and planning that went into the other phases of the study. The project team members must figure out who should be "sold," and in what order. They must be sure that they neither overlook someone important nor make random presentations to anyone who may seem interested. They must think through the strategy of gaining acceptance and inducing action. Or, stated another way, they must develop their presentation in terms of the interests, knowledge, and receptivity of their audience.

A well-planned program for putting across a major systems change usually consists of two steps: (1) testing the details of the plan with frontline supervisors and gaining their support, and (2) making a formal presentation to the executives involved and getting a final decision.

TESTING RECOMMENDATIONS AND GAINING FRONTLINE SUPPORT

The first step spans both the development and the presentation phases of the study. It is a transitional step, with the dual purpose of solidifying tentative recommendations and winning the agreement of those who must put them into practice.

Working from the bottom up by first winning the agreement of operating personnel is one of the most important "musts" in the whole systems improvement program. No matter how thorough an investigation the team has made, or how conscientious operating personnel have been about "telling all," the analysts will find that some significant facts, conditions, requirements, or limitations are always overlooked until tentative recommendations are discussed. If the analysts first check their thinking against the knowledge and experience of frontline personnel, they can uncover these concealed points without embarrassment to themselves while the plan is still fluid. Otherwise these points have a habit of arising later, at a time when they may reflect unfavorably on the thoroughness of the study and, therefore, on the soundness of the conclusions reached.

A second reason for starting at the bottom is that in reaching decisions on matters of system and procedure, top-level executives tend to rely heavily on the opinions of their operating supervisors. The paperwork of any organization consists of many details completely foreign to the average executive. It is not difficult, therefore, for an unwilling or unconvinced supervisor to raise objections that effectively close the mouths of the uninformed. On the other hand, the experienced analyst knows that department and division heads seldom turn down systems proposals on which their frontline supervisors are completely sold.

A final reason for this preliminary review is that it generally helps the analyst anticipate possible points of resistance in later presentations. Subordinates often reflect, or at least have a familiarity with, the attitudes of their superiors. A good many systems proposals have been put across only because some clerk or unit head forewarned the analyst by saying, "I think this is a good idea, but if I know the boss, he'll turn it down because . . ." By discovering these probable obstacles in advance, the analyst has time to formulate the best strategy for getting around them or for minimizing their effect on the final result.

In general, the proposed changes should be checked with at least one person in each of the smallest working groups affected. This may not be practicable, of course, where the new system is to be installed in many similar units within a multiple-branch or multiple-plant business. Even there, however, the analyst will be wise to check the tentative plan with one or two units besides those that served as guinea pigs during the earlier phases of the study.

How detailed the preparation for this initial clearance should be depends on the importance of the contemplated changes. For example, if a basically different system is being proposed, the analysts should try to obtain general agreement on the fundamentals of the plan before reviewing all its operating details. On the other hand, if the framework of the existing system is being retained, they should be prepared to discuss the recommended changes in complete detail during the initial review. For such a detailed discussion, they may find it helpful either to prepare a flowchart of the revised system or to mark the contemplated changes on the chart of the present system. In this way, the "before" and "after" routines can be visualized and compared more readily by operating personnel. The analysts' kit of materials for the presentation should also, of course, include all the important findings on which conclusions were based.

This is the juncture in the systems project at which the project team's skill in handling the human relations phases of its work will begin to reveal itself. How successful the members have been in gaining the confidence of frontline personnel during the earlier parts of the study will have a real bearing on how their recommendations are received. How they present those recommendations to operating personnel will tell the rest of the story. Here are some specific pointers on handling this second part of the job.

1. Approach the initial presentation with an open mind. Think of it more as an opportunity to *test* your ideas and less as a challenge to put them across. Be ready to modify them as weaknesses or better ways become evident.

2. Make the operator feel that you are presenting your material not as final recommendations but as *ideas for discussion.* Even on points that are well settled in your own mind, it is often wise not to give the impression that you have reached definite conclusions.

3. With the possible exception of higher-level policy or organizational matters, discuss all the proposed changes openly. Do not be mysterious about them. Tell the operator frankly what you are trying to do. By disclosing your own thinking, you will encourage the other person to exchange ideas with you.

4. Do not lose any opportunity to convince even a clerk that what you are doing is sound and beneficial to the company and therefore to her. Acquaint her with the anticipated results of the study so that she will see the changes affecting her as part of a broader purpose.

5. When objections arise—as they surely will—be sure, first, to hear the operating supervisor's arguments fully and to learn exactly why he or she thinks a given proposal will not work. Where necessary, set the point aside for further fact-finding or analysis. Then if you are still convinced that the suggestion should be adopted, follow these steps.

    a. Try to persuade the supervisor that the proposal will work—or at least that it is *worth a trial.*

    b. If the supervisor still does not agree, concede the point, provided it is not a major one and will not seriously interfere with the overall plan. Compromises on small points will often help you win the larger ones. If you are reluctant as a matter of principle to make such compromises, you need only remember how frequently operating managers or supervisors must adjust their ways to your ideas.

    c. If agreement cannot be reached on a major point, explain as tactfully as possible that because the point has an important bearing on the whole program, you do not believe it should be abandoned without reference to a higher authority. At the same time, assure the frontline supervisor that you will present all the pros and cons of the case clearly and impartially. Write out the supervisor's reasons for nonconcurrence and review them with him or her to ensure that they are a fair and complete statement of that person's position. (No matter how carefully this kind of situation is handled, a certain loss of goodwill will inevitably result. For that reason, a showdown should be resorted to only when important gains are at stake.)

6. Finally, try to make the operating supervisor feel some responsibility for the success of the program. Ask for help and advice in putting it across. Get his or her suggestions on how to slant the presentation at higher organization levels. Ask for ideas on tackling the installation or on coordinating the new procedures more smoothly with related routines. Tell the supervisor that you would like to make the final presentation a product of your joint thinking.

GETTING EXECUTIVE APPROVAL

Once the proposal has been cleared with frontline personnel, the analyst should seek final executive approval as soon as possible so that the momentum gained during the program development phase can be carried through to performance results.

How high in the organizational structure the proposal should be taken depends on the number of departments affected and the importance of the changes to be made. As a minimum, the head of each department that participates in the system should, of course, give approval. If the recommendations call for major changes in systems, a substantial additional investment in computers or other equipment, or a realignment of responsibilities among departments, the proposal should be carried up at least to the first executive having common authority over the departments affected. If the recommendations require changes in organizational structure or general policies, or if they will cause the dislocation of large numbers of workers, they will presumably be submitted to a general-management executive for final approval.

The most important rule in preparing for the formal presentation is that both form and content must be geared to (1) the background and interests of its recipients, and (2) the specific purpose of the presentation.

Identify the Audience

To communicate effectively, whether orally or in writing, analysts must first identify their audience and then estimate how it will receive, understand, and accept the message. Who will be the key recipients of the presentation? What is their frame of reference? How many of them are aware of the problem covered by the study? How much does each person know about it? How important is the problem in relation to the other problems confronting that executive? How much interest does he or she have in matters of systems and procedures? What arguments will have the most appeal? What questions is the executive likely to want answered? How much factual information should be presented to convince her or him?

In developing answers to these kinds of questions, analysts usually find that their audience is not a homogeneous group. Even though the executives may be intimately concerned with the matter under study, they invariably differ in their understanding of it, their attitudes toward it, their personal involvement, their ways of thinking, and their power to act. When analysts sense that such diversity does, in fact, exist among their audience, they must attune the form and content of the presentation to the "receiving apparatus" of the person or persons who are in a position to reach final decisions on the recommendations.

Define the Purpose of the Presentation

The substance and form of the staff's presentations—whether oral or written—will also vary substantially depending on the purposes sought. Although all such communications assist in getting action, they do it to different degrees and in different ways. That is, they may aim primarily at persuading, confirming, testing, or merely informing.

Persuading   Whenever analysts use a formal communication as the primary means of persuading executives to take action, they must first determine what kind of decision they want to obtain. For example, are they seeking approval to pilot-test the recommendations, or an immediate commitment to company-wide implementation?

Definition of their objective, coupled with analysis of the interests and attitudes of key recipients, will enable them to plan more intelligently the structure of the presentation and the type and amount of supporting evidence to be marshaled.

Confirming   Often a written report or memorandum, and sometimes an oral presentation, confirm facts already discussed and recommendations already agreed on. Since they do not need to convince, they differ from a communication designed for persuasion in that the argument is usually not so full, the supporting evidence is not presented in the same detail, and the reasoning behind recommendations and conclusions is not so elaborate. In addition, the confirming report will usually mention agreements that have been reached and may enumerate steps already taken to implement decisions.

Even though a confirming report may not be detailed, it should be complete. Confirmation is itself a part of the total process of securing action. An effectively written confirming report has these values:

1. It reaffirms the benefits of the systems project and strengthens executives' conviction that the decisions they have made are sound.

2. By providing a clear statement of results to be achieved, the confirming report establishes a basis for measuring implementation progress and accomplishment.

3. It helps to lock in the benefits realized—that is, to keep the department or unit from retrogressing or the benefits from being eroded with the passage of time.

Testing In some circumstances, the chief purpose of the project team's written or oral presentation may be to test the *adequacy of its work.* That is, the presentation may be intended to help the team members determine whether they have covered the ground sufficiently, whether their evidence is convincing, whether their conclusions are sound, whether their recommendations are clear, and whether they have overlooked important alternatives.

Another closely related purpose may be to test the *understanding* of the executive personnel concerned with the study. In other words, the presentation may help team members determine whether the agreements they believe they have reached with various executives as the study has progressed are real or superficial. And it may help them learn whether key recipients fully grasp the concepts underlying the recommendations and recognize the consequences of those recommendations.

Informing The final purpose of a written or oral presentation may be simply to inform the audience. This purpose comes into play when decisions have been reached on systems recommendations and the analysts must help communicate those decisions and the resulting action plans to the persons who will participate in the implementation. Ordinarily this type of presentation does not include detailed findings or conclusions, except as they are essential to an understanding of the decisions reached. Rather, such presentations concentrate on describing the agreed-to changes and the approach to installing them and realizing the expected benefits.

Select the Form of Presentation

The third consideration to be weighed is whether the final presentation should be oral or written. Oral presentations may be made to executives individually or in groups. Written presentations may take the form of memorandums, comprehensive formal reports, or proposed standard operating instructions.

In general, the best results come from making a presentation orally to a small group of two or three people. This type of presentation, skillfully conducted, elicits more warmth, understanding, and enthusiasm than even the best of written reports. It is also more flexible, offering alert analysts an opportunity to sense executive reactions and to respond accordingly by revising their emphasis, deleting or adding material, or changing the sequence of topics. For example, if the analysts sense great interest in a particular point, they can extend the discussion of it; or if they sense a lack of conviction, they can reemphasize their arguments or present additional evidence. Finally, an oral presentation provides a forum in which analysts can answer questions and clarify misunderstandings.

Under some circumstances, large conferences are a quicker means of obtaining many persons' approval or of reconciling many points of view. Such meetings, however, should be used sparingly, for too often they inhibit action rather than stimulate it. Also, they are difficult for the analyst to direct and control, they tend to crystallize opposition, and they encourage delay and "further study." For the most part, these complications arise because the analyst is not able to relate the presentation to each individual's interests and problems.

Although oral presentations are generally recommended, they are not always practicable or possible. Some type of written proposal must, of course, be submitted whenever the persons to be consulted are widely scattered geographically. Reports for this purpose should not only cover the recommended changes in detail but also anticipate and answer objections that might otherwise be answered verbally. Written presentations should also be used whenever

1. The recommended program is so extensive that it requires much supporting evidence and careful point-by-point deliberation.

2. The analysts must state their position clearly and fully because they have reason to believe that their conclusions will be challenged. Under these circumstances, the written word protects against mistaken recollections or personal interpretations of oral presentations and the discussions that accompany them.

In still other instances, a memorandum or set of written operating instructions will be sufficient when many elements of the proposal were discussed with executives and tentatively agreed upon during the survey and design stages. Here the written presentation simply brings all parts of the plan together for final executive authorization.

Even when recommendations are to be presented orally, the points to be covered should still be put in writing beforehand, much as if the presentation were to be made in report form. Preparing a written proposal forces analysts to convert general ideas into specific recommendations. It helps them organize the presentation more logically, and it serves as an invaluable aid in drafting the installation instructions.

Plan the Content of the Presentation

Whether oral or written, the formal presentation should usually cover the eight subjects described below. The amount of emphasis given to each, however, will vary depending on the recipients and on the purpose of the presentation.

1. Statement of the problem. The introductory section should describe the nature, cause, and significance of the problem. In addition, it may include comments on how the study originated, who requested it, and what its objectives were.

2. Scope of the project. The presentation should block out the breadth, depth, and limitations of the analysis—for example, the issues dealt with, functions or organization units surveyed, factors considered, and experiments conducted. This subject should be covered only in as much detail as is necessary to build confidence in the quality of the analysis, and, therefore, in the conclusions and recommendations that flow from it.

3. Summary of findings and conclusions. This section of the presentation covers the major deductions the analysts drew from the evidence and used as the basis for their recommendations.

4. Summary of recommendations. Next, the analysts should outline the principal changes proposed in each of the following areas.

    a. Organization

    b. Policies

    c. Information requirements

    d. Systems and procedures

    e. Methods and equipment

    f. Personnel

5. Dissenting points of view. The presentation should indicate whether operating personnel agree with the proposal. If their full approval was not obtained, the disputed points should be specified, together with an impartial statement of the facts and opinions supporting both sides. Even where opposing arguments have not surfaced, analysts should recognize that they may arise in the mind of the listener or reader. The presentation should therefore acknowledge and deal effectively with possible counterarguments, objections, or criticisms.

6. Anticipated benefits. These should be broken down into tangible and intangible benefits. For each expected tangible benefit (say, reduction in direct or indirect costs, reduction in inventory investment), the analysts should estimate its dollar value and explain how that estimate was made. For each intangible benefit, they should make an effort to highlight its significance. For example, what is the true competitive importance of the improved customer service that the recommended systems changes will bring about?

7. Cost and effects of changeover. The presentation should outline the measurable costs of converting to the new system. These might include the costs of purchasing or renting additional equipment, revising computer programs, retraining personnel, using additional space, printing new forms, making layout changes, converting records, or running the new and old systems in parallel during a pilot test. In addition, if the proposed improvement will require personnel changes, their nature should also be described.

8. Action program. Finally, the analysts should list the specific steps necessary for putting the recommendations into effect. This is one of the most important parts of any final presentation, for by charting a clear path to the desired goal, analysts make it easy for the executive to take action. Where some immediate executive authorization is desirable as a preface to action, the necessary memorandums, organization announcements, standard operating instructions, or other official notices should be prepared in advance so that the executive can be asked to sign them as a conclusion to the presentation.

Achieve the Desired Impact

If the analyst is to succeed in making people want to do what he wants them to do, he cannot rely solely on the soundness of his recommendations. They must be packaged attractively. They must be presented in a way that generates enthusiasm—that appeals to the imagination as well as to reason.

To have this sort of impact, the analyst's presentation must meet the critical tests of *clarity* and *persuasiveness.*

Clarity. Here are some suggestions which will help analysts ensure that their presentations—whether written or oral—are clear and easily understood.

1. Pay attention to structure. The first requirement to be met in achieving clarity is that the *structure* of the material must be instantly evident. Readers or listeners must be able to quickly perceive the relationship of the pieces to each other—to answer the question, "How does this piece relate to the whole?" They must also have the feeling at every point that they know where they are, where they have come from, and where they are being taken.

2. To meet this test, first write out a story line for the presentation—a brief two- or three-page statement of the key recommendations and the findings and conclusions that support them. Using this story line, plus the eight subjects listed in the section "plan the content of the presentation," write out all the headings of varying importance (major, intermediate, and minor) that you will use in the presentation. These headings constitute your outline. Next check the soundness of the outline: (*a*) Are all headings of the same value coordinate with each other? (*b*) Is every minor heading a logical subdivision of the intermediate or major heading under which you have placed it?

3. In the presentation, highlight only the major points and make them easy to grasp. Boil down, refine, and predigest the information. Instead of taking readers or listeners through the recommended system step by step, give them a concise picture of the principal differences between the old and the new. Then capsule all the proposed changes in terms of their significance. On a clerical cost reduction study, for example, summarize the percentage of existing activities, tasks, records, and reports that will be eliminated.

4. If the presentation takes the form of a written report, follow these guidelines.

    a. As a lead-in to a series of sections preceded by coordinate headings, insert a list of the points you are going to cover. An example can be found in the list of ten points on page 16 and the corresponding sections that follow the list in Chap. 2.

    b. Begin each paragraph with a topic sentence that states the single idea amplified in that paragraph.

    c. Use connectives or transitional words and phrases liberally to tie the whole report together and help the reader understand the relationship between its pieces. Often a phrase, clause, or sentence placed at the beginning of a new paragraph will meet this need—for example, "Notwithstanding this difficulty," "To achieve these benefits," "In light of these findings." In addition, use connectives within a paragraph to highlight relationships among sentences. Typically a single word or short phrase will suffice—"therefore," "however," "on the other hand," "finally."

    d. Choose your words carefully. Be sure they say *exactly what you mean.* Avoid management jargon, clichés, circumlocutions, vague words, and other evidences of failure to "think it through."

    e. As you write, fix your mind on a specific key reader. By doing so you are more likely to write clearly and to the point.

    f. When you have finished the first draft, challenge every sentence. Does it really say something? Is it precise or fuzzy? Is its meaning clear, or is it just a bunch of words strung together? When a long sentence is stripped down to subject, verb, and object, does it make sense and add value to the paragraph?

5. If, in contrast to a written report, the presentation is to be in oral form, observe these guidelines.

    a. Organize and outline the material as carefully as you would for a written report, and in much the same way.

    b. Try to limit the presentation to the amount of material that can be covered in about 30 minutes. If it must be longer, incorporate illustrative material or case examples at intervals to change the pace and give the listeners a chance to relax mentally.

    c. At an early point in the preparation, decide on the type of visual medium, if any, that you will use—for example, overhead projector, 35 millimeter slides, flip-over easel. The choice among these media should be based on considerations such as the following:

      (1) *Purpose and tone of the presentation.* Finished materials such as multicolored slides would be out of place in an informal progress report.

      (2) *Executive preference.* Some executives like polished formal presentations; others prefer grease pencil and pad.

      (3) *Audience size.* An audience of more than eight or ten people usually requires charts or text material so large that projected images are virtually mandatory.

    d. In preparing visual text material, remember the limited purposes it should be designed to serve. Its first aim is to highlight for the audience the structure of your argument. Its second is to keep your oral presentation on the track and prevent you from overlooking important points. In meeting these purposes, analysts are prone to make their visual texts too long and far too self-contained. The result is an oral presentation that often consists of little more than an analyst reading the visual material aloud. To avoid this pitfall, be sure to limit visual material to a very much stripped-down outline of the salient points, supplemented by high-impact graphs or charts. In other words, never make the visual text so complete that the audience can get the full message just by reading the material. If you are to stay in command of the presentation, your verbal amplification of the visual material must predominate.

    e. Remember that in a visual presentation—unlike a written report—listeners cannot flip back through the pages and review what has gone before. They must rely on their memory of what they have heard and seen. Whenever your presentation is long or deals with many subjects, take special steps to help listeners know where they are at any point in the presentation. Following are three useful techniques:

    (1) At appropriate points in your presentation summarize the material that you have just covered in order to place it in context with the subjects that have been previously discussed and those that are to come.

    (2) As an added piece of visual material, have in view at all times an easel sheet containing the presentation agenda. After you complete any point, this material will enable you to show the listeners where they have come from and where they are being taken.

    (3) In covering a long list of points, try to group them into two or three meaningful categories that the audience can remember even if it cannot keep all the individual points continuously in mind.

f. In preparing exhibits to supplement the visual text, remember that the audience must be able to grasp the significance of a chart or a graph at once. In a written report, readers can study charts and graphs at their own pace; they have ample opportunity to examine the scale, time periods, units, and sources of information. But charts and graphs that require close study detract from an oral presentation. They force viewers to divide their attention between looking and listening.

g. Finally, rehearse, rehearse, rehearse to ensure that you have (1) tested the material for clarity and persuasiveness, and (2) developed ease and fluency in delivering it. For this purpose, solitary practice is helpful but not sufficient. You will also want to make a dry run of the presentation before other team members who can criticize both the spoken words and the visual aids.

Persuasiveness   How effectively a presentation convinces the executive group to act on the project team's recommendations will depend to a great extent on the foregoing characteristics of the presentation itself. The very act of communicating findings, conclusions, and recommendations so clearly that they can be quickly grasped will, by itself, help to build executive confidence in the quality of the entire systems project. But beyond achieving clarity, analysts need to put some drama into their presentations. They must remember that to the average executive, systems and procedures are detailed, impersonal, and without much appeal. For that reason, real thought and skill are needed to create a presentation that awakens interest and generates enthusiasm.

One of the surest ways of dramatizing a systems proposal is through the intelligent and imaginative use of visual presentation techniques. Attention-getting charts, graphs, and pictures have been used widely and with outstanding results by advertising agencies in making program presentations. Since systems analysts must also be sellers of ideas, they should be alert to the opportunities for applying the same techniques.

Visual aids need not be elaborate to be forceful. They must only convey an

OPERATIONS	BRANCH SALES OFFICES	SALES HEADQUARTERS DEPARTMENTS	CREDIT DEPARTMENT
1. EDIT AND COMPLETE CUSTOMER'S PURCHASE ORDER	①		
2. TYPE PRINTED FORM COVERING THE SALE	② SALES ORDER ③ ACKNOWLEDGMENT		
3. ASSIGN SERIAL NUMBER TO THE FORM	④ SALES ORDER	⑪ FACTORY ORDER	
4. PROOFREAD TYPED FORM AGAINST SOURCE DOCUMENT	⑤ ⑥		
5. CHECK FORM FOR COMPLETENESS		⑫ SALES ORDER ㉗ FACTORY ORDER	
6. REGISTER OR LIST INDIVIDUAL ORDERS	⑦	⑬ CHRONOLOGICAL ㉘ BY SHIPPING DATE	
7. SUMMARIZE ORDERS	⑧	⑭	
8. MAINTAIN OPEN ORDER FILE	⑨	㉚ CHRONOLOGICAL ㉛ BY SHIPPING DATE	
9. CHECK CUSTOMER'S CREDIT			⑰
10. PRICE ORDERS		⑮	
11. ESTABLISH SHIPPING DATE OR PRIORITY		⑯	
12. ASSIGN ORDER TO PLANT			
13. POST TRANSACTION TO CUSTOMER RECORD	⑩ ORDER (54) SHIPMENT	㉙ ORDER (62) INVOICE	⑱ ORDER (73) PAYMENT
14. ENTER RAILROAD ROUTING ON ORDER			
15. COMPARE FACTORY ORDER WITH SALES ORDER	㉝		
16. MAIL ORDER ACKNOWLEDGMENT TO CUSTOMER	㉞		
17. CHECK PRICES	(59) ON INVOICE	㊽ AFTER SHIPMENT (61) ON INVOICE	
18. LIST SHIPMENTS	(55)		
19. SUMMARIZE SHIPMENTS	(56)		
20. MAINTAIN FILE OF SHIPPED ORDERS AND SHIPPING NOTICES	(57) (58)		
21. MAINTAIN FILE OF INVOICES	(60)		

FIG. 35 *Simple visual aids can dramatize duplication of clerical activities.*

ORDER DEPARTMENT	TRAFFIC DEPARTMENT	INVOICING DEPARTMENT	REGIONAL SALES OFFICES	PLANT SHIPPING DEPARTMENTS	ACCOUNTS RECEIVABLE DEPARTMENT	COMPUTER DEPARTMENT	NUMBER OF TIMES PERFORMED
							1
(22) FACTORY ORDER		(50) INVOICE		(42) LOADING TICKET (45) SHIPPING NOTICE			6
		(49)		(44) SHIPPING NOTICE			4
(23)		(51)		(43) (46)			6
(20) SALES ORDER							3
(19)		(35)		(39)			6
(24)			(37)				4
(25) (26)		(36)	(38)	(40)			8
							1
							1
							1
(21)							1
					(68) INVOICE (74) PAYMENT		8
	(32)			(41)			2
							1
							1
							3
				(47)		(70)	3
			(67)			(71)	3
(63) (64)	(65)	(52)		(66)			7
		(53)			(69)	(72)	4
						TOTAL	74

important set of facts or a fundamental idea simply and clearly. As an illustration, Fig. 35 shows a chart that was used in one company to bring out vividly the extent of clerical duplication in the processing of a sales order. The major operations in the process are listed at the left; the points at which each operation was then being performed are indicated by the circles; and the number of different times each operation was performed appears at the right. The numbers within the various circles designate the sequence of operations.

After using the chart to point up significant facts about the existing system, the analyst overlaid it with a transparent sheet containing colored stickers that had been positioned to cover up the operations recommended for elimination. This technique not only facilitated comparison of the "before" and "after" procedures but also dramatized the number of possibilities for eliminating work.

Another useful technique for arousing interest is to express savings or benefits in fresh, new terms. For example, "Adoption of these recommendations will save the cost of preparing, handling, and filing half a million pieces of paper a year." Or, "Now it takes us 14 days to deliver an order to one of our customers. Adoption of the recommended changes will give us a strong competitive advantage by cutting our average delivery time to 8 days without significantly increasing any of our costs."

One point of caution about using such techniques to enliven the presentation: They must not be obviously dramatic, forced, or artificial. If they are, they are more likely to provoke resentment or irritation than to foster reception. Neither must they be so novel that they divert attention from the substance of the presentation. Above all, analysts must be careful that in their effort to glamorize the results of their work, they do not overstate their case. Nothing will destroy confidence in their integrity more quickly than unsupported claims, sweeping pronouncements, or exaggeration of any kind.

SIXTEEN

Installing Approved Systems and Effecting Change

No other phase of sytems work is so consistently haphazard and imperfect as that of putting approved plans into operation. It avails little, of course, that the plans are workable if they are not made to work. And yet, in our experience, more systems failures result from muddled, slipshod installations than from any fundamental defects in the system itself.

There are two principal causes of this condition:

1. Systems analysts tend to feel that they have accomplished their mission when the line organization accepts their recommendations. By the time they reach this point, they have been through the often long and difficult processes of fact-finding, analysis, development, and persuasion. Once the victory is won, the tedium of implementation seems anticlimactic to them. They are more eager to meet the challenge of new problems than to buckle down to the prosaic task of thrashing out operating details. And since systems installation is more closely related to the operating function than to the planning or staff function, analysts are inclined to believe that responsibility for making the changeover cannot be separated from responsibility for operating the new system once it is in effect.

Line supervisors, on the other hand, are concerned chiefly with day-to-day operations. They do not usually have the time or the background to carry out the approved program by themselves. Analysts tend to forget that the plan they understand so well may still be only a skeleton outline to the operator. And because of these limitations, an earnest beginning is too often followed by halfway measures, imprudent compromises, loss of enthusiasm, and finally a bogged-down program.

2. Soundly conceived systems and procedures may be doomed to failure because analysts do not fully comprehend either the character or the magnitude of the job required to achieve and maintain the end-result benefits for which the new or revised system was designed.

Consider first the *character* of the implementation task. Analysts often lose sight of the fact that the success of their work depends on effecting two distinct kinds of changes—*tangible* change and *intangible* change.

Tangible change, as the words imply, consists of systemic or mechanistic change. It may, for example, take the form of installing (1) a new EDP system of inventory control, (2) a new career-path planning and performance evaluation system, or (3) a new sales order processing system (like the one described in the hypothetical installation plan later in this chapter).

Intangible change, on the other hand, consists of change in the way a company is managed or its components are supervised. Achieving this sort of change typically requires the development of new skills, new attitudes and perceptions, new approaches to decision making, or other new forms of behavior. Examples of intangible change are

1. Improvement in the skills of the frontline supervisors responsible for directing and controlling the day-to-day activities of the clerical force. (For a list of the specific skills that might be developed or sharpened in this process, refer to pages 248 and 249.)

2. Modification of management decision-making or control behavior as a result of new management information systems.

3. Change in the way sales representatives spend their time because of a significant reduction in their paperwork or the development of new call-planning and control tools.

As for the *magnitude* of the implementation task, analysts tend to underestimate the time and preparation required. It is a safe generalization to say that the installation of most systems changes takes more time and hard work than all the previous phases of the project combined. On a major study where the survey, analysis, development, and presentation take 2 months, the installation may take 6. It matters little that the new system, forms, records, and reports have been developed in complete detail, for these are still only a blueprint—a picture of the ultimate destination.

As the previous chapter noted, installation means getting from where you are to where you want to be. And the task of bridging that gap is often much greater than is generally realized. Forms and equipment must be ordered and their availability coordinated. Operating instructions must be written and tested, and personnel must be indoctrinated in them. The new systems must be refined and detailed points of operating mechanics settled. The cutover must be planned; records must be converted; and preparatory changes must be made in related systems and procedures. These are typical of the steps that must be planned and carried out in advance if just the *tangible* changes required are to be made without confusion, disruption of service, or a waste of the effort that went into the whole systems study. In addition, as will be discussed later, achievement of the *intangible* changes can increase the magnitude of the implementation task almost exponentially.

The material in this chapter is concerned primarily with tangible change—that is, with implementing major changes in transaction-processing or other operating systems. One of the sections that follows, however, does treat the problem of effecting intangible change—that is, change in the skills, attitudes, and behavior of managerial and operating personnel.

FIXING RESPONSIBILITY FOR THE INSTALLATION

As already implied, the analysts who developed the new or revised system should play a leading role in its installation. The nature of their participation, however, should change in midstream between the task of planning the installation and that of actually carrying it out. During the initial step of planning the changeover, they should retain the initiative. Even though the participation of line personnel will be greater at this point than in any previous part of the study, the analysts should continue to spearhead the program. In this way, installation planning not only benefits from their singular knowledge of the new system but also provides the occasion for them to transmit their knowledge to the operating supervisor. It is the transition step between the analysts' leadership of the project and the assumption of that leadership by the line organization.

Responsibility for actually making the installation belongs, of course, to the line organization. During the changeover, the analysts may continue to keep a close watch over progress and results and to confer regularly with operating supervisors. But they have no direct authority over performance of the work except at the specific request of a supervisor, who may consider it expedient to delegate such authority temporarily. In this event, the analysts are accountable to that supervisor for the duration of their assignment, and their authority over the installation extends no further than the supervisor's area of responsibility.

PLANNING THE INSTALLATION

The steps required in planning a major systems installation are fundamentally the same as those that were required in planning the study itself (see Chap. 8). That is, the total job must be broken down into its components, and each component must be assigned and scheduled.

Timing the Changeover

As a beginning point in installation planning, the analysts should reach decisions on three related questions concerning the time at which the cutover to the new system will be made.

1. Must the new system be inaugurated at a particular time? The need to have the changes in place by a set date generally arises when the activity

affected by the changes should be kept consistent throughout the calendar or fiscal year—as in the case of the accounting system, for example. Where an inflexible deadline must be met for these or other reasons, the dangers of confusion, breakdown, or excessive installation expense are greatest. Because the available time is invariably too short, the preparatory work must be planned and scheduled with extreme care and its progress followed just as closely.

2. Should the entire system be installed at the same time, or should it be installed in parts? This question is closely related to the first, since changes that must meet a specific deadline must also be installed wholly at that time. Where such a requirement does not exist, however, the analyst usually has a choice between concurrent and successive installation.

If the changes are not of major importance or are not numerous, they can usually be made simultaneously without risk of serious disruption. But if they are complex, they should ordinarily be handled progressively, so that the installation is not jeopardized by too rapid a rate of change. Obviously if too many adjustments are undertaken simultaneously—either at one or at several places throughout the organization—the supervisory force cannot coordinate them properly, nor can the work force assimilate them smoothly.

One simple but effective way of spreading the installation load is to separate changes requiring the elimination or combination of work from those requiring new methods of operation. Usually changes of the first type can be made quickly and at a time when the installation crew is ordering forms and equipment, writing job instructions, and taking other preparatory steps. In this way much of the underbrush is cleared away, and part of the operating personnel is freed to help with the changes still to be made.

Where the installation is undertaken one step at a time, the analyst must be careful that it does not spread over so long a period that momentum is lost or that the disruption or hardships accompanying the transition are unnecessarily prolonged. Another danger to guard against is the temporary loss of parallelism between related activities or transactions. These dangers are real, but they can be avoided. They do not argue against progressive installation so much as they emphasize the need for skillful installation planning.

3. At what point in the installation of the new system should the existing routines be discontinued? As a protection against unanticipated problems or hidden weaknesses in the new system, should both the new and the old be operated on a parallel basis for a given period? If not, should some *parts* of the old system be temporarily retained? Decisions on these points should be made early in the planning process, for they will materially affect the nature of the preparatory work, particularly the selection and training of personnel.

In general, parallel installations should be avoided wherever possible, for they always mean added expense and often create delays in the processing of source documents. Again, skillful planning should reduce the hazards of installation and the need for clinging to the old routines. Trial installations are often advisable, but they can be confined to a small sample of the work so that a full-blown parallel operation is avoided.

Planning the Preparatory Work

When the questions of timing the changeover have been resolved, the next step is to determine what preparatory work needs to be done before the new system can be put into effect. This is essentially a job of listing all the measures that must be taken to equip and train operating personnel for carrying out the new routine. Usually analysts allow too little time for this task, simply because they neglect to think it through fully when developing the installation plan. They either overlook important parts of the job or fail to break it down into sufficient detail to make realistic scheduling possible.

Inexperienced analysts can minimize these dangers by reviewing the approved system step by step and by asking the following questions about *each* of the *new* operations.

1. Is this operation described fully enough to cover every contingency, or does it need further testing, refinement, or amplification?

2. What new equipment, forms, supplies, work aids, or other clerical tools are required to carry it out?

3. How much training will the workers need to perform the operation satisfactorily?

4. How much indoctrination must frontline supervisors be given?

5. If persons outside the company (dealers, distributors, independent agents) are affected by the new system, how much and what kind of indoctrination should they be given?

6. What kinds of control tools will be necessary to monitor installation progress and results achieved?

Fixing Responsibility and Scheduling the Work

Having determined specifically *what* work needs to be done, the persons responsible for the installation should complete their plans by deciding *who* will carry out each step and *when* it should be started and finished. The analyst must let the operating supervisors make most of these decisions, for the installation timetable will be difficult enough to meet without risking the danger that supervisors will feel no responsibility for having set it up. The analyst can be of real help to supervisors, however, by urging that they keep the following requirements in mind when assigning responsibilities and preparing schedules.

1. The key individuals on whom the success of the installation will depend should be relieved of enough other responsibilities to ensure that they have adequate time for the job.

2. Where no inflexible deadline must be met, the actual cutover should be scheduled to take place when it will cause the least confusion. However well operations are planned and controlled, many departments have slack and peak periods during the month. These variables should be carefully studied before setting the changeover date.

3. The timetable must be *realistic*. It must make allowances for obstacles,

problems, delays, and unanticipated preparatory work. Above all, it must recognize the need for "staying in business" during the transition period.

Once the schedule has been drafted, analysts should make sure that every person participating in the installation receives a copy. In addition, they will do well to make the key executives who are interested in the program aware of the estimated time required to complete the changeover. Persons who have never participated in a major systems installation seldom realize the magnitude of the preparatory task. Unless they are told in advance what to expect, they are likely to become impatient over lack of results and to regard subsequent explanations as impromptu excuses.

One final suggestion on the matter of fixing responsibility: On major system changes involving several departments and extending over a long period (for example, conversion of many existing EDP applications to a next-generation computer), consideration should be given to setting up a steering committee of the operating managers principally concerned. As a backup to the full-time installation team members, this committee can fill an important need by facilitating interdepartmental coordination, revising priorities, and maintaining higher level support of a long and difficult job.

Revising the Schedules

As the preparatory work progresses, the need for additional preliminary steps often becomes evident. These steps should be incorporated in the formal installation schedule immediately so that they will not be forgotten.

If the preparatory work has not been completed when the installation date approaches, and if no deadline has to be met, the changeover should be postponed and the remaining work rescheduled. Starting the installation under a handicap usually prolongs the changeover and causes extra expense, excessive overtime, and undue strain.

One word of caution, however: Every effort should be made to avoid repeated postponements. At best, they dampen enthusiasm for the change and delay attainment of the benefits of the study. At worst, they raise doubts about the practicability of the new or revised system and cause operators to suspect that it has been poorly planned. Clearly the best protection against these hazards lies in developing the installation plan with great care, in following its progress closely, and in taking whatever measures are necessary to keep it on schedule.

A Hypothetical Installation Plan

The following hypothetical plan suggests a format for recording the installation steps. It illustrates the amount of detail required if the plan is to serve as a useful guide.

Plan for Installing a New Order Processing System

Action step	Responsibility of	Scheduled dates
1. Hold a series of meetings with headquarters personnel to outline new system and the steps by which it will be put into effect.	Holmes Bates	5/2–5/4 5/2–5/3
2. Prepare and send to regional managers a brief outline of approved system changes affecting branch and regional offices.	Bates	5/4

3. Eliminate immediately the following records, files, and clerical activities of the order department:
  a. Customer sales record card
  b. Commission record
  c. Duplicate file of open orders maintained by order schedulers
  d. Manifest of orders released to plants
  e. Duplicate editing of orders received from branch offices
  f. Checking of unit prices on invoices to customers and maintenance of an invoice file
  g. Preparation of daily summary of shipments by product class and sales region
	Holmes
	Gardiner
		5/5
		5/5–5/13

4. Determine the number of clerks who will be made available by the work eliminations listed in step 3 and arrange for their transfer.	Holmes Gardiner	5/6 5/6–5/13
5. Contact various suppliers of duplicating equipment, witness demonstrations, and visit companies using the equipment to determine the type and model that will best meet requirements; issue necessary purchase requisitions for the equipment and operating supplies.	Bates	5/5–5/20
6. Procure order schedule boards and sorting equipment for the order department.	Bates	5/9–5/13
7. Place order for visible-reference panels and strips.	Bates	5/9–5/13

8. Discuss tentative designs of following new or revised forms with various forms printers; clear any changes with operating supervisors concerned; release final designs to purchasing department for ordering:
  a. Order-invoice duplicating master for factory shipments
  b. Order-invoice snap-out set for shipments from branch warehouses
  c. Customer master information card
  d. Customer sales-statistical card
	Bates	5/16–5/27

9. Pending availability of new order-invoice duplicating master for factory shipments, have a supply of facsimile forms printed for use by branch offices in place of the present order forms.	Bates	5/20
10. Order window envelopes and binders for new invoice form.	Bates	5/27
11. To ensure uniformity of order preparation, make a survey of the sizes of typeface and the condition of typewriters in branch offices, order and billing departments, and plant shipping departments. Plan	Bates	5/31–6/10

Plan for Installing a New Order Processing System (cont'd)

Action step	Responsibility of	Scheduled dates
necessary shifts of equipment and initiate purchase of any new equipment required.		
12. Work with equipment suppliers to develop details of procedure for transmitting urgent orders to factories by wire. Place orders for equipment and forms and arrange for training of personnel to operate equipment.	Bates	5/31–6/10
13. Develop order department personnel program.	Holmes	5/9–5/11
a. On basis of revised plan of organization and division of work, prepare for the salary committee a definition and evaluation of each new or revised job.	Gardiner	5/9–5/20
b. Determine the number of persons required in each job.		
c. Reappraise qualifications of the present personnel in light of increased responsibilities of the job.		
d. Arrange any necessary personnel transfers.		
e. Select and train new personnel required.		
14. Draft an organization announcement listing new responsibilities of the order department.	Holmes	5/11
15. Instruct mail-room personnel to route all purchase orders, branch sales orders, and correspondence relating to specific orders or shipments to order department instead of to product managers.	Holmes	5/12
16. Notify branch managers that their phone or wire inquiries about specific orders or shipments should be made to the order department instead of to product managers.	Holmes Halverson	5/12
17. Discontinue product managers' maintenance of the following records:	Holmes Halverson	5/13
a. Open- and closed-order files		
b. Order-backlog record		
c. Daily and cumulative record of plant production		
18. Develop policy covering assignment of shipping priorities.	Holmes	5/16
19. Prepare instructions governing determination of plant from which orders should be shipped.	Holmes	5/17
20. Write detailed job instructions for each of the following positions in the new order department:	Holmes	5/18–5/20
a. Order scheduler		
b. Order clerk		
c. Statistical clerk		
21. Review these instructions with order department personnel, make any necessary modifications or additions, and issue instructions in final form.	Holmes Gardiner	5/23–5/25
22. Make approved layout changes in the order department and move the credit and traffic departments into the adjacent area.	Holmes Gardiner Wade	5/26–5/27 5/26–6/4 5/30–6/4
a. Prepare final layout drawings.	Callahan	5/30–6/4

Plan for Installing a New Order Processing System (cont'd)

Action step	Responsibility of	Scheduled dates
b. Reach agreement with building superintendent on tagging of furniture and equipment and scheduling of partition and telephone changes.		
c. Initiate moving and rearrangement authorization.		
23. Work out details of tying brokers' orders into the new procedure.	Holmes	5/31–6/3
24. Develop and carry out plan for transfer of customers' special packing, shipping, and invoicing instructions from headquarters to branch offices.	Holmes Gardiner	5/31–6/3 6/3–6/17
25. Expand the number of classifications into which order data are summarized for production planning and shipment scheduling.	Holmes Ryerson	6/6–6/10 6/10
26. Obtain from the credit manager a list of preferred credit risks and type this information on Linedex panel strips for use by order department clerks in checking orders.	Holmes Wade Gardiner	6/6–6/10 6/8–6/17 6/13–6/24
27. Work with credit manager on development of standard collection form letters and a standard mailing schedule.	Holmes Wade	6/6–6/10 6/6–6/17

28. Draft standard practice instructions covering the following procedures and clear them with operating supervisors:
  a. Processing of orders and invoices covering factory shipments. This should include
    (1) Distribution of order-invoice master
    (2) Information to be added at each point
    (3) Copies to be duplicated at each point and their distribution
  b. Processing of order-change notices
	Holmes
	Bates
		6/13–6/14
		6/13–6/24

29. Investigate any suggested revisions, points of disagreement, or areas of possible difficulty brought out by comments received on the draft of standard practice instructions in step 28. Revise instructions, if necessary, and issue them in final form.	Holmes Bates	6/23–6/24 6/20–6/24
30. Make a study of mailing time via airmail and first-class mail between each of the plants and each branch office. Issue instructions to each plant specifying which service to use in mailing shipping notices to each branch.	Holmes	6/15–6/24
31. Set up precalculated charts for billing department pricing clerks showing invoice amounts for varying quantities of each product.	Holmes Mitchell	6/15–6/17 6/15–6/24
32. Draft a procedure instructions manual covering all operations of branch sales offices.	Holmes	6/20–6/22 6/27–6/28
33. Analyze computer department's report schedule and determine the steps necessary to speed up monthly reports of shipments by branch office, product class, and customer.	Bates Holmes	6/27–7/15 6/29–7/6
34. Develop plans and make arrangements for installing	Holmes	7/7–7/8

Plan for Installing a New Order Processing System (cont'd)

Action step	Responsibility of	Scheduled dates
new procedures on a trial basis at the Philadelphia branch office.	Halverson	7/8
35. Develop standard unit time allowances for each clerical operation in branch sales offices.	Holmes Bates	7/11–7/15

36. Make pilot installation at Philadelphia branch according to the following sequence:
  a. Explain new procedures to office personnel.
  b. Eliminate activities and operations not required under new procedures.
  c. Convert existing records to new forms.
  d. Make test runs of the new order processing procedure using facsimiles of factory and warehouse order-invoice forms.
	Holmes
	Bates
		7/18–7/22
		7/18–7/29

37. On the basis of this pilot installation:
  a. Revise branch office procedure instructions manual.
  b. Refine standard time allowances for branch office clerical operations.
  c. Determine required clerical complement for each branch under the new procedure by applying standard time allowances to the branch's known work volume.
	Holmes
	Bates
		8/1–8/5
		8/1–8/12

38. Develop installation program for other branch offices, including:
  a. Detailed list of all steps that must be taken to change over to the new procedures
  b. Timing of each step
	Holmes	8/8–8/12

39. Arrange with representative of duplicating equipment supplier to train machine operators for order and billing departments.	Bates	8/15
40. Train a representative of each regional sales manager to assist branches in that territory in installing new procedures.	Bates	8/15–8/19
41. Send each branch manager a copy of the procedure instructions manual and the program for installation of new procedures.	Holmes Halverson	8/17
42. Issue memorandum of instructions to outside warehouses covering details of the new order-invoice procedure.	Holmes	8/15
43. Give each order scheduler a copy of order preparation instructions issued to the branch offices.	Holmes	8/16

44. One week before adoption of new order-invoice forms, issue instructions on the following points to order, credit, traffic, plant shipping, and billing departments:
  a. Cutover dates
  b. Handling of in-process orders that were originated under old procedure
  c. Potential problem areas requiring special supervisory attention during transition period
	Holmes	8/18–8/22

Plan for Installing a New Order Processing System (cont'd)

Action step	Responsibility of	Scheduled dates
45. Visit each branch office to review progress of installation, to assist with any problems, and to reach agreement with branch manager on required complement of clerical personnel.	Representatives of regional managers	8/22–9/16
46. During first month that the new forms are in effect, make a 100 percent audit of: *a.* Invoices covering warehouse shipments *b.* Order-invoice masters covering factory shipments	Bates Mitchell Gardiner	9/1–9/30

DOING THE PREPARATORY WORK

The preceding section indicated the nature of the preparatory work required on major systems installations. To supplement it, this section explains more fully the various types of work to be done and the objectives to be achieved in carrying out this part of the preinstallation job.

Preparing Personnel for the Changes

In setting the stage for the changeover, the analyst should first recognize the personnel problems involved and deal with them fairly, openly, and without hesitation. Decisions must be reached at the outset on what employee changes are to be made and how they will be handled. Here are some of the typical questions to be answered:

1. How many workers will be displaced by the new system? Who will they be?

2. Can some of them be transferred to other jobs without in the long run jeopardizing the benefits of the study? If so, which workers should be transferred, and to what jobs?

3. How many must be released? Who will they be?

4. What will be done to minimize the hardships of their release? What will the company's policy be on termination pay? How can the company help them find jobs elsewhere?

5. How many employees must be trained to perform work requiring new skills? Will the evaluation of their jobs be affected?

Next, every effort must be made to develop favorable employee attitudes toward the change before it is installed. Once decisions are reached on a major systems problem, the human relations aspect tends to be sidetracked by the technical aspect. Although the analyst may have done a skillful job in promoting understanding and interest at earlier stages of the study, he must remember that the employees' greatest concern is over the decisions finally reached and how those decisions will affect them. If they must await the actual changeover before learning what those decisions are, the gains already made in winning their acceptance are likely to be lost unnecessarily. Even where some employees will be displaced by the change, analysts must recognize that there is less risk and greater profit in disclosing their plans promptly and openly than in seeking to "protect" the work force until they are ready to move.

In the first place, rumor flourishes where knowledge and understanding are missing. And under these circumstances, the imagined consequences of a contemplated change are inevitably much worse than the actual consequences turn out to be. Thus, the analyst should. prevent needless uncertainty and unrest by a frank and early discussion of the facts.

A second reason for outlining the proposed course of action in advance is that employees then have an opportunity to think about it, discuss it, and become accustomed to it. Having made the necessary mental adjustment by the time the installation is begun, they are likely to take part in the changeover with greater acceptance and enthusiasm.

Finally, notwithstanding the effects of a change, when employees feel that nothing has been hidden from them, then are more likely to believe that they have been treated fairly. Even where some releases are necessary, skillful handling of the situation can convert an apparent liability to employee morale into an asset. A forthright acknowledgement of the problems, plus a clear explanation of what will be done to minimize the hardships of displacement, can have an incalculably favorable effect on employee attitudes throughout the whole organization.

Most of what should be covered in an initial explanation to the work force has already been implied. It is not necessary to describe the new system in detail, since that will be done later during the employee training phase of the preparatory work. The objective at this point should be to outline

1. The major features of the new system
2. The reasons why the changes are being made
3. The installation plan and timetable
4. The effect of the new system on personnel requirements
5. The steps the company has taken and will take in the interests of employees affected by the change

Testing the New System

A second important preparatory requirement is to make sure that the new system meets the test of actual operating conditions. This can be done (1) by putting a sample of actual work through the planned operations, or (2) by converting fully to the new system in one of several similar units of an organization while the other units continue operating under the old procedure—a trial operation that is usually referred to as a pilot installation.

Preliminary testing by either of these methods has many values, the net effect of which is to take the panic out of a major installation. The most obvious advantage, of course, is that testing enables the analyst to settle many minor details previously overlooked and to make certain that the new system will meet every operating contingency. Testing also serves as a means of checking the completeness and intelligibility of new forms and written instructions. In addition, where extensive changes are to be made in many similar groups throughout the company, a pilot installation provides the best opportunity for training the installation crew and determining the amount of worker retraining required before the full-scale installation is carried out. On changes that are limited to one area, the testing process can easily be combined with the worker training program by using either actual work or transcripts of it. A further advantage is that performance of the new routine under real or simulated conditions permits analysts to refine unit-time allowances for each operation so that they can determine clerical labor savings more precisely. Testing also frequently points the way to additional shortcuts and improvements that can be incorporated in the system before its inauguration. Finally—and this is perhaps of greatest importance—a successful test run supplies the confidence needed to ensure a quick, smooth installation.

So much for the benefits of pretesting. Now consider some practical aspects of carrying it out. Here are three rules that will help analysts get the best results from this part of the preparatory work.

1. Treat the preliminary tryout just as you would the actual installation. That is, conduct it as a tightly controlled experiment, not as a casual, unplanned exercise. Be thoroughly prepared before starting. Make sure that everyone knows in advance what is expected of him. Assemble all the necessary working tools, supplies, instructions, and forms or duplicated facsimiles. Simulate actual working conditions as closely as possible. Unless these precautions are taken, many of the important points that the test should bring to light will never be surfaced.

2. If the test is being made by the sample method, be sure that the transactions represent a cross section of the varying conditions that will have to be met when the new or revised system is in actual operation. Also include a number of the more difficult, nonstandard cases. Too often, in designing a new system or procedure, analysts tend to concentrate on the normal routine without paying enough attention to that small part of the work requiring special handling.

3. Be circumspect in selecting the unit in which a pilot installation is to be made. Try to pick a group that has a reputation for open-mindedness and willingness to adopt new ideas. The other units will be watching the experiment closely, and their attitude toward the change will be very much influenced by the response of those who first install it. The importance of this point was underscored by Irwin and Langham, who wrote: "Begin change in an area where the chances of success and benefits are greatest. Change is more acceptable if it follows a series of successful projects than if it follows failure. People like to play on a winning team."[1]

Obtaining Equipment and Rearranging Facilities

If the program calls for changes or additions in office equipment, the project recommendations often will not have gone beyond specifying the type of

[1]Patrick H. Irwin and Frank W. Langham, Jr., "The Change Seekers," *Harvard Business Review,* January–February 1966, p. 81.

equipment needed. Where this is true, an important part of the preparatory work will be to decide the exact make, model, capacity, and quantity of equipment to be obtained.

If more or less floor space will be required under the new procedure, if the location of a department is to be changed, or if equipment and machines are to be rearranged, as many of these changes as possible should be made before the installation so that quarters and facilities are ready in advance. The preparation of physical facilities should also include the efficient arrangement of such apparently minor items as work tables, racks, and cabinets, for these details often play an important role in ensuring smooth installations.

Developing Installation Controls

At this juncture, the analyst should design the ad hoc records and reports that will be used to measure progress against the installation plan and timetable, and to evaluate the results achieved. In most cases, the first part of this control task—measuring the work performed against the action plan and timetable—is performed reasonably well. But the second part—evaluating the benefits realized—typically goes by default. The analyst should, therefore, give special attention to this aspect of the control task by designing reports that will focus everyone's attention on the planned end results. For example, on a clerical cost reduction study, this report would presumably show the cumulative reduction achieved in number of workers; on a service improvement study, the reduction in the average as well as in the range of order-filling times by individual branch offices.

Preparing Written Operating Instructions

One of the most demanding tasks in launching a new or substantially revised system is translating a general plan of action into a form and language that are readily understood by the persons upon whom success of the plan depends. Too often analysts assume that they can meet this requirement simply by drafting standard practice instructions covering the new system. Although such instructions are the first prerequisite, they are still only part of the job. They tell only what operations are to be performed *after* the installation is completed. It is like giving workers a new and unfamiliar goal without explaining how to reach it.

On a major changeover, therefore, instructions on the steps to be followed *before* and *during* the installation are as important as those describing the ultimate system. Specifically, the following subjects should be covered:

1. The preparatory work to be done by operating personnel.
2. The procedures to be followed during the transition period. If thoroughly prepared, these instructions are often more elaborate and detailed than those covering the new system itself. They should include such matters as the time to discontinue each part of the old system, the method of handling

transactions that were started under the old system and are still in process when the cutover is begun, the provision of special checks and controls during the critical phase of the conversion, and so on.

3. The system and organization structure that will be in effect when the installation is finished.

4. The effect of the new system on departments that will not participate in it directly. For example, changes in a cost accounting or payroll function may be of concern to a factory supervisor even though he or she plays no part in performing the function.

In preparing such instructions, systems analysts will do well to err on the side of imparting too much rather than too little information. They must remember that because of their long exposure to the problem, their minds supply many of the minor details with which others are not familiar. A set of instructions that seems complete and understandable to them may have many gaps to the operating personnel.

If the system involves many departments, the analysts can enhance the usefulness of written instructions by preparing them in separable sections and given each work unit only the material with which it is concerned. In some instances, they will want to go even beyond this point and prepare detailed breakdowns of individuals jobs.

Finally, analysts should be sure to distribute copies of the instructions to *everyone* affected by them, not just to the department heads concerned. This need is often overlooked, with the result that relatively little written information reaches the great group of clerical personnel upon which management must depend for efficient execution of its policies and plans.

Training the Personnel

In most cases, written instructions alone cannot satisfactorily indoctrinate workers in a new or substantially revised system. Although such material forms the basis for indoctrination, it is still only a guide, a reference device, or a set of rules. If the instructions are to be properly executed, they must usually be supplemented by some sort of follow-up, encouragement, or training of both supervisors and workers. This indoctrination might consist of formal classes, group discussions, demonstrations, on-the-job coaching, or practice sessions during and after the testing of the system.

The amount of training required will be slight in some cases and great in others, but the need will be present in practically all. Where simple changes are to be made, the process of education will be simple; ordinarily a brief group meeting will suffice. But where many changes or radical ones are to be made, a more elaborate and carefully planned training program of several weeks' duration may be needed. This is particularly true when the projects entail the introduction of office equipment that requires specialized skills of the operators.

A word is needed here about responsibility for the training program. In general, the workers should be trained by their own supervisor—who in turn should be indoctrinated in the new system by the analyst. This arrangement not only reflects the supervisor's inherent responsibility for worker training, but also equips the supervisor to make better decisions both during the installation period and afterwards. Exceptions to this rule may occur when a tight deadline must be met or when highly mechanized operations are to be adopted. Under these conditions, the supervisor would presumably delegate much of the training job to outsiders, including the analyst and a representative of the equipment manufacturer.

Whether the training required is complex or simple, an important point to remember is that it should not be confined solely to the mechanics of the work. Enthusiastic and intelligent performance depends just as much on employees' understanding of the broad aspects as on their grasp of technicalities. For this reason, the analyst or supervisor should take care to familiarize each worker with the objectives and principles of the system and with the manner in which his or her job fits into the whole routine. A scheduling clerk, for example, should be told how the anticipated changes in a production control system will help shorten the production cycle and reduce inventories. Given this broad view of the whole, the clerk is better able to understand the effect of his or her errors or delays on subsequent parts of the system.

Besides being fully instructed in the work to be performed, each operator should have a period of actual practice, accompanied by discussion. The amount of time given to this part of the training varies, of course, depending on the degree of change, the type of equipment to be used, etc.

Revising Related Systems

Any required revisions in auxiliary routines that feed into the main procedure should be planned and carried out before the key changes are installed. Making these revisions ahead of time spreads the work and gives the analyst a chance to iron out minor difficulties before the larger job is undertaken.

The same principle can often be applied to activities that follow or depend on the system to be installed. For example, sometimes the forms of various reports can be changed beforehand even though they will not be used until the new plan is actually operating. Similarly, an allied procedure which depends upon these reports may also be revised to fit in with the report changes.

Making Preparatory Changeovers

As a conclusion to this section, it should be emphasized that any work that can be done before the installation should be placed on the preparatory schedule and not left for the installation schedule. Study will show that many tasks fall in this category. Records may be transferred or converted in advance, partial routines may be established, control systems set up, files rearranged, and so on.

MAKING THE INSTALLATION

Assuming the intrinsic soundness of the new or revised system, the success of the project will be assured if the preparatory work has been done with skill, foresight, and thoroughness and if the installation is made under calm, capable supervision. The following paragraphs give some suggestions and cautions on this final step of putting the system plan into operation.

Keep on Schedule

Since unexpected problems and troubles frequently develop, the actual installation seldom proceeds as smoothly as planned. And when these difficulties do arise, they tend to divert attention from the original plan and to provide excuses for delay. To keep from reaching a standstill, analysts must watch the progress of the work vigilantly, especially during the initial stages. Any lag in schedule that they permit to go unremedied is likely to throw the whole schedule off and to create serious difficulties at later stages. The analysts should not, however, let their zeal for keeping on schedule force them into beginning a new step until any preceding steps on which it depends have been satisfactorily completed.

Guard against Omissions

In the stress of installation there is a tendency to make do without certain features of the original plan, to ignore parts of the instructions, to leave apparently minor things until later, or to get along without certain facilities. Such lapses should be guarded against, for in all likelihood these items were included to facilitate the whole program. Their omission only begets problems later on.

Be Wary of Emergency Decisions

One of the most common contingencies during a large-scale installation is the need for issuing supplementary orders when the regular instructions do not cover a particular situation. Obviously such action should not be taken without thoroughly considering all sides of the problem, even though the work in question must be temporarily set aside.

To ensure that any such emergencies are handled quickly and wisely, analysts will do well, in staffing the installation organization, to station qualified supervisors or advisers at strategic points. Needless to say, the persons selected must have a full knowledge of the principles and objectives of the system and an ability to visualize the effects of their decisions on other parts of the plan.

Whenever an emergency decision is reached in response to some unanticipated condition, it should be made known to everyone affected. In addition, the change should be recorded on a master copy of the written operating instructions so that it will not be lost sight of and the new system will be kept continuously up to date.

Make In-Process Changes

Closely related to the need for issuing supplementary instructions is that of polishing or refining the original system design as operating experience reveals opportunities for further improvement. These opportunities should be exploited immediately if doing so will not materially slow down the conversion. If it will, they should be set aside until the more important parts of the system are being performed satisfactorily.

Occasionally, despite preliminary testing, an installation strikes such a serious obstruction that the system must be basically changed. Should this happen, the installation work should *not* be stopped immediately, for it will often be found that the problem is not so grave as originally thought and that much of the work can be salvaged.

Evaluate and Communicate Progress

Be sure to prepare, promptly and frequently, the special reports that were designed to measure installation progress and results. These reports will, of course, serve as a major control tool to those in charge of the installation program. But equally important, they should be communicated to those who are participating in the installation or are affected by it. This step assumes particular significance on long installations, for such progress reports are one of the principal means of maintaining momentum and enthusiasm and creating receptivity to further change. In addition, where a number of similar units (e.g., regional sales offices) are participating in the installation, regular reports of results achieved can stimulate a constructive spirit of competition. Finally, regular reports on progress and results should be made to the executives who gave final approval to proceed with the installation. This feedback is needed not only to reassure these executives about the soundness of their decision, but also to enlist their aid in encouraging installation participants and commending superior performance among them.

Leave a Work Program Behind

At the end of the formal installation work, the analyst and the supervisor should talk over the progress made and the soft spots still remaining in the system—the procedures that will need special watching, the operations that may bog down because of delays outside the system, and so on. Once they have agreed on possible trouble spots, they should jointly prepare a written work program of the critical steps to be taken before the first scheduled follow-up review. This program will serve as a guide to the supervisor during the interval and as a priority checklist to the analyst when he returns.

Summary

Here in summary form are ten guideposts that will help to keep the installation team out of trouble.

1. Be ready before starting.
2. Keep up to schedule.
3. Avoid rash emergency decisions.
4. Anticipate and eliminate crises.
5. Do not let minor kinks in the plan dampen enthusiasm or confidence.

6. Keep all phases of the changeover coordinated by informing executives and supervisors promptly of any changes in the new system or in the plan for installing it.

7. Prevent dissension among the personnel.

8. Do not require continuous or excessive overtime work of the installation crew. If the changeover is falling seriously behind schedule because of lack of personnel, get some extra temporary help.

9. Avoid disruption of service.

10. Do not sacrifice thoroughness for speed.

EFFECTING CHANGE IN SKILLS AND BEHAVIOR

Although the previous section deals primarily with tangible change, many of the guidelines and ground rules it contains also apply to the process of effecting intangible change—that is, change in the skills, perceptions, and motivations of the persons who must participate in a new or revised system. In addition to the guidelines already given, however, other special steps must be taken if the necessary changes in human behavior are to be achieved and sustained.

The first prerequisite to success in effecting these intangible changes is a realistic understanding of what these other special steps are. That is, what must be done to eliminate old operating habits and develop new ones? In response to this question, two points need to be emphasized:

1. Bringing about desired changes in a person's behavior requires two distinct but related achievements. The first is to create a clear *understanding* of what kind of change is needed and why. The second is to develop *skill* in doing what is wanted as well as *motivation* to do so.

2. These two end results—understanding and skill—are ordinarily *not* achieved by the same means or the same set of steps. Understanding can be instilled by the traditional processes of communication and education—for example, a clear written exposition, classroom lectures, group discussions, and the like. But essential as these processes may be, they are by themselves far from sufficient. They lack the elements of *demonstration, practice, correction,* and *reinforcement* that are needed to motivate people and build their skills. To rely on communication and education alone is much like trying to learn how to play golf by reading a book or attending a series of lectures on golfing. Again, such formal devices are useful in promoting understanding—for example, an appreciation of the importance of keeping one's head down and left arm straight in swinging a golf club. But as every golfer knows, and as others will readily understand, intellectual comprehension can never build skill. That can come only from practice and coaching.

As this analogy suggests, effecting change in people's behavior is much more difficult than effecting tangible, systemic change. In addition, we must recognize that although systems staff members can play a *lead role* in installing systemic change, they can only play a supporting role in bringing about behavioral change (particularly among management personnel). Useful as their facilitating role may be, it is clearly subordinate to the role of top management, which has primary responsibility whenever the behavior of significant numbers of people throughout the organization needs to be reshaped.

What the Systems Staff Can Do

The systems staff's facilitating role can take many forms. The following three kinds of steps are basic in most situations.

1. Help top management understand fully the new skills that must be developed and the behavior that must be changed at each point in the organization structure affected by the new or revised system. A given executive, for example, may see the job of bringing about an organizational change as one of developing new job descriptions and statements of reporting relationships, rather than one of getting some people to release authority and others to accept and use it. To call attention to these differences between form and substance, the systems staff should prepare a memorandum devoted exclusively to a description of (*a*) how a given executive's subordinates must change (in terms of what they should and should not do under the new system), and (*b*) what difficulties, shortfalls, and retrogressions are likely to be encountered in effecting the change. By forewarning management on "what may and what may not happen" during the early stages of change, the systems staff will help to eliminate unpleasant surprises that might otherwise lead to loss of confidence in the new system or the approach to its implementation.

2. Develop the content of the required communication and training programs. This material might include (*a*) written statements explaining the type of change needed and why, and (*b*) narrative and visual presentations to be used in formal classes, group discussions, or on-the-job coaching sessions.

3. Design a performance measurement or feedback system that will tell top management what progress is being made in bringing about the desired behavioral changes.

What Top Management Must Do

To play its lead role in effecting needed changes in behavior throughout the organization, top management must (1) create a climate in which change flourishes, and (2) visibly support the changes needed on individual systems projects.

Create the management climate   Chapter 6 described the kind of management climate that is needed to stimulate a continuum of constructive change in any organization. (See "What 'Management Support' *Does* Mean"—especially pages 89 through 91.) What bears emphasis here is this: The readiness with which changes will be brought about, whenever needed, in the skills of key managers and in the end results on which they focus their efforts, depends on top management's effectiveness in creating the proper climate—*in instilling a sense of competitive urgency* throughout the organization.

Facilitate change on individual projects   In addition to establishing the proper climate for change within the enterprise, top management must take the following steps to bring about the behavioral changes that are required to ensure the success of individual systems projects:

1. Give top-level endorsement to the program by issuing the communications announcing the changes to be made. Although these documents will often be drafted by the systems staff, they should clearly come from an executive who has authority over all the organizational components affected by the new or revised system.

2. Set an example for the rest of the organization by making the changes required in top management's own behavior. For example: In the implementation of a new strategic planning system or a new information system for performance measurement and control, down-the-line managers can have no stronger inducement to adopt new approaches than top management's own moves to apply these tools steadfastly and well.

3. Regularly use the special reports that the systems staff has designed for measuring group and individual performance in effecting the desired changes. By introducing these reports in management group discussions or drawing on them for the individual coaching and encouragement of subordinates, management will strongly reinforce the determination of others to develop new skills.

4. Finally, give visible recognition and rewards to the individuals who develop and apply the desired new skills most quickly and effectively.

MONITORING THE INSTALLED SYSTEM

Even when the new system is in operation and the installation is considered complete, the analysts' job is not over. One of their important continuing responsibilities is a systematic follow-up to ensure not only that the gains originally planned are realized, but also that they are maintained or extended.

To this end, the systems team should schedule a thorough checkup shortly after the installation to see that the change has been stabilized, and carry out a longer-range program of periodic audits to ensure that the new system continues to operate properly.

Installation Follow-Up

The initial follow-up should be thought of as continuing until all phases of the new system are fully in effect and running satisfactorily. Where the changeover is relatively simple, this review may be considered part of the installation work and made as the final step in that process. Where widespread changes have been implemented, however, a certain period of operation is required before the new system is shaken down and ready for review. Thereafter, continuing follow-up may be needed over a relatively long time before the desired results are fully achieved.

Follow-up objectives   Analysts should develop their plans for the follow-up around three objectives:

1. Ensure that all parts of the new system have been put into operation and that the written instructions are being adhered to.

2. Make any further modifications or refinements for which operating experience reveals the need.

3. Measure the results achieved against those originally predicted and take whatever action is necessary to get the maximum benefits from the new system.

Points to be checked   Following is a checklist of the major points to be covered in reviewing the newly installed system. The thoroughness of the check will vary from point to point in any one review. If the follow-up check shows that the anticipated benefits are in fact being realized at, or close to, the estimated operating cost, the other points in the checklist need not be pursued exhaustively. But if shortfalls are being experienced in the areas of benefits and costs, then detailed operating audits of all aspects of the installation should be made.

1. Achievement of objectives. Study the overall functioning of the system to see whether it is achieving the objectives for which it was designed. Has the quality of service to customers been improved to the degree expected? By what amount have inventories been reduced? How much additional money has been made available for investment through acceleration of cash flow? How do the clerical expense reductions actually realized compare with the savings estimate that was made as part of the project proposal?

Whenever the project analysts find that benefits so far realized fall significantly short of the plan, they must discover the underlying reasons and work out with the appropriate executives or supervisors a concrete program and timetable for overcoming the performance slippage.

2. Operating costs. How does the cost of the new or revised system compare with the cost that was estimated when recommendations were submitted?

3. Operating instructions. Check the written operating instructions on the new system in sufficient detail to determine whether they are being followed. If not, find out why. Is it indifference, misunderstanding, failure to receive copies of forms or instructions, insufficient follow-up or coaching by the supervisor, incompleteness of the instructions, or what? Having determined the cause, decide whether further training, better supervision, or other corrective measures are needed, or whether the variation is justifiable and should be incorporated in a revised set of instructions.

During this part of the review, ask those who are engaged in the work for their suggestions on further refinement or improvement.

4. Elimination of old routines. Make sure that no parts of the superseded system are being continued unintentionally. Where two systems have purposely been operated in parallel, make whatever tests or comparisons are necessary to satisfy the executives and supervisors concerned that the new system can safely replace the old.

5. *Clerical performance.* Check the quantity and quality of work produced against the performance estimates made in the original proposal. If performance is not up to standard, work with the supervisor to develop a plan for increasing production or realigning the work force. Where the written instructions call for the maintenance of clerical production records, check to see that these are being properly kept and used.

Also, as part of this step, determine whether controls over the accuracy of the work exist and are being used. These devices are of particular importance during and just following the installation, for they frequently provide the only reliable means of detecting errors in performance or failure to follow the new system.

6. *Facilities.* Make sure that all the prescribed facilities, supplies, and work aids are available and that they are being used as planned. Check the layout of the office against the original diagrams and determine whether the anticipated reduction in movement or backtracking has been realized.

Programming of improvement   Following each such review of the installation, the analyst should prepare jointly with the line supervisor a definite program of steps to be taken and results to be accomplished before the next review. This process should be repeated until no changeover work remains and the new system is considered satisfactory. When subsequent reviews are called for, the analyst's first step in each case should be to make sure that everything on the previous improvement program has been done.

Subsequent Operating Audits

The study of a given system should not be regarded as finished when the changes have been fully installed. Rather, that point should be viewed as the beginning of a development process that will continue indefinitely. Inefficient or unnecessary steps will creep back into the system. New personnel, not fully aware of the purpose behind the activity, will take over its performance. The basic requirements to be met will change. New equipment or techniques will open the door to further improvements. For all these reasons, the postinstallation reviews should be followed by regularly scheduled audits to make certain that the system is still operating effectively and can continue to do so.

Such a continuing review should be a responsibility of the systems staff, for in this way the staff is given an instrument to influence the final effectiveness of the organization's systems improvement program. Frequently, however, the staff will find that it can multiply its efforts by calling on the internal auditing staff to make the actual review. Where it does, both groups should work closely together in scheduling the reviews, in developing a detailed checklist of points to be covered, and in reviewing the results of each audit.

Appendix

One mark of the superior systems analyst is an ability to communicate recommendations in a convincing and well-thought-out way. The two examples below illustrate the communication skills of two analysts on the systems staff of a life insurance company. The analysts had been asked to describe their concept of a totally new position in the marketing department—a position that they, in fact, had recommended and that top management had agreed to create. The reader will have little trouble identifying the description that made the new position "come to life."

Example A   Director of Financial Analysis and Control—Marketing Department

SCOPE AND OBJECTIVES

To assure that the necessary procedures for budgetary control are properly developed and maintained; to direct the preparation and interpretation of consolidated budgetary control reports; and to control administrative procedures of a financial nature within the Agency Department.

Additionally—and of extreme importance—to create, stimulate, and integrate throughout the Agency Department the concepts of profitable volume as related to marketing programs.

RESPONSIBILITIES

1. To develop mechanisms for budgetary planning; to execute and control procedures developed; and to evalute performance reports

2. To help ensure the financial validity of home office staff, regional, and branch office special, annual, and long-term plans

3. To prepare analytical reports on field and home office expense performance, with emphasis on budget deviation

4. To prepare present and projected profitability analyses of individual offices

5. To conduct a continuous study of the effectiveness of compensation programs

AUTHORITY

As a home office staff member, the director of financial analysis and control exercises direct authority over personnel within his function.

RELATIONSHIPS

1. He is accountable to the vice president, administrative services, for the fulfillment of his functional responsibilities.

2. He assists and works with line sales management, providing guidance in budgeting procedures and in analytical studies of a financial nature; with the director of marketing research, providing financial assistance and recommendations as a complement to studies; with the director of sales promotion, providing guidance in budget preparation and financial comment on promotional programs; with the director of product management, providing financial assistance in the preparation of market programs; and with other staffs, providing financial support for their functional specialty.

3. He works with representatives of the Actuarial Department in profitability studies involving establishment of actual product cost.

4. He works with the Controller's Department on questions of budgeting procedures.

Example B   DIRECTOR OF FINANCIAL ANALYSIS AND CONTROL—MARKETING DEPARTMENT

BASIC MISSION

1. Provide a focal point within the Marketing Department for thinking on the "economics" of marketing.

2. Ensure that the company is getting the maximum possible return from its marketing dollar through continuous scrutiny of expenses and investment in field expansion.

3. Instill a greater degree of cost consciousness in the Marketing Department.

4. Give meaning to and make explicit a concept of "profitable" volume in all areas of the company's marketing operations.

SPECIFIC RESPONSIBILITIES

(A) *Financial and Cost Analysis*

1. Continuously analyze each element of cost incurred by the field offices of the company's Marketing Department (branch offices, group and brokerage offices, and regional offices). On the basis of this analysis

    a. Develop long-term trends of *unit* costs for each cost element

    b. Develop the basis for making *meaningful* comparisons of costs among like field units (for example, comparisons of costs among branch offices)

    c. Identify the elements of cost to which special study or cost reduction effort should be applied

2. Continuously evaluate the relationship between costs incurred and results achieved on alternative approaches to sales expansion (for example, opening additional branch offices, increasing the number of apprentice agents, modifying the apprentice agents' contract, significantly changing the level of sales promotion or service to agents). This responsibility includes

    a. Evaluating the results achieved (when possible) and the costs of each alternative method of expanding sales

    b. Developing recommendations for increasing the "payout" of each method of achieving sales expansion (for example, recommending changes in the branch manager's compensation penalty for apprentice agent failures)

3. Evaluate the overall "profitability" of individual offices or channels of distribution. This responsibility is the equivalent of distribution cost analysis in make-and-sell businesses and includes

   a. Developing yardsticks for different products—the contribution they generate over and above the Marketing Department costs of selling and servicing them

   b. Developing yardsticks for selling outlets—the contribution they generate over and above the cost of operating them, taking into account the impact of such factors as sales volume and product mix, persistency, size of policies, size of office, etc.

   c. Making break-even analyses of marginal offices

   d. For all offices, identifying the factors that have the greatest "profit"-improvement potential, as a guide to sales planning

4. Undertake both requested and self-initiated special studies of an economic or financial nature. For example:

   a. Establishing projected cost and sales value relationships for various volumes and product mixes for new types of selling outlets

   b. Analyzing all group insurance offices to determine which are economically justifiable and which should be discontinued

   c. Evaluating expense trends of comparable competitive life insurance companies and reviewing the company's own trends in this light

   d. Establishing Marketing Department costs for proposed new products

5. Work closely with the Marketing Research Department to ensure that financial considerations are properly taken into account in the projects it carries out. In addition, institute special studies with the department to find the answers to such questions as

   a. What are the consequences of housing agents away from branch offices in urban areas, rural areas, shopping centers, and key market areas?

   b. To what extent do agents use company-furnished leads, and what are the sales results from these leads versus the cost of providing them?

   c. What is the sales value and cost of various services rendered to agents?

(B) *Standards of Performance*

1. Establish, install, and continuously refine all *nonselling* performance standards, including

   a. Unit-time standards for regional service centers to help control costs

   b. Quality-of-service standards in regional service centers to help ensure that service costs are not reduced to the detriment of quality

   c. Clerical personnel standards (number of clerks) for selling outlets

   d. Expense standards for regional service centers and branch offices

2. Develop yardsticks for measuring the relationship between costs incurred and results achieved for alternative methods of expanding sales.

(C) *Sales Planning*

Assist Marketing Department management in the *whole sales planning* process by furnishing it with data for deciding between major marketing alternatives and for developing annual operating plans and budgets.

1. For decisions on major marketing alternatives, this responsibility includes

   a. Analyzing and bringing to management's attention the financial and cost implications and the projected payout of alternatives (for example, changes in compensation plans, opening of additional branch offices)

    b. Working with those responsible for the "field marketing laboratory" program in setting up the mechanics for each "laboratory" project to ensure that costs incurred and results achieved are properly identified and related to each other.
2. For the development of annual operating plans, this responsibility includes:
    a. Exercising functional responsibility for developing, refining, and installing field sales planning and budgeting procedures to be used by marketing planners to establish projected sales volumes, personnel development, and operating expense of each field region and each selling outlet (including group insurance offices)
    b. Developing expense budgeting procedures for central service offices and Home Office Marketing Department units
    c. Ensuring that field and Home Office Marketing personnel understand the purpose and application of the procedures underlying the planning and budgeting process
    d. Helping the Marketing Vice President establish a realistic timetable for carrying out the steps involved in each year's planning process
    e. Ensuring the financial validity of annual and longer-term plans developed by the selling outlets and field regions—for example, making sure that goals specifying the number of apprentice agent candidates to be matured are costed out properly, and that projected relationships between costs and results for planned volumes and mix of sales are realistic
    f. Consolidating the sales plans (including expense budgets) of the field regions and the Home Office Marketing Department sections into a plan for the Marketing Department as a whole

(D) *Measurement of Performance*
    1. Develop and install a structure of management control reports and the necessary underlying procedures for reporting actual performance against (*a*) annual and longer-term plans, and (*b*) unit performance standards. These reports will include nondollar summaries of agency development performance, personnel development performance, clerical productivity, etc., as well as dollar summaries of volume, premium income, and expenses.
    2. Analyze the recurring control reports as issued, interpret the significance of variances from plans or standards, and identify for management situations requiring its attention.
    3. Achieve—through the consistent use of such performance reports and reviews—a greater consciousness of "profitability" and cost at all levels within the Marketing Department.

(E) *Methods and Procedures Improvement*
Through the use of a small staff of outstanding systems analysts:
    1. Continuously analyze and improve systems, methods, and procedures of a nonselling nature in
    a. Branch offices and other selling outlets
    b. Regional service centers
    c. Home Office Marketing Department activities
    2. Search other fields and industries for successful procedural concepts and techniques applicable to the insurance industry.

Index

Index

Accounting-systems approach to systems analysis, 45
Activities, clerical: causes of complexity in, 81–82
    centralization versus decentralization, 210
    combination opportunities, 224–226
    opportunities for savings in, 23–24
Administrative costs (*see* Costs, administrative)
Analysis:
    of functions, 83–84
    of potential computer applications, steps in, 300–306
    of systems (*see* Systems analysis)
Attitudes, defensive, forestalling, 141–142
Authority, failure to delegate, 209–210

Backlogs as index of systems problem, 112
Bottlenecks, causes of, 233
Budgets as administrative cost control mechanism, 42
Business Week, 9, 273

Cash management, sample improvement project, 196–201
    improvement opportunities, 198–201
    information to gather, 196–198
Centralization of clerical functions, advantages of, 210
Charts:
    advantages of, 160
    Gantt, 139, 309
    organization, value in systems study, 150, 161–162
    PERT, 309
    work-distribution: illustrated, 164–165
    use in systems study, 162
    (*See also* Flowcharts)
Chief executive (*see* Top management)
Clerical activities (*see* Activities, clerical)
Clerical costs (*see* Costs, clerical)
Code simplification, 227
Communications as means of gaining support for systems improvement, 88–91
Computer applications, 297–307
    benefits of, estimating, 302–303
    costs of, estimating, 303–304

Computer applications (*Cont.*):
  definition of project objectives and
    scope, 301
  evaluation of potential, 300–306
  examples of successful, 299–300
  feasibility of, determining, 304–306
  identification of alternatives, 301
  planning of, 297–306
  responsibility for identifying, 300
  (*See also* EDP systems)
Computers:
  economic payoff, 27
  as factor in MIS growth, 271
  as method of carrying out systems,
    293
  number installed, 27
  as operating expenditure, 27
  place in systems improvement
    program, 46–47
  specialists, assignment of responsibility
    to, 296–297
Control:
  information needed for, 278–284
  relation of systems to, 10, 12
  reports for, characteristics, 285–286
  of staff's overall performance, 113–
    119
  of systems improvement project:
    control boards, 115
    Gantt chart, illustrated, 139
Control procedures as index of systems
  problem, 112
Controller as leader of systems
  development effort, 62–64
Controls:
  analysis of, 215–217
  over EDP project progress, 309
Coordination, relation of systems to,
  12
Cordiner, Ralph, 9
Cost reduction studies, statement of goals
  for, 96
Cost savings, clerical, as contribution to
  profit, 25–26
Costs:
  administrative: control through
    budgets, 42
    guides to study of, 208–249
    illusions about, 42
    indirect factors affecting, 81

Costs, administrative (*Cont.*):
    opportunities for cost reduction, 23–
      26
    steps in cutting, 47
  clerical: guides to cutting, 208–249
    duplication of work, eliminating,
      217–226
    external factors, modifying, 208–
      212
    necessary work, simplifying, 226–
      230
    need for work, evaluating, 212–
      217
    routine, mechanizing, 234–235
    timing of operations, changing,
      230–234
    opportunities for cost reduction, 23–
      26
  direct: effect of poor procedures on,
    17–19
    reduction of, through systems
      improvement, 17
    information to gather on, 157–
      158
  overhead, opportunities for cost
    reduction, 23–26
Customer:
  complaints of, as index of systems
    problem, 113
  service to: conduct of typical study,
    181–189
    improvement opportunities, 188–
      189
    information to gather, 182–188
  improvement of, through systems
    study, 19–20

Decentralization of clerical functions,
  advantages of, 210
Delegation of authority, lack of, effect on
  systems, 209–210
Department heads (*see* Line
  management)
Diversification of work, arguments for,
  247
Duplication:
  of function, 218
  of information, 220–226
    illustrated, 222

Duplication (*Cont.*):
    of operations, 219–220
        illustrated, 324–325
    of purpose, 218–219
    of systems development effort, 58

EDP (electronic data-processing) systems:
    control: of in-process changes, 309–
        310
        of project progress, 309
    evaluation of, 295–296
    project plan, elements of, 308–309
    responsibility for, assignment of, 296–
        297
    selection of projects, guides to, 306–
        307
    standards for, 296
    (*See also* Computer applications;
        Computers)
Education of executives and supervisors,
    top management's objectives in, 88–
    91
Employees (*see* Personnel; Work force,
    clerical)
Equipment, mechanical (*see* Mechani-
    zation of office routines)
Executives:
    education of, objectives, 88–91
    lack of profit consciousness in, 40–42
    (*See also* Top management)

Fact-finding:
    checklists of facts to gather, 146–160
        details of existing system, 150–157
        effectiveness of system, 158–159
        external factors, effect on system,
            159–160
        flow of work, 154–155
        objectives and requirements of
            procedure, 146–147
        operating costs, 157–158
        organization and personnel, 147–
            149
        policies, 149–150
        quality of work, 156
        quantity of work performed, 153–
            154
        reports, forms, records, 156–157

Fact-finding (*Cont.*):
    importance to systems improvement
        project, 145–146
    methods of obtaining facts, 176–178
    suggestions on recording facts, 160–
        176
        functional organization charts, 161
        systems flowcharts, 162, 166–175
        task lists, 162–163
        work-distribution charts, 162, 164–
            165
        working papers, 175–176
Files:
    information to gather on, 157
    open, symbol for, 168
    permanent, symbol for, 168
    suitability for posting methods used,
        245
    temporary, symbol for, 168
Filing, aids to, 245–246
Filing sequence, factors to consider, 246
Filing systems:
    indexing of, 245
    requirements for, 244–246
Fire-fighting approach to systems
    analysis, 44–45
Fish, Lounsbury S., 21
Flow of work:
    as approach to systems study, 74–76
    delays in, elimination of, 233–234
    effect of office layout on, 175, 239–240
    information to gather on, 154–155
    irregularities in, cause and effect of,
        154–155
    symbols for, 168
    use of layout flowchart to illustrate,
        175, 240
Flowcharts:
    columnar or horizontal: advantages of,
        171
        description of, 174
        illustrated, 172–173
        limitations of, 171
    layout: illustrated, 175
        preparation of, 174
    skeleton or vertical: advantages of,
        168
        construction suggestions, 169–171
        illustrated, 166–167
        symbols in, 168–169

Form letters, use of, to simplify work, 227
Forms:
  combination of, 221–224
  control, as approach to systems
      improvement program, 45–46
  design of, standards for, 241–244
  information to gather on, 156–157
  instructions for, illustrated, 243
  logical groups of, 221–223
  methods of producing, 223–224
  performance evaluation for systems
      staff, illustrated, 121–124
  symbols for, 168
  unnecessary, analysis of, 214–215
  use of, in project planning, 136–141
      sample forms, 137–140
Functions:
  analysis of, in systems study, 151
      criteria to apply, 79–81
  clerical (*see* Activities, clerical)
  division of, sample questions, 148
  duplication of, 218
  unnecessary, examples of, 213

Gantt chart, 139, 309
General Electric Corporation, 9
General-management systems:
  computer applications for, 299–300
  definition of, 14
  improvement of, 250–269
  scope of program for, 111
  strategic analysis and decision-making
      process, study of, 250–268
  types of, 268–269
Gerstner, Louis, 265
Golub, Harvey, 302, 307–308, 310
Graphic presentations (*see* Charts)

Hertz, David, 27–28
Holden, Paul E., 21
Human relations, importance in systems
    improvement effort, 87–91, 315–
    316

Information, management, steps in
    improving, 47–48

Inspection:
  definition of, 168
  symbol for, 168
Installation of approved systems, 327–349
  cautions on, 327–329, 343–345
  follow-up of, 347–349
  hypothetical installation plan, 333–337
  operating audits, 349
  planning, 329–332
      changeover schedule, 329–330
      preparatory work, 331
      timetable for, 331–332
  preparatory work, 337–342
      with affected personnel, 337–338
      development of installation controls,
          340
      physical arrangements, 339–340
      for testing new system, 338–339
      for training personnel, 341–342
  responsibility for, 329, 331–332
  suggestions for effecting change in
      skills and behavior, 345–347
  staff's role, 346
  top management's role, 347
  written instructions for, points to
      include, 340–341
Interviews with operating personnel,
    guides for, 177–178
Inventories, reduction of, 19
Inventory management, sample
      improvement project, 189–196
  benefits, 190–191
  improvement opportunities, 193–
      196
  information to gather, 192–193
  symptoms of poor management,
      191–192

Jones, Reginald, 9

Labor shortages, clerical, systems as
    answer to, 27
Layout:
  of department, example of
      improvement, 241–242
  of office: objectives of study, 239–241
      revision of, 239–241

Line management:
  ability to meet change, 38–39
  attitude toward systems improvement
    staff, 43–44
  awareness of systems improvement
    opportunities, 40
  education of, objectives of, 88–91
  guides for managing clerical
    employees, 247–249
  lack of profit consciousness in, 40–41
  limitations of, in systems improvement
    program, 59
  responsibility for interdepartmental
    systems: benefits, 57
    disadvantages, 57–59
    effect on development program, 59–
      62
  responsibility for intradepartmental
    systems, 54–55
  role in installing approved systems, 329
  as source of project ideas, 111–112
  as systems review committee, 60

McKinsey & Company, 27
McNair, Malcolm, 74
Management (*see* Line management; Top
    management)
Management information systems (*see*
    MIS)
Management support, meaning of, 87–91
Management tools, 3
Mechanization of office routines, 234–
    239
  cost considerations, 236–237
  effect on labor force, 239
  flexibility possible, 238
  quality considerations, 238
  speed of performance, 237–238
Methods:
  clerical, survey of, what to include,
    152–154
    (*See also* Mechanization of office
      routines)
  mechanization of, 234–239
    evaluation of, 236–239
    principles governing, 235
  new: demonstration of, 107
    evaluation of, during study, 83–84

Michigan State Housing Development
    Authority, 201
  case study, 201–205
MIS (management information systems)
  approach to improvement, 276–292
  characteristics of, 284–288
    control reports, 285–286
    overall design, 284–285
    planning reports, 285
  data banks, place in, 274, 276–278
    relation to planning and control
      process, illustrated, 277
  evaluation of existing reports, 288–290
  information needed for planning and
    control, 276–284
    financial information, 278
    key factors for success, 279–284
      associated activities, illustrated,
        282–283
      examples of, 281–284
      identification of, 279–284
  interest in, reasons, 270–271
  problems with, 272–276
  report design, guidelines on, 290–291
  top-management approach to, 276–
    292
Morale of employees, systems as stimulus
    to, 30–31

Occupations, clerical, growth of, 22
    (*See also* Activities, clerical)
Operational systems:
  computer applications for, 299
  definition of, 13
  improvement of, 180–205
    cash management system, 196–201
    customer service, 181–189
    inventory management, 189–196
    public agency operations, 201–205
  scope of program for, example, 109–
    110
  subdivisions, 13–14
Operations:
  clerical: backtracking in, 231
    frequency, examples of reduction in,
      232–233
    mechanization versus manual
      methods, 234–239

Operations, clerical (*Cont.*):
    sequence of, suggested changes in,
      230–232
    simplification of, checklist, 226–230
    (*See also* Mechanization of office
      routines)
    productive: definition of, 168
      symbol for, 168
Organization structure:
    effect on systems, 208–210
    fact-finding on, value of, 149
    multiple enterprise, location of systems
      staff, 66–68
    single enterprise, location of systems
      responsibility, 56–66
Overhead expense, contribution of
    clerical cost to, 26
    (*See also* Costs)
Overorganization as index of systems
    problems, 208–209
Overtime as index of systems problems,
    112

Personnel:
    as index of systems problems, 112
    information to gather on, 147–149
    interviews with, guides for, 177–178
    operating: participation in systems
      improvement project, 102–104
      relationships with systems staff,
      100–107
PERT chart, 309
Planning:
    information needed for, 278–284
    reports for, characteristics, 285
    strategic (*see* Strategic planning)
Plans, execution of, through systems, 11–
    12
Policies:
    adjustments in, to reduce volume of
      work, 211
    analysis of, in systems study, 210–211
    effect on cost of clerical activity, 81–82
    implementation of, 11–12, 29
    information to gather on, 149–150
    relation of procedures to, 11–12
    for systems improvement program,
      50–51

Procedures as components of systems, 9–
    10
Protection, desire for, as barrier to
    systems improvement, 42
Public agency operations, sample im-
    provement project, 201–205
    improvement opportunities, 204–
      205
    information to gather, 203–204
    symptoms of poor operating
      performance, 202–203
Purchasing function, elements to be
    planned and controlled, example,
    281–284

Recommended systems:
    installation of (*see* Installation of
      approved systems)
    presentation of, 316–326
      objectives of, 317–318
      oral, 318–319, 322
      points to cover, 319–320
      rules in preparing, 320–326
      use of visual aid, 323–325
      written, 318–319, 321
    review with line management: guides
      to, 314–316
      need for, 313–314
Records:
    information to gather on, 157
    symbols for, 168
    unnecessary, analysis of, 214–215
Reports:
    as control and measure of systems
      staff, 117–118
    control: characteristics of MIS, 285–
      286
      illustrated, 289
    design of, guidelines, 290–291
    difference between planning and
      control, 286–289
    evaluation of, 288–290
    as index of systems problems, 112–
      113
    information to gather on, 156
    planning: characteristics of MIS,
      285
      illustrated, 287

Reports (*Cont.*):
   preparation of, reducing frequency, 233
   symbols for, 168
   unnecessary, analysis of, 214–215
Responsibility:
   for interdepartmental systems, 56–66
      controller, 62–64
      department heads, 56–59
      divided, effect on studies, 57–59
      general-management executive, 64–65
      principles, 65
   for intradepartmental systems, 54–55
   lack of clarity in, results, 209
   for strategic planning, 251–252
   of systems staff, 93–95
   of top management, for relationships with systems staff, 91–93

Sales order processing, typical procedure, 75–76
Sloan, Alfred P., 15
Smith, E. Everett, 39
Smith, Hubert L., 21
Standard practice instructions:
   as line management responsibility, 56
   manuals, misconceptions about, 45
Strategic planning:
   definition of, 9
   examples of corporate strategic moves, 252–253
   importance of, 252
   process, 257–268
      analysis of alternatives, 261–264
      assessment of strengths and weaknesses, 260
      contingency planning, 264–265
      decision making, 266–268
      definition of organization's objectives, 258
      development of guidelines and assumptions, 258–259
      estimation of strategic gap, 260–261
      evaluation of businesses, 259–260
      testing of alternatives, 265–266
   role of systems in, 253–255

Strategic planning (*Cont.*):
   system, design of, 255–268
      factors influencing, 255–257
   as top-management responsibility, 251–252
Supervisors (*see* Line management)
Symbols, illustrations of, 168–170
Systems:
   computer-based: development of, 293–310
      need for commitment to, 294
   definition of, 9–10
   external constraints affecting, 78–79
   hidden waste in, 40
   industry experiences, 4–9
      brokerage, 4
      insurance, 6
      petroleum, 7–9
   installation of approved (*see* Installation of approved systems)
   interdepartmental: definition of, 53–54
      responsibility for, 56–68
   intradepartmental: definition of, 53
      responsibility for, 54–55
   management information (*see* MIS)
   need for, examples, 4–9
   objectives, analyst's appraisal of, 77–78
   role in business, 10–11
   and top management, 11–12
   types of, 13–14, 53–54
      (*See also* EDP systems; General-management systems; Operational systems; Strategic planning, system)
   as vehicle for policy execution, 29
   (*See also* Computer applications; types, *above*)
Systems analysis:
   approaches to avoid, 44–48
   of clerical procedures, 23–24
   evaluation of alternatives, 83–84
   top-management approach to, 72–85
   value of, 15–34
Systems analyst, cautions for, 84–85
   (*See also* Systems staff)
Systems improvement:
   answer to labor shortage, 27

Systems improvement (*Cont.*):
  barriers to, 37–48
  benefits: to clerical performance, 26
    to employee morale, 30–31
    to production personnel, 16–19
    to top management, 21
  effect on customer service, 19–20
  functional organization as barrier to,
    42–43
  impact on computer operations, 27–29
  as means of coping with complexity,
    31–32
  responsibility for: in multiple-enter-
    prise organization, 66–68
    in single-enterprise organization,
    56–66
  role in facilitating change, 32–34
  as staff function, 51–53
  as stimulus to teamwork, 29
Systems improvement program:
  management relationship to staff, 91–
    95
  management support of, 86–91
  need for flexibility, 109
    special staff, 51–53
  objectives of, 50, 109
  organizational requirements, 51–54
  performance standards, 95–100
  policies, 50–51
  projects to include: assignment of
    priorities, 111–113
    sources of project ideas, 111–113
  scope, 109–110
  staff for (*see* Systems staff)
  steps, 49
Systems improvement projects:
  control: of new work undertaken, 114
    of work progress, 114–115
  evaluation of alternative solutions, 99
  guides in developing recommenda-
    tions, 99–100
    to fact-finding, 96–99
  information to gather, 145–160
    (*See also* Fact-finding)
  for line operating functions, 180–205
  measurement of results, 115–119
  modification of, 179
  monitoring performance of, 144

Systems improvement projects (*Cont.*):
  origin of, 108–109
  participation of operating personnel,
    102–104
    benefits, 102–103
    cautions, 103–104
    example, 103
  planning of, 128–141
    assignment of responsibilities, 136
    benefits, 128
    completion timetable, 136
    definition of problem or objective,
      130–133
    development of fact-finding plan,
      133–134
    identification of issues, 133
    prerequisites for, 128
    steps to gain acceptance of results,
      135–136
    use of forms, illustrated, 137–140
  for reduction of administrative costs
    (*see* Costs, clerical, guides to cut-
    ting)
  sample projects, 181–205
    cash management, 196–201
    customer service study, 181–189
    inventory management study, 189–
      196
    public agency operations, 201–205
  setting of project goals, 95–96
  staff-initiated, balance and sequence
    of, 114
  steps to secure support, 141–144
  for strategic analysis and decision-
    making process, 250–268
  (*See also* Recommended systems)
Systems staff:
  acceptance of, 43–44
  approach to individual studies, steps,
    74–84
  authority of, 68
  criteria for selection, 69–71
  development of project program,
    108–113
  guides to gaining confidence of
    operating personnel, 104–107
  location of, 66–69
  mission of, 68

Systems staff (*Cont.*):
  nature of task, 88
  need for top-management viewpoint,
      73–74
  performance: control of, 113–119
      evaluation of, 120–124
      report on results achieved, 118
      standards, 95–100
  personal conduct, rules for, 105–107
  place in multiple-enterprise
      organization, 66–68
  qualifications of, 69–71
  relationships: with operating
      personnel, 100–107
      with top management, 93–95
  responsibilities: recognition of
      operating personnel's
      contribution, 104
      reporting, 56–68
      sample position description, 70
      to top management, 93–95
      for understanding of operating
      personnel, 101–102
  role in effecting change in skills and
      behavior, 346
  technical competence as means of
      building confidence, 105–106
  as technical specialists, approaches
      followed, 44–48
  training, 119–120

Top management:
  attitude toward systems, 38
  responsibilities, 11–12
      for strategic planning, 251–252
      to systems staff, 91–93
  role in installing approved systems,
      346–347
  use of systems staff, 91–92
Top-management approach:
  to MIS studies, 276–292
  to systems analysis, 73–85

Transportation:
  definition of, 168
  symbol for, 168

U.S. Air Force Systems Command, 5
U.S. Department of Defense, 31–32
  planning, programming, and
      budgeting system, 31–32
Urwick, L., 29

Value-cost relationships:
  in clerical operations, 24
  in computer-based systems, 304–306
  examination, as test of system, 79–81
  examples of application, 80–81
Volume of work:
  control of, examples of savings, 82–
      83
  effect: of outside departments on,
      159–160
      of policies on, 210–211
  information to be gathered, 153–154
  reduction through policy changes,
      210–211

Wall Street Journal, 39
Work elimination:
  examples of, 80–83
  test for, 79–80
Work force, clerical: division of labor,
      guides for, 246–247
  management of, 246–249
  need to balance with workload,
      249
  shortages in, 27
  size of, 22
  tasks of supervisor, 248

Young, Richard, 294